Neighbourhood Renewal & Housing Markets

Community engagement in the US & UK

Edited by

Harris Beider

Centre for Urban and Regional Studies
University of Birmingham, UK

Blackwell Publishing

Editorial offices:
Blackwell Publishing Ltd, 9600 Garsington Road, Oxford OX4 2DQ, UK
Tel: +44 (0)1865 776868
Blackwell Publishing Inc., 350 Main Street, Malden, MA 02148-5020, USA
Tel: +1 781 388 8250
Blackwell Publishing Asia Pty Ltd, 550 Swanston Street, Carlton, Victoria 3053, Australia
Tel: +61 (0)3 8359 1011

First published 2007 by Blackwell Publishing Ltd

ISBN: 978-1-4051-3410-1

Library of Congress Cataloging-in-Publication Data
Neighbourhood renewal & housing markets : community engagement in the US & UK / Harris Beider, editor.
p. cm.
Includes bibliographical references and index.
ISBN: 978-1-4051-3410-1 (pbk. : alk. paper)
1. Urban renewal–United States. 2. Urban renewal–Great Britain. 3. Community development, Urban–United States. 4. Community development, Urban–Great Britain. 5. Housing policy–United States. 6. Housing policy–Great Britain. 7. Urban policy–United States. 8. Urban policy–Great Britain. I. Beider, Harris. II. Title: Neighbourhood renewal and housing markets.

HT175.N3973 2006
307.3'4160973–dc22
2006015835

A catalogue record for this title is available from the British Library

Set in 10/12.5 pt Trump Mediaeval
by Graphicraft Limited, Hong Kong
Printed and bound in Singapore
by COS Printers Pte Ltd

The publisher's policy is to use permanent paper from mills that operate a sustainable forestry policy, and which has been manufactured from pulp processed using acid-free and elementary chlorine-free practices. Furthermore, the publisher ensures that the text paper and cover board used have met acceptable environmental accreditation standards.

For further information on Blackwell Publishing, visit our website:
www.blackwellpublishing.com

RICS Research

The Royal Institution of Chartered Surveyors is the mark of property professionalism worldwide, promoting best practice, regulation and consumer protection for business and the community. It is the home of property related knowledge and is an impartial advisor to governments and global organisations. It is committed to the promotion of research in support of the efficient and effective operation of land and property markets worldwide.

Real Estate Issues

Real Estate Issues is an international book series presenting the latest thinking into how real estate markets operate. The books have a strong theoretical basis – providing the underpinning for the development of new ideas.

The books are inclusive in nature, drawing both upon established techniques for real estate market analysis and on those from other academic disciplines as appropriate. The series embraces a comparative approach, allowing theory and practice to be put forward and tested for their applicability and relevance to the understanding of new situations. It does not seek to impose solutions, but rather provides a more effective means by which solutions can be found. It will not make any presumptions as to the importance of real estate markets but will uncover and present, through the clarity of the thinking, the real significance of the operation of real estate markets.

Books in the series

Adams & Watkins *Greenfields, Brownfields & Housing Development*
Adams, Watkins & White *Planning, Public Policy & Property Markets*
Allen, Barlow, Léal, Maloutas & Padovani *Housing & Welfare in Southern Europe*
Ball *Markets & Institutions in Real Estate & Construction*
Barras *Building Cycles & Urban Development*
Beider *Neighbourhood Renewal & Housing Markets*
Ben-Shahar, Leung & Ong *Mortgage Markets Worldwide*
Buitelaar *The Cost of Land Use Decisions*
Couch, Dangschat, Leontidou & Petschel-Held *Urban Sprawl*
Couch, Fraser & Percy *Urban Regeneration in Europe*
Dixon, McAllister, Marston & Snow *Real Estate & the New Economy*
Evans *Economics & Land Use Planning*
Evans *Economics, Real Estate & the Supply of Land*
Guy & Henneberry *Development & Developers*
Jones & Murie *The Right to Buy*
Leece *Economics of the Mortgage Market*
Lizieri *Towers of Capital*
McGough & Tsolacos *Real Estate Market Analysis & Forecasting*
Monk & Whitehead *Affordable Housing and the Property Market*
Newell & Sieracki *Property Investment & Finance*
O'Sullivan & Gibb *Housing Economics & Public Policy*
Seabrooke, Kent & How *International Real Estate*
Tiwari & White *Real Estate & Finance in the New Economic World*
Wellings *British Housebuilders: History & Analysis*

Contents

Preface

The world is becoming a smaller place. Countries are changing rapidly and no more so as a result of immigration, both old and new, and the impact on neighbourhoods and housing markets. Alongside demographic drivers is the continuing importance attached to policy issues on making equality of opportunity happen in practice rather than developed as a theory, addressing problems of racism and segregation and gaining the trust of people when confidence in institutions has reached an all-time low.

These policy and political debates are happening in different places and different contexts but have a clear resonance in the US and UK. Shaping neighbourhoods of opportunity and aspiration, addressing the problems of entrenched segregation and restricted choice and revitalizing communities are not in themselves new. However, both governments in recent times have grappled with these issues. So it was in this context that the Centre for Urban and Regional Studies (CURS) decided to convene a symposium on Housing, Neighbourhoods and Communities in the US and UK at the University of Birmingham, England, in late June 2004. The event brought together academics, policy makers and practitioners to share ideas, issues and trends on housing, neighbourhoods and communities in the US and UK. Through invited papers and presentations we wanted to explore critical themes, such as convergence of the role of the state, the impact of changing housing markets, conflict and displacement of communities and local organizational responses to these local, regional and national changes.

Since 2004 planning has taken place in preparation for this book. Original papers have been substantially revised to reflect the fast moving and fluid policy environment on housing, communities and 'race' in both countries. New papers have been commissioned to fill gaps from those presented at the original symposium. Discussion between participants has sharpened the focus of the book.

This is not designed to be a dry and esoteric academic volume. The issues are too important to be discussed in such an abstract way. Moreover many of the participants at the Birmingham symposium and contributors to the book have interesting career trajectories. We have been campaigners, advocates, lawyers and housing practitioners. This, combined with our academic training, means that the chapters that unfold are laced with the objectives of using knowledge to generate an informed debate that leads to public policy change.

Complex and deep-seated problems require practical and radical solutions. If we are successful in starting this journey then preparing this book would have been time well spent indeed.

Acknowledgements

Many people have made this book a reality. In early 2004 I took my proposal to convene a symposium to Phillip Leather who was then Director of Urban and Regional Studies. We discussed how the event could help to develop new ideas and thinking on critical comparative areas. Special thanks should be given to the CURS administrative staff for the conscientious way that they have coped with transatlantic correspondence. In particular, the efforts of Jane Simpson and Yvonne Harley are much appreciated.

CURS has been an encouraging place to embark on an academic career, and a number of people have helped. David Mullins was instrumental in bringing me to the department and has been a constant source of support and advice. He has struck just the right balance of encouragement and constructive criticism in the search for the highest standards. Lisa Goodson has been generous in allocating time to discuss new ideas, concepts and themes, as well as providing many laughs. Other colleagues who have helped include Austin Barber, Ed Ferrari, Stephen Hall, Ricky Joseph, Peter Lee, Jane Lutz, Alan Murie, Jenny Phillimore, Rob Rowlands, Andy Tice and Chris Watson.

Outside CURS a number of people have also encouraged my work. Kusminder Chahal has been my colleague and close friend for nearly ten years. During that time we have spent many occasions intellectually dissecting 'race' and politics, getting angry and frustrated with the slow pace of change but at the same time having fun. Danielle Walker and Katherine Knox from the Joseph Rowntree Foundation provided a grant to review UK and US approaches to 'race' and housing that sharpened some of the ideas in this book. More recently, Malcolm Harrison, David Robinson, Sarah Blandy, Steve Gayle and Glyn Robbins have all listened and shaped ideas.

My understanding and knowledge of 'race', housing and neigbourhoods in the United States have been deepened with the help of a number of people. Nicole Marwell has shared ideas and views on the role of community organizations and politics on my numerous visits to Columbia University and has always been a good colleague and friend. Jerilyn Perine demonstrates boundless knowledge on housing policy and practice in the US and has been an incredible host in New York since the time when she was Housing Commissioner. Conversations with Lucille McEwen have helped to deepen my understanding of the role of faith organizations in regeneration and renewal. Friends and colleagues at the Urban Institute in Washington DC have provided much needed space and rigour in comparing UK and US approaches to 'race' and housing. Sue Popkin has become a mentor and true friend. Marge Turner and Lynette Rawlins have provided a forum to develop some of the ideas contained in this book.

Friends in Birmingham and London have lived through this project. Thanks to Ben and Nicola for providing food and sustenance to keep me going; to Alec and Nic for providing me with a London 'home'. Lastly, this book could not have been achieved without the love and support of my wife Sarah and our children Jemima, Alys, Jess and Tom. My kids in particular have had to put up with a lot over the years and have turned out in spite of this to be fantastic, humorous and well adjusted people. I suppose the reason is that Sarah has been an incredible mother and partner.

Harris Beider

Dedication

To Sarah, sometimes my critic, always my friend and forever my soul mate.

Contributors

Austin Barber
CURS, School of Public Policy, University of Birmingham, UK
Austin Barber is a Lecturer in urban development and planning at the Centre for Urban and Regional Studies in the University of Birmingham. His main research interests include city centre regeneration, city living and gentrification, inter-city competition and cooperation and comparative urban development in Britain, Europe and North America. In recent years, he has written numerous studies for local and regional agencies around these themes as well as for academic audiences. Prior to his academic career, Austin worked as a newspaper journalist for several years in London and Birmingham, covering urban development, economic and planning affairs from a day-to-day perspective.

Harris Beider
CURS, School of Public Policy, University of Birmingham, UK
Harris Beider was appointed as Lecturer at the Centre for Urban and Regional Studies in the University of Birmingham in April 2005 after being an Honorary Lecturer since September 2001. Harris is also an independent management consultant specializing in diversity, housing and organizational change. At the Centre for Urban and Regional Studies his main research interests are focused on housing, race and neighbourhoods. His major recent projects include *Involving Black and Minority Communities in Housing Investment Decisions* (published by Office of the Deputy Prime Minister in 2004) and *Black and Minority Ethnic Communities in the Eastern Corridor: Aspirations, Neighbourhood 'Choice' and Tenure* (Birmingham City Council, 2006). He is currently editing a special issue of *Housing Studies* on 'Rethinking Race and Residence'. Prior to working at CURS Harris was Executive Director of the Federation of Black Housing Organisations (1997–2001) and Founding Director of People for Action (1995–1997). At the former, Harris was involved in helping to shape national policy on black and minority ethnic issues. Harris has been an advisor to the Prime Minister's Social Exclusion Unit and Visiting International Scholar at Columbia University, New York City.

Pauline Card
School of City and Regional Planning, Cardiff University, UK
Pauline Card is a Lecturer in Housing at the School of City and Regional Planning. Her main research interests are housing policy, particularly in relation to social inclusion/exclusion; housing management with an emphasis on the management of anti-social behaviour and access routes

and allocation processes; the governance of housing; and housing and regeneration with particular interest in the role of housing organisations in community regeneration. Recent work includes 'The role of housing stock transfer organisations in neighbourhood regeneration: exploring the relationship between regeneration, "new localism" and social networks' (with Jane Mudd, 2006) in the journal *Housing Studies*; and 'Governing anti-social behaviour – inclusion or exclusion?' (2001) in *Two Steps Forward: Housing Policy into the New Millennium* (edited by A. Marsh and D. Cowan, published by Policy Press, Bristol).

James DeFilippis
Department of Black and Hispanic Studies, Baruch College, City University of New York, USA
James DeFilippis is Assistant Professor in the Department of Black and Hispanic Studies at Baruch College, CUNY, New York. His work focuses primarily on issues of community development, but has also dealt with more specific housing policy issues and has recently broadened to include looking more explicitly at issues of immigration in community development and urban labour markets. He is the author of many publications on housing, community development and the politics of urban development, including: *The Community Development Reader* (with S. Saegert, forthcoming); *Unmaking Goliath: Community Control in the Face of Global Capital* (2004), named the best book on urban politics in 2004 by the American Political Science Association; 'The Emancipatory Community? Place, Politics and Collective Action in Cities' (with P. North, 2004) in *The Emancipatory City? Paradoxes and Possibilities* (edited by Loretta Lees); *Keeping the Doors Open: HUD-Subsidized Housing in New York City* (2003); and 'The myth of social capital in community development', *Housing Policy Debate* (2001).

Xavier de Souza Briggs
Department of Urban Studies and Planning, Massachusetts Institute of Technology, USA
Xavier de Souza Briggs is Associate Professor of Sociology and Urban Planning at MIT. His main research interests are race/ethnicity, inequality and democracy in cities. Currently he is working on the effects of housing mobility and neighbourhood context on low-income families and children in the US; local civic process ('democracy as problem-solving') in Brazil, India, South Africa and the US; and interracial networks (bridging social capital) in America. In 2005, Brookings published his book *The Geography of Opportunity: Race and Housing Choice in Metropolitan America*. He is Founder and Director of The Community Problem-Solving Project @ MIT (www.community-problem-solving.net) and was former acting Assistant Secretary for Policy Development and Research at the US Department

of Housing and Urban Development in the Clinton Administration (1998–1999).

Ed Ferrari

CURS, School of Public Policy, University of Birmingham, UK

Ed Ferrari is a Lecturer at the Centre for Urban and Regional Studies at the University of Birmingham. Following an education in town and regional planning, Ed worked in local government where he was involved in the development of strategic research and GIS capacity for Birmingham City Council's Housing Department. The themes of using GIS and modelling techniques for housing strategy were continued in a PhD undertaken at the University of Sheffield. Since then, Ed has developed research interests in the analysis of housing markets, mobility in the social rented sector, application of GIS to housing research, and use of secondary data-sets for policy research. He has been involved in a wide range of research and consultancy projects for local government, central government and voluntary sector clients. Recently, he has been very closely involved with the development of the evidence base for, and the evaluation of, Housing Market Renewal (HMR) in England.

Rick Groves

CURS, School of Public Policy, University of Birmingham, UK

Rick Groves is Head of Department, Centre for Urban and Regional Studies, at the University of Birmingham. He has undertaken a wide range of research and consultancy into various aspects of housing policy and urban development both in the UK and overseas. In the UK he has undertaken commissions for ODPM (and its predecessors), the Joseph Rowntree Foundation, ESRC and a variety of other organizations and authorities. Recent publications include *Implementing New Powers for Private Sector Housing Renewal* (with Sian Sankey, published by the Joseph Rowntree Foundation) and *Neighbourhoods that Work: A Study of the Bournville Estate, Birmingham* (with Alan Middleton, Alan Murie and Kevin Broughton, 2003, published by Policy Press).

Stephen Hall

CURS, School of Public Policy, University of Birmingham, UK

Stephen Hall has been a Lecturer in Urban and Regional Economic Development at CURS since September 1995. His principal research interest is in area regeneration policy in England. Stephen has undertaken research on area regeneration and neighbourhood renewal for the Office of the Deputy Prime Minister, Joseph Rowntree Foundation, Local Government Association and the Housing Corporation, and has wide experience of evaluation of local regeneration projects in different parts of England. Stephen has undertaken research into partnerships, housing and employment aspects of area regeneration policy in Britain and Europe (particularly France) for the European Union, Housing Corporation, British Academy, British Council and Centre National des Recherches Scientifiques. He was Visiting Lecturer at the Université Blaise Pascal, Clermont Ferrand, France, in 2004. Recent publications include *Restructuring Large-scale Housing Estates in European Cities: Good Practices and New Visions for Sustainable Neighbourhoods and Cities* (with R. van Kempen, K. Dekker and I. Tosics, editors, published 2005 by Policy Press).

Peter Lee
CURS, School of Public Policy, University of Birmingham, UK
Peter Lee is a Senior Lecturer in Urban and Regional Studies at CURS, having joined the department in 1994. While at CURS Peter has completed research and consultancy projects for a number of clients including the Office of the Deputy Prime Minister (ODPM), the European Commission, the Joseph Rowntree Foundation, the Housing Corporation and the Economic and Social Research Council (ESRC), as well as a number of individual local authorities and housing associations. Peter's research interests have gravitated towards analysis of housing markets and neighbourhood change to explain geographies of social exclusion and social cohesion. Peter's recent research has focused on the function and trajectory of neighbourhoods, regional and local housing strategy development and housing market restructuring in the context of social and economic change.

Lucille McEwen
Harlem Congregations for Community Improvement, Inc., New York, USA
Lucille McEwen is President and Chief Executive Officer of Harlem Congregations for Community Improvement. Founded in 1986, HCCI is a coalition of more than 90 inter-faith congregations that has implemented a comprehensive portfolio of programs to provide affordable housing and safer streets; offers opportunities for individuals and groups to become and remain economically independent; increases understanding of and access to health care; and provides substantive educational programs for adults and young people. Through alliances with other communities and organizations, elected officials and local residents, HCCI has also helped reduce crime in the community; increase public sanitation; and preserve and transform open space.

David Mullins
CURS, School of Public Policy, University of Birmingham, UK
David Mullins has been based at CURS since 1989, and is Professor of Housing Policy and Programme Director for the Public Service MBA (Housing Stream). His research interests include the governance, management and

regulation of housing organizations and the role of the non-profit sector in public policy access to social housing, equality and diversity. His books include *Housing Policy in the UK* (with Alan Murie) published in 2006 and *Housing and Public Policy* (with Alex Marsh) in 1998. David has published in *Housing Studies, Policy and Politics, Public Administration, Public Money and Management, Voluntas* and *The International Journal of Non-Profit Marketing*. He has undertaken research for the Office of the Deputy Prime Minister, Housing Corporation and Audit Commission and many individual social landlords. He is a coordinator of the European Network for Housing Research Study Group on Institutional and Organisational Change in Social Housing. Before joining CURS he worked in local authority housing for 10 years and also for the Housing Corporation.

Alan Murie
CURS, School of Public Policy, University of Birmingham, UK
Alan Murie is Professor of Urban and Regional Studies at the Centre for Urban and Regional Studies, University of Birmingham. He was Director of CURS for the period 1995–2000. He is currently Head of the School of Public Policy at the University of Birmingham. Previous posts include Professor of Planning and Housing, Heriot-Watt University, Edinburgh (1990–94) and Lecturer, Senior Lecturer and Reader in Urban Studies, School for Advanced Urban Studies, University of Bristol. He has also held visiting professorships at the Universities of Utrecht and Amsterdam and City University Hong Kong. Alan Murie has more than 30 years' experience of research on housing and related topics. His research work has related to the changing housing situation and housing policy in the UK and other countries. This has referred in particular to privatization of public housing, the residualization of social rented housing, differences within the home ownership sector and changing demand for housing in the UK. He is the author of key texts related to housing including *Housing Policy in the UK* (with David Mullins, 2006, published by Palgrave Macmillan) and was the founder editor of the journal *Housing Studies*. He is a Fellow of the Royal Society and an elected member of the Academy of Learned Societies in the Social Sciences.

Jerilyn Perine
BLOCK BY BLOCK, LLC, New York, USA
Jerilyn Perine is President of BLOCK BY BLOCK, LLC which is focused on affordable housing development, community preservation and redevelopment of former industrial sites in cities in the US and abroad. From 2000 to 2004 she was Commissioner of the New York City Department of Housing Preservation and Development and has more than 25 years of experience in housing and community development in New York City's neighbourhoods.

Jerilyn was appointed by both Mayor Rudolph Giuliani and Mayor Michael Bloomberg to lead America's largest municipal housing agency with more than 2500 employees and an annual operating and capital budget of $800 million. As Commissioner, Jerilyn was the author of Mayor Bloomberg's New Housing Marketplace Plan that provided $3 billion over 5 years to preserve and create over 65,000 units of affordable housing. Under Mayor Giuliani she designed and oversaw the management and operation of programs designed to return a significant inventory of tax-foreclosed residential property to local, private ownership. She was a member of the International Brownfield Exchange between 1998 and 2002 and has lectured at conferences and universities in the US, Germany, the UK and Australia.

Susan Popkin
Metropolitan Housing and Communities, The Urban Institute, Washington DC, USA
Sue Popkin is a Principal Research Associate in The Urban Institute's Metropolitan Housing and Communities Policy Center, Washington. A nationally recognized expert on assisted housing and mobility, Sue directs The Urban Institute's 'Roof Over Their Heads' research initiative, which examines the impact of radical changes in public housing policy over the past decade. A second key area is the impact of neighbourhood environments on outcomes for public housing families. A third focus is evaluating strategies for promoting mobility and choice for assisted housing residents. This body of research includes the HOPE VI Panel Study, the first large-scale, systematic look at outcomes for public housing families; the HOPE VI Retrospective Tracking Study; the Ida B. Wells Relocation Fall-Out Study; the CHA Relocation Assessment; and the Three-City Study of Moving to Opportunity, which builds on her work on the MTO Interim Evaluation. Previously, she directed the only national study of public housing desegregation, and was the Project Director for a series of studies on the Gautreaux Housing Desegregation programme, which assisted public housing residents with moving to housing in integrated communities in Chicago and surrounding suburbs. Sue is the lead author of *The Hidden War: Crime and the Tragedy of Public Housing in Chicago*, and the author of numerous papers and book chapters on housing and poverty-related issues.

David Robinson
Centre for Regional Economic and Social Research (CRESR), Sheffield Hallam University, UK
David Robinson is Professor of Housing and Public Policy at the Centre for Regional Economic and Social Research (CRESR), Sheffield Hallam University, where he leads the housing research team. David has researched extensively around the interface of housing, social justice and cohesion. He

has a long-standing interest in efforts to understand the complex interplay of factors reproducing racialised inequalities within the housing system and efforts to manage the consequences of residential settlement patterns for community relations and cohesion. David's recent project work includes separate studies of the housing pathways and the neighbourhood experiences of new immigration for the Joseph Rowntree Foundation and analysis of the contribution of housing management to the objectives of community cohesion for the Chartered Institute of Housing. David has produced numerous academic outputs and policy reports. Recent examples include the report *Neighbourhood Experiences of New Immigration* (2006) and his editorship of a special edition of *Housing Studies* focusing on the transforming of social housing in the UK (2006).

Rob Rowlands
CURS, School of Public Policy, University of Birmingham, UK
Rob Rowlands is a Lecturer at the Centre for Urban and Regional Studies. His background is in housing and planning and he has worked as a housing officer. Rob has developed a number of research interests based around housing. He is interested in the notion of community and its application to public policy. In particular he has developed research around public participation in policy development and is interested in the role of 'community' in the development and regeneration of residential areas. He has been involved in a number of research projects focused on communities and regeneration. Recent publications include 'Place making in theory and practice: collaborative planning, power, governance and large estates' (with S. Hall, 2004) in *Restructuring Large Scale Housing Estates in European Cities* (edited by R. van Kempen, K. Dekker, S. Hall and I. Tosics, published by the University of Utrecht), 'Commodification, normalisation and intervention: cultural, social and symbolic capital in housing consumption and governance' (with J. Flint, 2003) in *Journal of Housing and the Built Environment* and *More than Tenure Mix: Developer and Purchaser Attitudes to New Housing Estates* for the Joseph Rowntree Foundation.

Michael H. Schill
UCLA School of Law, University of California, Los Angeles, USA
Michael Schill is Dean and Professor of Law at UCLA School of Law. He is a national expert on real estate and housing policy, deregulation, finance and discrimination. He has written or edited three books and over 40 articles on various aspects of housing, real estate and property law. He is an active member of a variety of public advisory councils, editorial boards and community organizations. Before joining the faculty of UCLA School of Law, Michael was the Wilf Family Professor in Property Law at New York University's School of Law and Professor of Urban Planning at NYU's

Robert F. Wagner Graduate School of Public Service. From 1994 to 2004, Michael served as the director of the Furman Center for Real Estate and Urban Policy. Prior to that, he was a tenured Professor of Law and Real Estate at the University of Pennsylvania. He has also been a visiting professor at Harvard Law School.

1

Introduction

Harris Beider

Background

This book is about cities, neighbourhoods and people in the US and UK and the impact of debates on housing, renewal, community engagement and 'race'. These are concerns that have been at the forefront of public policy in both countries for many years. While there is a long-established dialogue about these issues between British and American academics and policy communities, there has been an absence of a coherent comparative approach[1]. This book is one of the first to try to bridge the gap at academic and policy levels.

In recent years the US and UK have been witness to substantial changes to national and local approaches to housing, community engagement and renewal within cities and neighbourhoods. The processes of privatization, deregulation and the changing structure of cities with the long-term decline of manufacturing industry have considerably moved the agenda on. In the UK, 'disturbances' in several northern cities and associated concerns about race and segregation were discussed in a series of high-profile reports and influenced subsequent policy. In 2000, the Sustainable Communities Plan signalled renewed interest in investment in housing. In the US, the new approaches being adopted to housing, with the Voucher Program, HOPE VI Program and the further move away from public housing, have contributed significantly to policy discourse and practice.

Changing patterns of race and residence, and new approaches to community involvement and empowerment are also of central interest and are themes that run throughout this book. There is an underlying debate about

[1] The most recent texts making a comparison between housing issues in the UK and the US are Karn and Wolman (1992) and Wolman (1975).

convergence and divergence of policy and research agendas across the two countries.

This book provides an important new contribution to these debates. It does not set out to provide a systematic or comprehensive comparison of approaches in the US and UK, but rather highlights specific issues and developments which are crucial to an understanding of residence, renewal and community empowerment in both countries. It presents original research material, a discussion of the transferability of experience between these and highlights of the issues emerging from the consideration of the two countries in parallel.

The material presented is research based but also directly engages with issues around policy and delivery. It is designed to reflect the concerns of academic, activist and policy communities on both sides of the Atlantic. Indeed, the book is enriched by the contributions from housing practitioners who manage the complex issues of providing housing to diverse and changing communities on a day-to-day basis.

The objective is to make a positive contribution to the debates on housing, ethnicity and community renewal. This is a challenging task when discussing these issues in one country, let alone comparing two housing systems. For this reason the first step is to consider the tradition of comparative housing studies, and specifically those that compare the US and UK. The current approach will be placed in a context that both acknowledges the tradition of comparative work but also seeks to make a departure towards the importance of origins of housing policies and their outcomes. After reviewing comparative housing approaches we will discuss the extent to which the UK and US have different housing typologies shaped by approaches to policy issues, housing affordability and racial segregation. The plan of the book and discussion of chapters recognizes the differences between the two countries. The chapters are analysed with reference to housing and welfare typologies, comparison between cities and neighbourhoods and an exploration of change agents within institutions.

Comparative approaches

There has been growing interest in comparing housing policies and systems across different countries (Wolman, 1975; Ball et al., 1988; Karn and Wolman, 1992; Power, 1993; Barlow and Duncan, 1994; Harloe, 1995; Doling, 1997; Kemeny and Lowe, 1998; Katz, 2004; Van Kempen et al., 2006).

The world has become a smaller place during the last 20 years, not only in terms of the ease and cost of travel and the increased interest this brings in different cultures, people and ideology, but also the advent of the internet, making communication between people from different parts of the world reduced to a click of a computer button. The technological revolution, together with the growing dominance of English as the lingua franca of the internet, has encouraged dissemination of ideas, policies and papers between academics and practitioners. People have come together in an unprecedented manner. Differences have been acknowledged, but it could be argued the emphasis has been to find common solutions to what appear to be common problems at a number of different levels. Most notably Sassen (1991) viewed globalization as shaping a new hierarchy of cities, neighbourhoods and people, driven by flows of international capital and the reconfiguration of urban economies.

Here global forces create a prosperous and rich workforce employed by transnational companies living in the wealthiest part of cities. A workforce, that is also internationalized but very different, services transport, health and catering infrastructure. These are poor migrants seeking a better deal in the rich cities. The cleaners, cab drivers and cooks that live in the fragile world of low skills and low wages are essential to make the global city function. Globalization not only helps to remove barriers for increased communication in policy analysis but also leads to income and spatial inequality (Sassen, 1991).

The globalization thesis has been subject to criticism. Hamnett questions the basis of Sassen's methodological perspective, process and outcomes (Hamnett, 1996). Specifically the view put forward is that Sassen at best neglects and at worst ignores the importance and legacy of the welfare state in European countries. In short, dual cities of rich and poor do not follow automatically from globalization. Polarization does exist within cities but, because of the welfare state and the existence of a system of benefits, this is cushioned in Europe. Further Hamnett suggests that Sassen and her followers are guilty of placing an American template laced with assumptions about the limited role of the welfare state and driven by immigration. The world and the concept of globalization are more complex than this.

The importance of political integration or ideological attachment between different countries helps us to understand the increase in comparative housing studies. It could be argued that the fall of the Soviet Union together with the rise of the US as the pre-eminent political and economic global power paved the way for a free market political and economic philosophy to 'let rip' across the world. In truth the election of the Thatcher government in the UK in 1979 followed by Reagan's presidential success in 1980 had already started the process that has been followed by successive administrations in both countries and replicated in many other parts of the world.

The characteristics of the new political ideas resulted in to a sharp reduction in the importance of and funding to the welfare state, a concerted attack on unionized labour and the dominant position of free market economics (Gamble, 1994). Social housing no longer featured among the main spending priorities of government but was, instead, increasingly depicted as part of a malaise associated with a bloated and inefficient welfare state and social disorder. Apart from the sharp cutbacks in public expenditure there are other aspects of ideological re-alignment that need to be considered. These include, for example, the economic management in both countries that prioritizes low interest rates, active promotion of home ownership as the preferred form of tenure for all income groups, demolition and selling of public sector housing stock and increasing dependence on private sector finance to support housing development and maintenance programmes.

The ideological processes that have become so dominant during the last 20 years had a profound impact on the direction of housing policy. It could be argued that the importance of low inflation, limiting public investment and reforming public management have shaped housing decisions on the importance of home ownership as a policy objective leading to sustainable communities, the increased role of private finance to build social housing, the growing professionalization of housing organizations to reflect the importance of markets as opposed to state intervention and the demolition of public sector housing estates and their replacement with mixed-tenure and mixed-income neighbourhoods.

One of the most interesting comparative accounts on US and UK housing policies was provided by Karn and Wolman (1992). Though now somewhat dated, and based on research during the 1970s and 1980s, it nevertheless provided a systematic comparison between the two systems that helps to contextualize this book. These rested on the importance of tax incentives given to home ownership in the US during the 1980s and the relatively small and marginalized role of public sector and assisted housing. In contrast the local authorities and, increasingly, housing associations in the UK, continued to develop and manage housing for low-income communities. Karn and Wolman commented that that quality of housing stock in the UK was not as good as in the US. This of course is less important because the Labour government's Decent Homes Standard (DHS) establishes a series of quality measures that all public sector housing stock should reach by 2010 (ODPM, 2003). It should also be noted that DHS applies to the private sector but to a lesser standard. In July 2002 a public service agreement (PSA) target was announced for housing in the private sector. This applied not to all dwellings but to those occupied by vulnerable households (defined as people on income support, housing benefit, council tax benefit, disabled persons tax credit, income based job seekers allowance, working families tax credit, attendance allowance, disability living allowance, industrial injuries, disablement benefit, war disablement pension). The target was to increase the proportion of vulnerable households in the private sector who live in homes that are in decent condition.

Housing typologies

It could be argued that divergence analysis owes much to, and indeed flows from, the influential work of Esping-Andersen (1990). Some have even suggested that this body of work has 'paradigm status' in that it has shifted theoretical thinking (Matznetter, 2001). Here, three typologies of welfare regime are grouped based on data on key social policy areas. These are *liberal*, which is characterized by the importance of the market, restricted public goods and emphasizing the role of the private sector; in contrast *social democratic* regimes are typically those that provide universal public goods and services, promote the importance of the state and the relatively small role of the market; finally the *conservative* regime is one which is characterized by the centrality of the family and relatively marginal role of the state and markets (Table 1.1). Based on comparative research of certain features of welfare states (excluding housing), Esping-Andersen was able to accommodate 18 capitalist countries into these three ideal types. Britain and the US were grouped under the liberal category.

The work of Esping-Andersen provides a systematic way of discussing social policy and practice in different countries. However, housing is not discussed as a key area of analysis. The focus was on social security and pension systems. Indeed there is only one housing reference in the index and that relates to home ownership. As Esping-Andersen recognizes, 'our

Table 1.1 Different types of welfare regime.

	Liberal	**Social democratic**	**Conservative**
Role of:	Marginal	Marginal	Central
Family	Central	Marginal	Marginal
Market	Marginal	Central	Subsidiary
State			
Welfare state:	Individual	Universal	Kinship
Dominant mode of solidarity			Corporatism
			Etatism
Dominant locus of solidarity	Market	State	Family
Degree of decommodification	Minimal	Maximum	High (for breadwinner)
Modal examples			Germany
	US	Sweden	Italy

Source: Esping-Andersen (1990).

study has clearly ignored many crucial policy arenas and institutional aspects of modern industrial capitalist societies' (1990, p. 221). It must also be recognized that the research was undertaken during the 1970s and 1980s.

Turning to the contemporary housing question, we might ask whether it was appropriate in the 1970s and 1980s to categorize the two housing systems in the US and UK as examples of a liberal welfare regime. We might also consider the trajectory of change in the housing systems since 1980 and ask whether they are closer to each other today than when Esping-Andersen published his seminal work in 1990. Private housing tenure has become increasingly dominant in both countries; finance capital to support owner occupation has become easier to access, even for the lowest-income groups; public sector housing has become more residual and stigmatized, and is undergoing transformation in the form of HOPE VI in the US and the 'Right to Buy' together with local authority stock transfer in the UK.

The different starting point and very different institutional context have led to many differences being preserved despite these somewhat common trajectories of change. In particular the continued existence of a social housing 'profession' with skills, knowledge and capacity has helped preserve a social housing sector of some size in the UK, whereas its more limited development in the US has restricted the ability to resist market mechanisms and the advance of real estate professions into the social realm. Nevertheless, it could be argued that these differences are minimal and that a common set of political assumptions about the limited role of the state, the importance of markets and discussion on active and responsible citizenship will eventually lead to similar outcomes for policies concerned with housing, communities and renewal of cities and neighbourhoods. Table 1.2 considers the evidence and attempts to build towards housing typologies for the US and UK.

Given the above, do US and UK housing systems show a move towards alignment? In formulating a response we need to take a step back and reflect briefly on the starting points, process and outcomes of housing policy. We have already discussed processes and outcomes but said little about starting points.

Social housing after 1945 in the UK was about large-scale volume building by local authorities. The context is important. World War II meant that many towns and cities had to be rebuilt. Decent quality housing was in short supply and out of reach of all but the very rich. In these circumstances, it was local authorities rather than the private sector that had the capacity, skills and political support to build large-scale public sector housing efficiently. Indeed the Labour government elected by a landslide in 1945 birthed the welfare state and was driven to create a vision of a post-war Britain very different from mass unemployment, squalor and

Table 1.2 Comparing US and UK public assisted housing systems.

US	UK
Public housing authorities, not-for-profit organizations and for-profit organizations build and manage housing for low-income communities	Local authorities and housing associations build and manage housing for low-income communities but also increasingly engaged in intermediate housing markets
Stigmatized public housing sector	Stigmatized public housing sector but much less so than US
Public and assisted housing accounts for about 5 percent of total housing	Public and assisted housing accounts for about 20 percent of total housing
Public housing demolition by HOPE VI	Public sector housing transfer by stock transfer; demolition of mainly private housing in the Housing Market Renewal Areas
Public housing tenants have to perform 8 hours of community service a month or face eviction	Local authority and social housing tenants have an opportunity to become involved in tenant participation
Emphasis on home ownership because seen as way to build sustainable communities	Emphasis on home ownership because seen as a way to build sustainable communities
Housing seen as entitlement, not a right	Housing increasingly viewed as an entitlement and not a right
Low levels of supported housing provided to vulnerable communities, e.g. ex-offenders	High levels of supported housing provided to vulnerable communities, e.g. ex-offenders, elderly
Public sector housing characterized by segregation by race and income	Public sector housing generally characterized by segregation by income but not race. There are exceptions to this, such as Tower Hamlets in east London that has high concentration of Bangladeshis in public sector and assisted housing (see Chapter 11)
Low levels of government spending on public and assisted housing	Relatively high levels of spending on social housing to address problems of affordability; real spending on housing fell 1992–2003, but increased thereafter
Voucher programmes provide opportunity to move from public sector housing	No voucher programmes but Housing Benefit a significant voucher-like subsidy for low-income households to rent privately. In addition options for social housing tenants under the Right to Buy legislation and Home buy
Intensive programmes offered to public housing tenants on basic housekeeping, credit repair and counselling on home ownership	Programmes on home ownership and support fragmented and small scale

destitution that had marked pre-war Britain. The political emphasis was to design and build quality housing for low- and middle-income communities. As Malpass and Murie document, social housing was an important policy and political priority with private tenure relatively marginal (Malpass and Murie, 1999). More recently Malpass argues that housing was not incorporated into the post-war welfare state and as early as 1953 had moved back to a more residual role (Malpass, 2005). Despite the subsequent increase in home ownership, cutbacks of public expenditure, and policies of 'Right to Buy' and local authority stock transfer, social housing continues to account for 18.5% of all dwellings in England during 2005 (Wilcox, 2005). It could be argued that the welfare state has shown a remarkable capacity to withstand disinvestment and ideological attack.

In the US the starting point was very different. Cities did not suffer bomb damage. By 1950 the majority of households were homeowners and the impact of a booming economy, rising prosperity and access to finance led to a home ownership reaching 61.9% by 1960 (US Census Bureau, 2001). In these circumstances, public housing authorities in the US could not play a critical role in meeting housing supply or housing management. Indeed, public assisted and public housing in the US accounted for just 3% of all housing units in 2001 (Ditch et al., 2001). The social housing sector has remained small. Although the processes have been similar to some extent, the underdeveloped welfare state and small social housing sector means that convergence between housing systems is not likely to take place.

Political ideology has been an important driver for change in recent years but it is even more important in the origins of social housing in the UK and US. There has been significant convergence but the two housing systems are also very different because of social housing legacy in the UK (Figure 1.1). It could be argued that in both the US and UK public sector housing is declining. However, it was never a serious competitor to private sector housing in the US compared to the dominant role it once had in the UK. Moreover, it continues to occupy an important political position despite the impact of housing stock transfer and Right to Buy.

The common set of policy processes during the Thatcher/Reagan era, and more latterly by Blair/Clinton and Blair/Bush II, sharpened the importance of earned rights as well as responsibilities in housing. These policies, together with actions such as HOPE VI and DHS, reforms of housing benefits and a residualized public and assisted housing sector in the US and UK, have still resulted in differences in access to, and size of, public housing. Policy and ideological pressures have led to the withering of already weak and small public sector housing in the US, whilst in the UK social housing still continues to play an important role in meeting housing supply. The reasons for the outcomes are related to the starting positions of public housing in both countries.

STARTING POINTS

- Role of welfare state
- Housing as a right or entitlement
- Building up institutional capacity
- Housing volume and size
- Promotion of home ownership

↘

PROCESSES

- Thatcher/Reagan
- Public sector under attack
- Specific initiatives
- Blair/Clinton
- HOPE VI/Decent Homes
- Large-scale voluntary transfer (LSVT) quasi public sector

↘

OUTCOMES

- Residualized housing sector in US and UK
- Differential size of social housing in US and UK
- Protected housing sector in UK/welfare legacy
- Public sector 'withering' in the US

Time →

Figure 1.1 Developing housing processes and typologies in the US and UK.

Policy comparisons

In this section we will compare responses to a common set of problems that afflict cities and neighbourhoods in the US and UK. The issues of growing racial polarization, housing quality and reviving cities and neighbourhoods have taxed governments in both countries. Again differences exist because of the policy context and responses will vary.

Patterns of residence, segregation and discrimination

Minority communities account for almost 25 percent of the population in the US compared to fewer than 8 percent of the population in the UK. In both countries these groups are largely based in urban areas (US Census Bureau, 2001; UK Census, 2001).

The ten major cities of residence for minority groups in the US were 'majority minority' cities, ranging from Detroit, where the white population formed only 10.5 percent of the overall population, to Phoenix where they formed 45 percent of the population. Some cities such as New York, Chicago and Los Angeles had relatively diverse populations. Much of the population growth in US cities during the 1990s was the result of

Latino, Caribbean and, to a much lesser extent, eastern European migration. It should be noted that this minority growth occurred in the context of a general population decline in most cities. More specifically, rapid population growth in minority communities coincided with slower population growth in white communities. For example, Detroit lost a fifth of its population between 1980 and 2000, especially whites, which left African-Americans accounting for over 80 percent of the city population. In most of the cities there was a high degree of racial segregation (US Census Bureau, 2001).

The situation in the UK is different, with no city having a majority minority population. Twenty-nine percent of London's population is drawn from minority communities. The next largest cities of minority residence are Birmingham and Manchester where they comprise 29 and 20 percent of the population respectively. This overall suggestion of clustering in certain cities tends to mask an increasing segregation of minorities, with many residing in certain neighbourhoods. For example, the inner London boroughs of Newham and Brent have 'majority minority' populations (61 and 55 percent respectively), whilst 90 percent of the people living in the outer London boroughs of Havering, Bromley, Bexley and Richmond are white. Birmingham and Manchester show similar patterns of concentration of minority groups. Moreover, concerns about increasing segregation were (and continue to be) an important issue in the UK in towns and cities such as Bradford, Burnley and Oldham (with minority populations ranging from 11 to 22 percent; UK Census, 2001).

Minorities are generally spatially concentrated in large cities within industrial regions but that the overall size of the minority population and the scale of ethnic segregation are much greater in the US than in the UK. However, both countries are becoming increasingly ethnically diverse and fragmented. The growth of the Latino communities in the US and Asian and African communities in the UK are making previous discussion of race relations in terms of a 'black–white' dichotomy less relevant to the challenges facing both countries. The US is seeing rapid 'latinization' of cities, which will have fundamental consequences for public policy. Similarly, the growth of dual heritage communities and the arrival of new migrant groups, both as asylum seekers and refugees and migrant workers (e.g. from southern and eastern Europe in the UK), are posing challenges in terms of shifting cultures and identities.

In the US, African-American and Latino households have lower rates of home ownership, higher rates of poor housing and are more likely to live in overcrowded households than their white counterparts (Schill, 2004). These groups continue to be polarized by a concentration of poverty and race. For example, African-Americans accounted for over 50 percent of people in poor neighbourhoods between 1970 and 1990 (this was also the period when the *actual* number of people living in poor neighbourhoods doubled) (Massey and Denton, 1993). The flight of whites from cities such as Detroit and Baltimore and the spiral of economic decline led to hypersegregation (a term based on an index of segregation of sixteen indicators first coined by Massey and Denton, 1993), lower tax yields and problems with local governance. Discrimination continues to operate in the private sector, with minorities losing out through practices such as 'racial steering' in up to 25 percent of transactions in the housing market (Schill, 2004).

In the UK we can also see problems of racial polarization in cities, while resources have moved away from inner cities to suburbs and the wider region. 'White flight' from urban centres to the suburbs shows no signs of abating. This is despite the impact of urban renaissance and attempts to increase the attractiveness of city centre living, and leaves surrounding 'inner ring' areas unaffected in all but the most buoyant housing markets, which appeal to a relatively narrow band of young to middle-aged, professional people. These factors give rise to concentrations of poverty where mainly minority groups are trapped and disconnected from jobs, schools and decent housing, and communities fragment. Neighbourhood policies have done little to alleviate this cycle of decline which impacts disproportionately on minority communities.

Some believe that there is an alternative way to address racial and income polarization (see Katz, 2004). This is based on de-concentrating poverty and racial segregation by innovative lettings policies, demolishing concentrations of social housing and replacing them with mixed communities, connection of people to jobs and active intervention of the private sector to increase employee support programmes. The analytical lens must be at a regional and national level and the policy tools need to be driven by housing and economic mobility programmes. Though this type of diagnostic and solution has been forcefully proposed in the US it also finds an echo in public policy in the UK. The Home Office's report into the disturbances in the north of England in 2001 (Home Office, 2002) argued as much in its conclusions on both housing association and especially local authority letting policies.

Importantly Karn and Wolman stressed the levels of racial segregation as key variants in the housing experience in the UK and US. Extreme levels of racial segregation, especially between blacks and whites, marked housing (and particularly public housing) in the US during the 1970s and 1980s (Karn and Wolman, 1992).

Segregation levels in the US continue to scar public policy debates on race, income and housing. For example, the average white person lives in a neighbourhood that is almost exclusively white, whereas the rates for African-Americans, Latinos and Asians is much higher in terms of contact with people from other minority groups (Briggs, 2005). Recently both

Government and its agencies in the UK have become concerned with patterns of racial segregation in some towns and cities. This view holds that increased immigration, growing diversity and fragmentation of communities masks deepening concentration and separation of communities that have been picked up by Government reports (Home Office, 2002). In September 2005, Trevor Phillips, the Chairman of the Commission for Racial Equality (CRE), in a speech to the Manchester Council for Community Relations, alarmingly stated that Britain was 'sleepwalking to segregation' (CRE, 2005). Phillips stated that America, in the aftermath of Hurricane Katrina, had persistent levels of racial segregation after years of equal opportunities legislation. This was manifest in schools, housing, politics and higher education. He concluded that America 'is a segregated society, in which the one truth that is self-evident is that people cannot and never will be equal. That is why, for all of us that care about racial equality and integration, America is not our dream, but our nightmare' (CRE, 2005, p. 8). Turning to the UK, Phillips stated that there was growing evidence that residential and educational segregation are nearing levels seen in US cities such as Chicago and Miami.

It is more accurate to state that the UK is not 'sleepwalking into segregation'. The reverse appears to be the case. Using the Index of Segregation[2], Peach states that between 1991 and 2001 there was decreasing or stable levels of segregation in English cities. Compared to a US city, such as Chicago where African-Americans lived in neighbourhoods that were typically 78 percent black, only 22 percent of wards in the UK had a population that was more than 50 percent from black and minority ethnic groups (Peach, 2005). Additional research has found that the number of ethnically mixed wards increased during the same period from 864 to 1070 (Simpson, 2005).

As Phillips' remarks in 2005 demonstrate, segregation has become more important in policy and political discourse even though the evidence has not substantiated it. Perhaps it is wiser to talk about ethnic enclaves rather than 'ghettos'. In addition the picture is more complex because some groups, especially Indians and to a lesser extent African-Caribbeans in the UK, have access and reach to a wider choice of housing and neighbourhoods and are starting to disperse across cities (CURS, 2005). This is not the case for poorer minorities and new migrants. Pakistanis and Bangladeshis continue to suffer disproportionately from poor housing, high unemployment and low income (ODPM, 2003). Between 1991 and 2001 these groups demonstrated the highest levels of population growth which manifested

---

[2] Measures the percentage of a minority population, which would have to change its location to copy the rest of the population of a city. It ranges from 0 (indicating no segregation) to 100 (total segregation).

in deepening residential concentration as economically mobile whites and other groups fled the inner housing markets of cities. For example, recently completed research shows that the white population in some parts of Birmingham fell by over 30 percent during this period, as whites moved out and Pakistanis and Bangladeshis moved in, followed in the 1990s by new migrants from Africa, eastern Europe and the Middle East (Beider and Goodson, 2005).

Growing ethnic segregation in the UK is found in poor-quality private sector housing. It should be noted that some of the worst housing in the country is located in the private as opposed to the public sector. The private sector has some 2.2 million households (10 percent of the all the housing stock) with more than 43 percent of the housing built before 1919 (Mullins and Murie, 2006). So much concern has been expressed about the private sector that the government is introducing a mandatory licensing scheme for houses of multiple occupation (HMOs); that is, housing that has three or more storeys and five or more residents drawn from more than one household (ODPM, 2005).

Discrimination in local authority housing by race has been widely documented and demonstrated (Rex and Moore, 1967; Rex and Tomlinson, 1979; Henderson and Karn, 1987). Public sector housing was conceived as high specification and spacious dwellings for the 'deserving' low- and middle-income classes. Newly arrived migrants from the Caribbean and Indian subcontinent were neither seen as deserving nor indeed qualified in terms of length of residency for this type of housing. In addition, the residualization of local authority housing stock and parallel rise of home ownership meant that minorities, like most aspiring home dwellers, viewed public housing as being a last choice rather than something that they may do at some point in their housing careers. The stigma was not just in terms of the quality and location of stock. Low-income and residualized white estates became effectively 'no go' areas for minority communities. They were perceived as being areas where harassment and racist abuse were commonplace. Moreover, research has shown that in some places there was widespread verbal and physical harassment that was having devastating consequences for minority households (Chahal and Julienne, 1999; Beider and Goodson, 2005).

Minority housing preferences have traditionally been towards private tenure in neighbourhoods populated by established black and minority communities. Discrimination in public sector housing markets together with the need to be close to places of work and community networks combined to lead to patterns of residential segregation in some towns and cities across England. This area of housing careers and pathways is picked up in Chapters 2 and 5.

More work needs to be done to map the housing and neighbourhood trajectories of minority communities in the UK and US. Whilst it is generally the case that minority communities continue to live in economically disadvantaged and concentrated communities, the pattern is by no means consistent for all minority ethnic groups. Indians and some African-Caribbean communities in the UK and Asian-Americans (used in the US context to describe people descended from the Far East) and Latinos in the US are slowly moving out of inner-city housing markets for a new life in contiguous areas. This appears to be driven by increased educational achievements, higher incomes and labour mobility. However, it is not always associated with dispersal of minority communities into wider society; some higher-income minority communities will choose to live a partially segregated life. For example, Chinese communities in the US continue to regard established areas of settlement as providing different types of support including social, financial and human capital. Moreover, these enclaves provide an essential 'stopping point' for movement to more suburban locations (Zhou and Lin, 2005). There are other minority groups such as Pakistanis and Bangladeshis (UK) and African-Americans (US) who remain locked in static or declining housing markets and have consequently experienced very little spatial mobility. It could be argued that these communities experience the greatest levels of social disadvantage and educational under-achievement, reducing their choices regarding housing, employment and neighbourhoods of residence.

Within the UK, dispersal is particularly evident among Indian communities (the largest ethnic group in the UK), which show signs of movement from neighbourhoods in London and the west Midlands towards settlement in the south-east and south-west. Between 1991 and 2001 the growth nodes for Indian communities were towards the south-east and areas contiguous to major conurbations and London. More than 40 percent of local authorities showed an above average increase in the Indian population. Most were located in the line between Bristol and the Wash (i.e. the area of southern England commonly associated with being the economic engine of the British economy) (see Chapter 2).

Taking this further, the areas seeing the most dramatic growth rates in minority communities in the UK were rural and semi-rural local authorities, such as North Dorset, Malvern Hills, North Shropshire and Ryedale, all of which experienced increases of between 100 and 300 percent in their minority populations, apparently reflecting both the dispersal of established communities and official migrant dispersal programmes. However, all were growing from a very low base and, whilst these figures may not point to a significant outward migration of minority communities from traditional areas of settlement, a pattern of change is nevertheless discernible (see Chapter 2). The housing and race equality challenges in the future may be located in towns and cities of secondary migration that have had little previous experience of meeting the needs of minority communities.

Policy makers have picked up the changing nature of immigration and settlement (Spencer, 2005). 'New' communities are people who have migrated to the UK since the 1990s. These groups are drawn from a diverse range of countries. Some groups have come from places that have seen significant conflict, such as Somalia, Afghanistan, Iraq, Congo and the Balkans. Political reasons are often just as important as economic reasons for coming to the UK. Following the establishment of the National Asylum Support Service (NASS), the past 5 years have seen significant numbers of asylum seekers and new communities arriving into major cities but also dispersed to towns and regions that have had little experience of immigration. Other groups have come from the European Union seeking better wages and employment. There has been significant immigration from the new Accession countries[3]; workers have been recruited by employers again located in areas where there has been very little experience of supporting migration, such as East Anglia and the south-west of England (Spencer, 2005).

The lessons about minority sprawl are mixed in the US. Similar patterns to the UK are slowly emerging. Latino and Asian-Americans – the two fastest growing minority groups – are moving into diverse neighbourhoods located in the suburbs, as are some middle-income African-Americans. The story is by no means consistent across the nation. For example, the geographical 'sprawl' of people is moving much faster in sunbelt cities located in the west than in the rustbelt cities of the east. Racial segregation and polarization remain uncomfortable features for far too many Americans (Briggs, 2005).

Segregation only becomes a problem when communities have a lack of choice and realistic options in employment and key public policy areas such as housing and education. Indeed it functions to support essential day-to-day activities, such as social interaction, shopping for specific goods and services as well as providing access to places of worship. There is a need for researchers to consider the importance of social capital allied to housing and neighbourhood choice in so-called segregated communities. They are not necessarily 'problem' areas but neighbourhoods that help people and cities to function (Beider, 2005).

These trends of movement of minority communities in the US and UK will have profound consequences for our understanding of delivering

---

[3] The Accession countries joined the European Union on 1 May 2004. They are the Czech Republic, Cyprus, Estonia, Hungary, Latvia, Lithuania, Malta, Poland, Slovakia and Slovenia.

public services, racial tolerance and the move towards a political acceptance of the importance of diversity.

Home ownership for some?

Housing tenure also shows some similar trends across the US and UK with home ownership a common aspiration among minorities as well as the wider population. The home ownership rate in the US reached 69 percent in 2003. This is the highest it has ever been and confirms the perception that this is a nation of homeowners and aspiring homeowners (US Commerce Department, 2004). However, the high rates mask disparities by race. For whites, home ownership is the tenure of choice of 75.1 percent, but it is significantly lower for African-Americans, at 48.4 percent. The fastest growing segment is Latinos but, at 47.4 percent, they still have amongst the lowest rates of ownership of all ethnic groups in the US. Overall, however, African-Americans and Latino groups have recently shown a surge in owner occupation that has outpaced even whites. Between 1993 and 2003 their respective rates increased by 6 percent and 7 percent compared to 5 percent for whites (Schill, 2004).

What are the key drivers for increasing numbers of minority groups opting for home ownership? First we can point to rising prosperity amongst some minority families allowing access to better quality housing. This is reinforced by statistics drawn from the 2000 US Census showing dispersal of African-American and Latino communities. Second is the opening up of mainstream lending to minority groups. Increasingly banks, pushed by US Government legislation such as the 1977 Community Investment Act, regard minority groups as an untapped and lucrative market rather than a bad risk. Third, and linked to the second point, is the growth of what has been called 'risk-based underwriting' from some financial institutions to low-income communities. However, the rise of home ownership amongst minorities in the US should not hide the fact that racial concentration and poverty have been features of American life for several decades.

The position in the UK is not dissimilar. Nearly 72 percent of the population is homeowners. Between 1971 and 2002 there was a 20 percent increase in home ownership with most of the surge associated with changes in the 1980s. This followed the Right to Buy legislation introduced by the Conservative government, which enabled tenants to buy their council houses from the local authority at discounted rates. Correspondingly the number of council tenants declined from 34 percent in 1981 to 14 percent in 2002 (ODPM, 2003). Apart from making it easier for tenants to become homeowners, the decline can also be linked to stigma surrounding council housing, and its perceived associations with high rates of crime, drugs and poverty. The residualization of council housing stock made it appear to be the housing of last resort. Financially home ownership was more advantageous because it provided opportunities to increase capital as well as use equity to improve living standards.

This has not always been the case for black and minority ethnic communities in the UK. Many immigrants arriving from the new Commonwealth in the 1950s and 1960s experienced various forms of financial exclusion that restricted their pathways into home ownership (Henderson and Karn, 1987). Institutional racism in lending policies and practices within the formal financial sector during this period is often referred to as one of the main barriers preventing many immigrant households from breaking into home ownership in Britain. In response many of these 'new' communities made use of self-help strategies and social networks to counter unresponsive financial lending markets. For example within the African-Caribbean communities there was widespread use of rotating savings and credit associations (Roscas) or 'pardners'. These informal savings and lending schemes enabled financially excluded communities to pool resources to raise sufficient funds for a deposit to secure their first house purchase (Senior, 1991; Joseph, 2006).

Problems of quality and affordability

One of the key drivers of housing change in Britain is the 2010 Decent Homes Standard. Recognizing that a significant proportion of public sector housing requires investment, the government has stated that social landlords need to review, repair or demolish housing units to meet this new standard. It is estimated that there are 1.6 million unfit homes in the UK. This equates to one in fifteen homes in the country falling below the standard (Leather and Morrison, 1997). As noted earlier in this chapter most unfit housing is located in the private sector but the DHS is weaker here than in the public sector.

It has been argued that minority communities are the least likely to be satisfied with their housing choices as well as tending to be overrepresented in some of the worst housing conditions, both in the private sector and in social housing (Madood et al., 1997). Of the longstanding minority communities, African-Caribbean and Bangladeshi communities are the most concentrated in local authority housing and have therefore been disproportionately affected by the backlog of disrepair in the council housing sector. Patterns of housing disadvantage also affect new migrants who are now well represented on some low-demand local authority estates as a result of the government's dispersal policy. Poor housing in the private sector compounds the situation. A recent report found that one out of six dwellings visited, and housing asylum seekers, was unfit for human habitation (Garvie, 2001). Problems surrounding status and tenure together with language barriers and a lack of information may result in many new migrants being forced into poor-quality private sector housing.

In contrast to the UK situation, housing quality in the US has never been better (Schill, 2004). Only 2 percent of the occupied housing units have severe physical problems (American Housing Survey, 2001). This is perhaps not surprising as 91 percent of US housing stock was built after 1920, but, where there are severe physical problems, these have a racial dimension. African-Americans and Latinos experience rates twice the average for living in housing in physical distress (at 3.5 and 3.2 percent respectively).

In the US, those renting homes are paying increasing amounts of disposable income to ensure that there is a roof over their heads. In 2001, 53 percent of renters paid out more than 30 percent of their income on housing costs (and 23 percent spent over half of their income) and 10 percent of homeowners paid more than half of their incomes to service mortgage debt. The problems of housing affordability cut across income and ethnic groups. They are most acute in growth cities such as New York, Houston and San Francisco. Much of the new demand for housing emanates from new migrants, especially those coming from Central and South America during the 1990s – many drawn because of low incomes and poor prospects in their native countries. By becoming a replacement force for jobs that established communities simply do not want to do (such as those in the service industry, construction and landscaping), new migrants are changing the social and demographic map of the country.

There are similar problems of housing affordability and low-paid and low-skilled jobs in the UK. This is most acute in the large cities and conurbations of London, Birmingham and Manchester. Here too, new migrant communities are inadvertently helping to prop up some parts of the economy where work is often low skilled and low paid (Sassen, 1991). Research highlights the problem of refugee communities in securing decent housing, and the associated high levels of homelessness, for example in the Midlands (Mullins and Rowlands, 2005). In contrast to the US, there are no obvious examples of cities actively encouraging minority communities, especially refugees and asylum seekers, although some housing organizations do regard new minority communities as an important part of their overall housing business strategy (such as Whitefriars Housing Group in Coventry). Here refugees and asylum seekers are viewed as a new customer group sustaining the demand for social housing (Mullins et al., 2004).

New urbanism and urban renaissance

This book is about cities and neighbourhoods. Specifically it is about a much bigger policy agenda based on renewal of cities as places to live and work. Known as new urbanism in the US and urban renaissance in the

UK, the guiding principles are to prevent the sprawl of cities and anonymous housing developments. The emphasis is on high design standards and well planned and integrated neighbourhoods, communities and regions. Moreover, the rise of cities is predicated on mixed retail, tenure and ethnically diverse neighbourhoods with space for people to come together. Participation, renewal and regeneration are some of the key themes running through this movement. The policy context is similar in both countries. There is a movement to demolish stigmatized public sector housing, deconcentrate poverty and support the development of sustainable and thriving communities.

The new lens focusing on regions rather than neighbourhoods is a radical shift for policy practitioners and politicians on both sides of the Atlantic. It was not so long ago that housing, renewal and neighbourhoods were at the centre of discussion and debate. Urban policies, such as subsidized housing, community reinvestment, community development, empowerment zones (recent US initiatives), task force, city challenge, single regeneration budget and social exclusion policies (recent UK initiatives), were being promoted by the policy and academic community as mechanisms to revitalize communities. Commentators now suggest that these well intentioned relatively small-scale initiatives have done little to reduce poverty or polarization of race and class. Indeed, they may have been counterproductive by stigmatizing poverty and race still further (Orfield, 1997; Katz, 2004).

The Urban Task Force in the UK was established in 1998 to identify the causes of decline in cities and set about a vision for renewal based on social well-being and regeneration. According to Richard Rogers, the chair of the Urban Task Force, there have been some successes in the regeneration of cities and neighbourhoods. People have started to return to live as well as work in cities. This has been based on a combination of the economy expanding and tenacious leadership from local authorities. Rogers also states that this has happened in other cities across England: in 1990 there were 90 people living in central Manchester, today there are 25,000 residents. Over the same period the population of central Liverpool has increased fourfold (Rogers, 2005).

However the Urban Task Force is critical of the performance of the government. It had failed to address a number of problems that are highlighted in this book such as income and racial inequality. New and subsidized housing developments are frequently built outside cities, attracting mainly white and middle-income families. This increases income and ethnic polarization and leads to pressures on public services such as schools, housing and the environment (Rogers, 2005).

Demolition of housing stock in established inner-city neighbourhoods is seen as a problem rather than a solution (see also Chapter 12). In addition, the ability of larger housing organizations to engage with communities and provide services at a local level is questioned: 'The trend towards even bigger housing associations driven by development ambitions is diluting commitment to community involvement that was the hallmark of good management. Many smaller associations perform better on this front, but funding drifts to large scale organisations' (Rogers, 2005, p. 11). Again this is a concern in Chapters 8, 9 and 10.

New urbanism and urban renaissance point to the centrality of cities, diverse communities and renewal as a key motor for economic success and well-being in the US and UK. They share similarities in both diagnosis and also solutions.

Plan of this book

This is not a book that seeks to make direct comparisons in relation to housing, race and community engagement in the US and UK. Rather the focus is on understanding that cities, neighbourhoods and institutional responses will be different. These are dependent on local and national policy priorities that will change within different contexts. Contextual differences can mean that apparently similar problems, such as racial polarization, the stigmatization of public and assisted housing and the future role of not-for-profit housing organizations, take on a different meaning in different cities and regions. Starting points for housing systems in the US and UK vary and, whilst there is some apparent political alignment in terms of ideology and policy practice, significant differences continue to be evident in outcomes at all levels.

Our concern is not about whether the UK and US are converging or diverging in terms of approaches to housing, neighbourhood renewal and community engagement. On balance this seems a futile debate. The reality, as the chapters in this book illustrate, is that responses to housing and social problems in different cities and neighbourhoods vary depending on institutional capacity, local conditions and national policy drivers (Table 1.3). There are lessons to be learnt about the way that New York City has addressed problems of housing decline and abandonment that are probably more relevant to London rather than smaller American cities. Similarly, housing associations in England may be able to develop meaningful dialogue with community development corporations operating in the east and west coasts of the US. In addition, we need to note the ideological approaches that were driving both countries under Blair/Clinton and are continuing under Blair/Bush II. Specifically, whilst policy approaches appear to be converging the context for housing, renewal and community engagement remain markedly different, shaped by the legacy of housing policies adopted immediately after World War II. The starting points for public and assisted

Table 1.3 Towards a framework: policy similarities, local differences and change agents.

Themes	**UK**	**US**
Policy similarities	Housing supply, segregation and revival of cities	
Local differences and drivers	Primary role of national government on: • Housing supply • Identifying segregation and community cohesion as problems • Addressing housing quality	Primary role of local government on: • Increasing opportunities for homeownership • Public sector housing reform and poverty rather than race acknowledged • Housing quality not regarded as a problem
Change agents	Local authorities Housing associations representing nearly 20 percent of housing stock	Local authorities Not-for-profit organizations representing less than 5 percent of housing stock

housing are very different. Though both form a liberal welfare regime, it is the legacy of the welfare state in the UK, as much as the adherence to the free market in the US, that continues to shape housing responses.

Policy and demographic change

The first section of the book attempts to map race, housing and communities in the US and UK. Peter Lee and Ed Ferrari suggest in Chapter 2 that changing patterns of residence and ethnicity in the UK have coincided with housing market change and restructuring. These are long-term trends that have been addressed by short-term housing interventions. A much longer view is required that considers minority population growth as a driver to economic and social renewal in cities across the UK. In Chapter 3, Xavier de Souza Briggs also discusses the long-term trends within neighbourhoods and communities that have led to growing polarization by both race and class. This has devastating effects on the sustainability of communities and neighbourhoods as minority communities are squeezed out of opportunities to access employment, decent housing and successful schools. These opening contributions provide an empirical and policy framework on housing, race and residence.

Changing cities and neighbourhoods

The second section drops down a level to analyse change in cities and neighbourhoods. Jerilyn Perine and Michael Schill, in Chapter 4, provide an overview of how a public housing department can intervene in private housing markets and have an impact on transforming neighbourhoods in America's largest city. Assessing the Ten Year Housing Plan that took place between 1987 and 1997 they show how cities and housing authorities can learn from the way that New York rebuilt and rehabilitated housing stock, supported neighbourhood renewal and engaged with diverse organizations. Though this is a positive chapter, showing how a city regenerated itself using housing policy and practice, it also illustrates the difficulties of working across departmental disciplines and the problems of consulting with a wide group of stakeholders. In Chapter 5, Austin Barber and his colleagues discuss the impact of changing welfare regimes on housing markets and the impact on black and minority ethnic communities in Birmingham. Whilst acknowledging that new patterns of spatial and ethnic segregation have appeared in the city, Barber also points out that some minority communities, by being excluded from public sector housing that has rapidly deteriorated, have actually received capital gains from private sector housing that can increase mobility and options. Moreover, new high-value housing markets created in the centre of the city have attracted ethnically diverse consumers, confounding fears that gentrification will lead to increased polarization.

In Chapter 6, Susan Popkin reviews public housing transformation in the US drawing material from a number of research sites, but especially Chicago. Accepting that the levels of investment funnelled into distressed communities using programmes such as HOPE VI have helped low-income communities, Popkin also counters that the underplaying of race and desegregation as an explicit policy goal has led to housing interventions having mixed results. The bold call is for the US to commence new narratives not just on housing, neighbourhood and city renewal but also on race equality. In the concluding chapter in this section, David Robinson critically reviews the policy importance and application of community cohesion as a mechanism to encourage ethnically mixed and successful neighbourhoods and cities in the UK. Turning the argument on its head, Robinson argues that the problem is not so much of segregated (and implicitly) minority communities but addressing the key issue of increasing housing choice for low-income groups. In this way policies and practice will change to meet the real challenge of low income and housing quality, rather than the normative and politically constructed problem of segregation and community cohesion.

The chapters in this section demonstrate that local authorities can play an important and critical role in addressing deep-seated social problems. In many cases they are the only agencies that have the combination of strategic overview, capacity and skills to understand and act on housing and social issues. In this respect, New York should be regarded as the exception rather than the rule for US cities. It is by far the largest local housing authority in the country and has the longest legacy of public sector and assisted housing. Moreover, the global importance of New York as a financial centre helped to rescue the city from a housing and fiscal crisis after the federal government had washed its hands of the problem. New York probably has more in common with London and other big cities in the UK rather than its US peers.

Institutions and change agents

The third section drops down another level to assess the role of housing organizations in managing evolving communities and policy agendas. Chapter 8, by Lucille McEwen, is based on the not-for-profit organization that she leads in Harlem. It provides a close-up view of how micro-housing strategies are conceived with the objective of increasing housing supply and choice in a community that is a political and cultural symbol for African-Americans across the US and further afield. Here housing development and management is something that should not be left to the private sector. Community-based organizations need to respond and grasp opportunities that changing housing policies may provide. David Mullins, in Chapter 9, also looks at a not-for-profit housing organization and assesses how it has adapted to changing welfare regimes during its 100-year history. Connecting welfare regime theory, national housing policy changes and local influences enables Mullins to demonstrate how institutions such as housing associations can be seen as both adapting and helping to shape the regimes within which they operate. More than this, the case study organization has shown a remarkable penchant for spotting emerging opportunities to fuel substantial growth during specific periods while maintaining local embeddedness and close linkages with local communities and stakeholders.

Harris Beider compares community development corporations in the US with black and minority ethnic housing associations in the UK in Chapter 10. Both appear to follow similar trajectories, starting off as advocacy organizations rooted in community work. He shows how they have had to change to meet the demands of housing policy. This has challenged and changed organizational culture and opened up a gap between communities and organizations that has led sponsors in government and its agencies to question their role.The examples drawn from the work of community organizations in the areas of renewal, regeneration and housing demonstrate the capacity of individuals to intervene positively in local neighbourhoods. Of course, they operate within frameworks established by local and national governments. This being said it should not diminish the powerful and entrepreneurial impact of progressive local organizations and their role in local renewal.

Conflict, participation and power

The importance of communities and organizations continues as a theme in the final section. Specifically the concern is about community empowerment and engagement. In Chapter 11, Stephen Hall assesses the success of area-based regeneration initiatives using evidence from two symbolic projects in London and suggests that these attempts are about property renewal rather than people engagement. Hall criticizes the 'bootstrapism' of much regeneration work; that is, expecting local communities to transform neighbourhoods that have been in long-term economic decline. This neglects the importance of decline in regions and cities and the disconnectedness that exists between housing, regeneration and regional economic planning. Community consultation and engagement in these circumstances are likely to have a marginal impact. Chapter 12, by James DeFilippis, places community organization and development at the heart of the debate. Arguing against the excessive marketization of renewal and regeneration that washes out poor communities and makes it easier for gentrification to take root, DeFilippis puts forward a radical view of community investment and control that secures neighbourhoods and builds from the bottom up. Rob Rowlands and Pauline Card take the themes of engagement and power a step further in Chapter 13. Contextualized within policies that decentralize decision making on housing and community development, the concern is whether this really amounts to empowerment and influence of low-income communities or simply is a means for government and housing organizations to tick a box marked consultation and get on with the task of property renewal.

The story of this book and the individual chapters does not amount to a dry academic debate between divergence and convergence theories. The real task is to observe the lessons that these contributions provide and absorb knowledge from different contexts, locations and circumstances. To the academic it is important to consider whether housing typologies overlain with policy discussion provide a good basis for analysing different systems. For policy makers there are lessons to be learnt about engaging with communities, increasing choice and addressing the problems of segregation. Finally for the activist the experiences and research evidence can be deployed critically to review actions that have often been taken for granted. In this way the book seeks to reach out to a range of people in different places, cities and countries and connect them to ideas and views that may contribute to shaping public policy agendas.

References

American Housing Survey for the United States (2001) US *Bureau of the Census*. Washington DC.

Ball, M., Harloe, M. and Martens, M. (1988) *Housing and Social Change in Europe and the USA*. Routledge, London.

Barlow, J. and Duncan, S. (1994) *Success and Failure in Housing Provision: European Systems Compared*. Pergamon, Oxford.

Beider, H. (2005) *Housing, Social Capital and Integration*. Paper presented to ENHR Conference, Reykjavik, 29 June–3 July 2005.

Beider, H. and Goodson, L. (2005) *Black and Minority Ethnic Communities in the Eastern Corridor: Aspirations, Neighbourhood 'Choice' and Tenure*. Birmingham City Council, Birmingham.

Briggs, X. (ed.) (2005) *Metro Dilemma: Race, Housing Choice and Opportunity in America*. Brookings Institution, New York.

Centre for Urban and Regional Studies (CURS) (2005) *West Midlands Regional Housing Strategy 2005: Delivering a Housing Vision for the West Midlands in the 21st Century – Pathways of Choice*. West Midlands Regional Assembly and West Midlands Regional Housing Board, Birmingham.

Chahal, K. and Julienne, L. (1999) *Racist Victimisation in the UK*. Joseph Rowntree Foundation, York.

Commission for Racial Equality (CRE) (2005) *After 7/7: Sleepwalking into Segregation*. Speech given by Trevor Phillips at Manchester Council for Community Relations, 22 September 2005.

Ditch, J., Lewis, A. and Wilcox, S. (2001) *Social Housing, Tenure and Housing Allowance: An International Review*. DWP In-house report 83, Department for Work and Pensions, London.

Doling, J. (1997) *Comparative Housing Policy: Government and Housing in Advanced Industrialized Countries*. Macmillan, Basingstoke.

Esping-Andersen, G. (1990) *The Three Worlds of Welfare Capitalism*. Princeton University Press, Princeton.

Esping-Andersen, G. (1999) *Social Foundations of Postindustrial Economies*. Oxford University Press, Oxford.

Gamble, A. (1994) *The Free Economy and the Strong State: the Politics of Thatcherism*. Macmillan, Basingstoke.

Garvie, E. (2001) *Far from Home*. Shelter, London.

Hamnett, C. (1996) Social polarisation, economic restructuring and welfare state regimes. *Urban Studies* **33** (8): 1407–30.

Harloe, M. (1995) *The People's Home?: Social Rented Housing in Europe & America*. Blackwell, Oxford.

Henderson, J. and Karn, V. (1987) *Race, Class and State Housing: Inequality and the Allocation of Public Housing in Britain*. Gower, Aldershot.

Home Office (2002) *Community Cohesion: A report of the independent reviewing team*. HMSO, London.

Joseph, R. (2006 forthcoming) *Housing Wealth and Accumulation: African Caribbean Experiences of Afro-Caribbean Households of Home Ownership in Birmingham and London in the Period 1950–1979*. Unpublished PhD Thesis: Centre for Urban & Regional Studies, University of Birmingham.

Karn, V. and Wolman, H. (1992) *Comparing Housing Systems: Housing Performance and Housing Policy in the United States and Britain*. Clarendon Press, Oxford.

Katz, B. (2004) *Neighbourhoods of Choice and Connection: The Evolution of American Neighbourhood Policy and What It Means for the United Kingdom*. Brookings Institution, Washington DC.

Kemeny, J. and Lowe, S. (1998) Schools of comparative housing research. From convergence to divergence. *Housing Studies* **13** (2): 161–76.

Leather, P. and Morrison, T. (1997) *The State of UK Housing: a Factfile on Dwelling Conditions*. Policy Press, Bristol.

Macpherson, W. (1999) *The Stephen Lawrence Inquiry: Report of an Inquiry by Sir William Macpherson of Cluny*, Cm 4262-1. London, HMSO.

Madood, T.T., Berthoud, R., Lakey, J., Nazroo, J., Smith, P., Virdee, S. and Beishon, S. (1997) *Ethnic Minorities in Britain: Diversity and Disadvantage*. Policy Studies Institute, London.

Malpass, P. (2005) *Housing and the Welfare State. The Development of Housing Policy in Britain*. Macmillan, Palgrave.

Malpass, P. and Murie, A. (1999) *Housing Policy and Practice*. 5th edition. Palgrave, Basingstoke.

Massey, D. and Denton, N. (1993) *American Apartheid: Segregation and the Making of the Underclass*. Harvard University Press, Cambridge, Mass.

Matznetter, W. (2001) *Housing Classes and Welfare Regimes. Making Sense of Welfare Regimes in Different Times and Places*. Paper presented at Housing Studies Association Conference, Cardiff.

Mullins, D. and Murie, A. (2006) *Housing Policy in the UK*. Palgrave Macmillan, Basingstoke.

Mullins, D. and Rowlands, R. (2005) *West Midlands Regional Housing Strategy*. WMRA, Birmingham.

Mullins, D., Beider, H. and Rowlands, R. (2004) *Empowering Communities, Improving Housing: Involving Black and Minority Ethnic Tenants and Communities*. ODPM, London.

Office of the Deputy Prime Minister (ODPM) (2003) *Housing and Black and Minority Ethnic Communities. Review of Evidence Base*. HMSO, London.

Office of the Deputy Prime Minister (ODPM) (2005) *Houses in Multiple Occupation (HMO): Licensing Scope and Schemes*. ODPM, London.

Orfield, M. (1997) *Metropolitics: A Regional Agenda for Community and Stability*. Brookings Institution, Washington DC.

Peach, C. (2005) Britain has enclaves, not ghettos. *The Guardian* 28 September 2005.

Power, A. (1993) *Hovels to High-Rise: State Housing in Europe since 1850*. Routledge, London.

Rex, J. and Moore, R. (1967) *Race, Community and Conflict. A Study of Sparkbrook*. Oxford University Press, London.

Rex, J. and Tomlinson, S. (1979) *Colonial Immigrants in a British City: a Class Analysis*. Routledge & Kegan Paul, London.

Rogers, R. (2005) *Towards a Strong Urban Renaissance. An Independent Report by Members of the Urban Task Force*. Urban Task Force, London.

Sassen, S. (1991) *The Global City: New York, London, Tokyo*. Princeton University Press, Princeton.

Schill, M. (2004) *Housing, Neighbourhoods and Communities: An American View*. Paper presented at Housing, Neighbourhoods and Communities in the US and UK Symposium, University of Birmingham, England, 29/30 June 2004.

Senior, O. (1991) *Working Miracles: Women's Lives in the English Speaking Caribbean*. James Currey, London.

Simpson, L. (2005) *Measuring Residential Segregation*. Paper presented at Census: Present and Future Conference at University of Leicester, 16/17 November 2005.

Spencer, S. (2005) *Integration of New Migrants: 'What Works?'* Unpublished report.

UK Census (2001) *Census 2001*. HMSO, London.

US Census Bureau (2001) US *Bureau of the Census*. Washington DC.

US Commerce Department (2004) US *Bureau of the Census*. Washington DC.

Van Kempen, R., Murie, A., Knorr-Siedow, T. and Tosics, I. (eds) (2006) *Regenerating Large Housing Estates in Europe. A Guide to Better Practice*. Utrecht University, Utrecht.

Wilcox, S. (2005) *UK Housing Review*. Chartered Institute of Housing and Council of Mortgage Lenders, Coventry.

Wolman, H. (1975) *Housing and Housing Policy in the US and the UK*. D.C. Heath, London.

Zhou, M. and Lin, M. (2005) Community transformation and the formation of ethnic capital: immigrant Chinese communities in the United States. *Journal of Chinese Overseas* **1** (2): 260–84.

Policy and Demographic Change

2

Changing Spatial, Ethnic and Tenure Patterns in England

Ed Ferrari and Peter Lee

Introduction

In this chapter, some of the embedded, long-term and structural patterns of change that have altered the shape of housing, neighbourhoods and communities are examined. The particular viewpoint taken is that of the housing market. The emergence in the UK of severe disparities in housing market (and neighbourhood) outcomes cannot, we argue, solely be seen in the context of sudden change caused by market shocks or other short-term stimuli. Rather, these outcomes have been the result of continuing long-term shifts in population and ownership structures around the country. In evidencing this argument, we focus principally on the changing spatial expression of ethnicity and housing tenure in England.

Arguably the main strategic concern among housing professionals in the UK recently has concerned the management of change. A sudden and severe discovery of widespread housing market weakness in the late 1990s led to the characterization of much of the post-industrial heartlands of the north and Midlands of England, the central belt of Scotland and south Wales as evidencing 'low demand' for housing (Lowe et al., 1998). After the initial shock the debate calmed and a more analytical interpretation of the phenomenon followed. Most commentators observed that the needs and aspirations of communities were changing and that the problem needed to be understood from a number of spatial perspectives: neighbourhood, city, sub-regional and regional housing markets (Murie et al., 1998). But, crucially, one of the aspects of change that appeared to mark out the 'low demand' phenomenon as distinct from previous housing and neighbourhood problems was that the risk of low demand affected both the public and private sector (Bramley and Pawson, 2002).

While, in the most extreme cases of low demand, neighbourhood change can occur rapidly – as expressed by abandoned and boarded-up properties – the underlying processes affecting neighbourhoods occur over a longer period of time. Patterns of economic development since World War II have resulted in chronic distortion of demand for housing between the regions of the north and the south. This has resulted in housing markets in England increasingly dichotomizing between the declining industrialized north and Midlands regions (generally characterized as low demand) and the over-heating in the advanced tertiary economies of the south-east and London (high demand and high house prices). However, low demand is not simply driven by differential speeds of economic development. De-industrialization, coupled with high levels of ethnic segregation in the north and Midlands, have resulted in a pattern of housing consumption that is highly skewed. This is manifested by high rates of overcrowding and high concentrations of black and minority ethnic communities in the inner cities. There is, conversely, an under-utilization of urban space in low-density suburbs and peripheral social housing estates, characterized mainly by white households. In the case of the peripheral estates, severe problems of social exclusion exist, which, with similarly deprived communities in the inner cities, bookend a comparatively affluent suburban 'middle class'.

These patterns of spatial residence and segregation reflect the historical immigration of a minority ethnic population to the industrial inner cities and the concurrent decentralization of mainly white households to the growing suburbs. The housing focus of the post-1945 welfare state was in providing good quality municipal housing with gardens in the suburbs, whilst immigrants settled in inner city private sector lodging houses and low-value owner-occupied housing (Rex and Moore, 1967; Lee and Murie, 1997). This had at least three implications for English cities. First, it established a pattern of residence in the industrial towns and cities that has largely endured to the present day. Second, it *de facto* assigned private housing tenure status to a cohort of immigrants who were not enfranchized by the growth of welfare-subsidized social housing. Third, as the inner city declined physically and environmentally, the problems of inner city deprivation, social unrest and the stigma of inner city failure became associated with high concentrations of minority ethnic communities. The speed and direction of changing demand for housing, identified in the mid-1990s, therefore, obscured the failure of the policy environment to recognize the underlying drivers of change and react appropriately.

This chapter is divided into two parts. We start the chapter by introducing the concept of changing demand and the policy responses. The primary purpose of this chapter, however, is to consider the longer-term trends affecting changing demand for housing. This is considered in the second part from the perspective of three aspects of urban geography:

- Housing choice (in this case, exemplified by housing tenure)
- Patterns of ethnic segregation
- The resulting underlying patterns of deprivation

Together these trends affect housing market trends, shape the 'path dependency' of households' journeys through urban space, and influence the composition of cities and neighbourhoods in England. The main empirical evidence that is presented is from a number of related studies of the UK's population census in 1981, 1991 and 2001. The focus is England because of the devolved nature of housing strategy and urban policy in the UK[1]. We conclude the chapter by suggesting that whilst government housing market renewal initiatives are welcome, there is a danger that short-term cyclical alterations in housing markets may obscure long-term trends and that the path dependency of urban space presents a more significant obstacle to the longer-term delivery of sustainable communities in the UK.

Changing demand for housing

During the late 1990s a major shift in housing demand throughout the Midlands and the north of England was identified (Murie et al., 1998; Bramley and Pawson, 2002; Lee and Nevin, 2003). The principal manifestation of this change in demand was a decline in demand for social housing measured by numbers on waiting lists, increased turnover of rented housing and the surfacing of significant numbers of abandoned properties in the north and Midlands. Bramley and Pawson (2002) and Murie et al. (1998) identified a number of micro-level (neighbourhood) and macro-level (regional and national) impacts that have an influence on whether an area will experience low or changing demand. These are summarized below.

[1] Similar processes of housing market and neighbourhood change have occurred in Scotland, Wales and Northern Ireland but the policy responses have been separate or different to those in England.

Micro-level factors:

- Reputation of the area
- The quality and condition of the housing stock
- Unpopular dwelling types: monolithic/systems built, maisonettes; smaller terraced housing stock with no parking facilities
- Anti-social behaviour: lettings policies and impact of Anti-Social Behaviour Orders

Macro-level factors:

- Residualization of council housing
- Aspirations and changing markets
- Age profile in parts of market is becoming skewed
- Portability of housing benefits
- Planning and estimates of population growth
- Economic growth
- Interest rates

The debate on low and changing demand was important because it broadened the analysis of housing's role in delivering wider urban policy goals. Conservative policies during the 1980s had viewed inner cities as a problem, and the housing policy response had followed a pattern of spatially targeting policies to improve the physical infrastructure of housing 'estates' (social housing projects). While approaches to delivery differed, such investment was typified in programmes such as *City Challenge*, *Estate Action* and the early New Labour *Estates Renewal Challenge Fund* (ERCF). While these programmes may have had some local success, dependent on partnering and governance arrangements (e.g. Davoudi and Healey, 1995), the fundamental approach was ineffective in tackling the root cause of inner-city decline. It was unable to counter the structural processes of land use, change and migration that have been defining features of the UK urban–rural relationship throughout the last century. That the outward migration of more affluent households from British cities has been matched by increasing levels of deprivation within the urban core has been an established characteristic of post-war urban geography (Power and Mumford, 1999; Champion, 2002). The debate therefore brought into sharper focus the dynamic between housing demand and aspirations, processes of migration impacting on community development and housing's interaction with the economy. The changing demand debate implicitly called for a re-evaluation of the role of older industrial cities, requiring an understanding of the potential function of housing as a mechanism for reversing migration trends.

Alongside the growing awareness of changing demand within housing markets, the 1997 New Labour Government created the Social Exclusion Unit (SEU) which commissioned a wide-ranging enquiry into the causes of social exclusion. This had an explicitly spatial focus and broadened the perspective on housing's role in reproducing patterns of deprivation. The

SEU's 'Policy Action Team' report on *Unpopular Housing* (DETR, 1999) acknowledged that housing supply side factors contributed to social exclusion through increased turnover and abandonment, whilst Lord Rogers' final report of the Urban Task Force (again commissioned by the New Labour government) called for an 'urban renaissance' in British towns and cities (UTF, 1999). These strands contributed significantly to the debate by implicitly recognizing the need for an underpinning strategy to retain affluent households, including those who aspire to move away, in order to reverse historical migration trends.

The academic and policy community was, therefore, articulating a set of conditions impacting on housing which was complex but which operated across a range of spatial scales, cut across housing tenures and was dependent on changing consumption patterns and household aspirations. But these contributions did not fully reflect the asymmetry between housing investment and the economy. Housing is unique in that investment (in 'bricks and mortar') is durable and its relationship to the (more fluid) economy changes over time. It does so in such a way that the shape of cities emerges in response to the varying synchronicities and disconnections of a number of economic phases and periods of housing development. Whilst housing investment may initially synchronize with the economy, over time it may impact negatively on subsequent consumption patterns and patterns of investment. In this context we identify four phases of development which illustrate the function of housing and the economy:

- **Production of primary manufacturing and raw materials.** During a period in which foundries and the production of iron, steel and heavy metal-based manufacturing developed rapidly, working class housing developed in these places to bring workers to the workplace.
- **Increased productivity and personal mobility.** In the second phase, increased productivity brought about by greater skill levels and new production processes led to increased differentiation in the workforce, greater personal mobility and an expansion of housing choice.
- **The maturation of housing markets.** In this phase, changes in the role of the economy to more service-led spheres results in a separation of housing and economic functions. The location of services is not strictly determined by the location of production of housing, therefore older housing no longer fits with its original economic function. Well connected urban areas provide an enormous range of residential choice for middle- and higher-income earners. Dual-income households and the polarization of household incomes sharpen these increasing distinctions between the location of housing and the functioning of the economy.
- **Housing as an investment.** Finally, the increased affluence and the aspiration for home ownership has led to a situation in which the life plans of

many households include a significant financial investment in housing. This may have become more pronounced in the UK as the uncertainties concerned with private pension provision, the state pension and the stock market mean that investment in housing is seen as one of the most secure options for a household. Households seek to maximize their investment in housing especially at certain stages in their career and they are concerned to invest in housing that they will see appreciate in value.

This phasing of development highlights the tension between flexible economic systems, the inflexibility of fixed housing assets and the inability of fixed assets to respond to the changing culture, lifestyles and aspirations of the new service classes. The implication of these patterns of housing investment for low demand in the UK were alluded to by Lee and Murie (2004):

- Changes in the function of the economy led to a loss of employment with long-term impacts on housing market investment.
- Over time an over-supply of lower priced housing – built to serve a previous economic mode – became noticeable.
- Policy interventions tended to deal in a piecemeal way with changes in markets by temporarily 'reinventing' the function of neighbourhoods through the intervention of social landlords and the private rented sector (e.g. through the housing of students) in previously owner-occupied housing areas.
- The industrial legacy meant that the impact of former industrial areas had a physical impact and a dampening effect on sales prices in parts of the housing stock.

The policy implications of this signalled a need for significant improvements in the housing choices available to households in urban areas in order to reverse migration trends and deliver urban renaissance. Furthermore, an emphasis on 'housing pathways' (Payne and Payne, 1977) re-emerged, which was a concern that the choice of housing available in an area was sufficiently wide (and would be in the future) that it could respond to demands from different household types at different stages in their housing career. The response to changing demand by central government was the launch of the Sustainable Communities Plan (ODPM, 2003a). This set out a long-term programme for delivering sustainable communities and urban renaissance by tackling housing supply and affordability issues in the south-east of England and low demand in the north and Midlands.

Tackling low demand was the remit of nine Housing Market Renewal Pathfinders newly created in the north and Midlands. Housing market renewal represented a new approach to regeneration in the UK[2]. The Pathfinders cut across local authority boundaries and demanded new governance arrangements and approaches to tackling demand issues across all housing tenures. As opposed to more tightly defined regeneration programmes of the past, such as the government's New Deal for Communities (each approximately 4000 households), the target area of the Pathfinders covered a much wider area, averaging just over 87,000 households[3]. This meant that the aims of housing market renewal would have a more strategic impact at the sub-regional and regional level. But, the resources committed to tackling low demand (£500 million) (ODPM, 2003b) were dwarfed by the commitment to delivering affordable and key worker housing in the south-east (£6 billion) (ODPM, 2003a). Despite this, HMR Pathfinders moved swiftly to remove some of the worst manifestations of low demand, remedying this through clearance and demolition of unpopular stock.

Three years on from its launch, recent housing market trends (specifically, a rapid increase in house prices across England and also spreading to 'low demand' areas) have raised a question over the long-term prospects of the market renewal programme. Continued house price growth has posed a threat to the longer-term strategic ambitions of housing market renewal, including the synchronization of housing choice and 'aspirations' reflected by a changing economy. This is a clear example of the fourth phase of synchronicity between the housing market and the economy. The increased emphasis on housing as an investment good has inexorably led to the disconnection of local housing markets from local populations. This is evidenced through a house price to gross annual earnings ratio that, at around 4.75, is at an historic high (Figure 2.1). Housing is increasingly being bought and sold for a number of secondary uses (such as buy-to-let, second homes or as part of a land portfolio). These uses do not necessitate occupation or any other connection to the locality on the part of the owner. Reform of pension provision in the UK is likely to galvanize this trend. Under new rules, Self-Invested Pension Plans (SIPPs) offer significant tax

---

[2] The Housing Market Renewal Fund (HMRF) is a grant available in selected local authority areas in the north and Midlands of England for the purposes of large-scale housing market restructuring. The delivery of projects is managed by nine 'Pathfinder' organizations that have been specially established as strategic partnerships between local government, social housing providers, and the private housing sector. See Lee and Nevin (2003) for the policy context leading to the establishment of the HMRF, and Cole and Nevin (2004) for a review of the Pathfinders' first 2 years of activity.

[3] The smallest Pathfinder is 'Urban Living' (in Birmingham and Sandwell) with 57,160 households. The largest is 'Transform South Yorkshire' (Sheffield, Rotherham, Barnsley and Doncaster) with 125,132 households. Data are calculated from ODPM (2004) and are based on the 2001 Census of population.

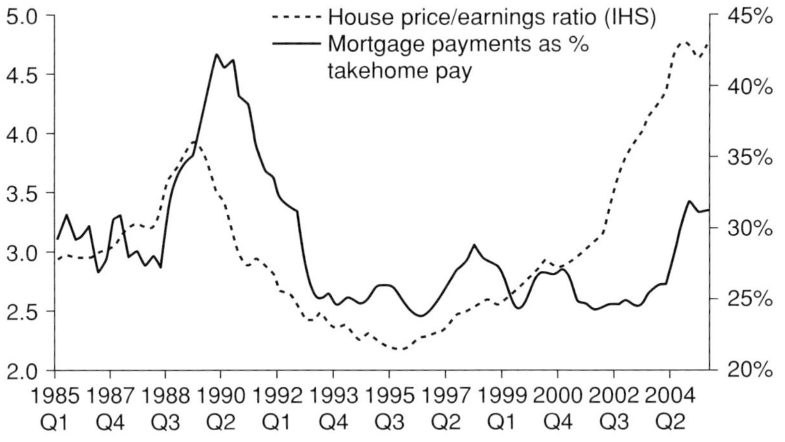

Figure 2.1 Average earnings and mortgage payments.
Source: Nationwide Building Society (2005).

breaks to those investing in property (and other investment commodities). Early criticism of these reforms has focused on the suspicion that they will disproportionately benefit the wealthy, and that they will lead to a surfeit of buy-to-let properties and holiday homes, effectively pricing lower income groups out of their local housing market.

However, while housing investment activity is undoubtedly growing, its popularity will be dependent on its performance as an investment compared with other commodities and markets. The cyclical nature of the housing market, together with economic cycles more generally, would suggest that housing's current relationship to the wider economy will not remain stable and that it is insufficient in totally explaining longer-term urban processes.

Socio-economic trends affecting sustainable communities

In the second part of this chapter we consider longer-term trends in spatial patterns of segregation, deprivation and the consumption of housing affecting housing markets in England.

Whilst changing demand has affected both the private and social sector, the concept of housing market renewal has coalesced around the idea that a transformational improvement in the 'urban offer' (including, but not only, housing) is required in low-demand areas. (This is reflected in some of the names that Pathfinder organizations have adopted: 'Transform', 'Renew', 'Urban Living', 'Elevate', etc.) Such improvements are seen as necessary in order to retain 'aspirant' households and prevent a further loss of population from the urban cores. The urban offer encapsulates the full range of facilities and services (including housing, transport, retail, crime reduction measures and neighbourhood management) that make up the urban experience. Improvement in the urban offer emphasizes the importance of regeneration of the physical housing stock, but is also key in the delivery of effective housing pathways to dissuade potential outward migrants.

An important component of an appropriate choice of urban housing includes provision for home ownership, and this has featured strongly in plans for housing market renewal. This is not simply a matter of increasing the volume of home ownership but also delivering the appropriate size, design and layout of housing for sale. One of the main objectives of housing market renewal, therefore, is to use public subsidy to create the right conditions for the private sector to build the type of houses that people want, whilst providing them in the tenure that they aspire to. Alongside this, a net reduction in the number of social housing units to re-balance the market is the objective in most low-demand areas. This approach parallels the New Labour government's view of home ownership as an aspiration for all and an instrument for increasing social cohesiveness, tackling social exclusion and enabling personal wealth accumulation. In the following section we consider the growth of home ownership and other trends in tenure change over the past 20 years. This will allow some assessment of the potential of further growth in home ownership.

Tenure change

Tenure change has been the explicit target of a number of successive housing policies, including the 'Right to Buy' (Jones and Murie, 2005). Where the objective has been to promote home ownership, the underlying aim is to increase individual wealth and reduce the burden on the welfare state through the use of housing and capital assets bound up in housing. This is consistent with a now dominant ideological belief in the 'property owning democracy' and the perceived primacy of ownership as a 'natural tenure' (Saunders, 1990).

The rise of the housing association sector

In line with changes in policy and ideology towards home ownership, the way that social housing has been provided has also changed. The growth of the housing association sector[4] has supplanted direct provision of social

---

[4] Housing associations are the main providers of social housing among a group of organizations legally referred to as 'registered social landlords' (RSLs). Other types of RSL include charitable bodies, tenants' management organizations and other not-profit landlords such as 'arm's length management organizations' (ALMOs), which are companies controlled by local authorities but which manage social housing on their behalf. The Housing Corporation, an executive agency of the Office of the Deputy Prime Minister, regulates all RSLs in England.

housing by local authorities. Whilst local authorities remain the largest social housing landlords, based on considerable historic stock holdings, practically no new 'council housing' is being built. Sales of such housing to sitting tenants, both through local authorities' discretionary powers and the statutory 'Right to Buy' introduced in the 1980 Housing Act, has skewed the profile of the remaining local authority stock. Local authority housing was effectively 'residualized' because untraditional, difficult to maintain, or poor condition properties and those in unpopular areas were understandably eschewed by purchasers in favour of larger, well built properties, such as those in inter-war garden suburb estates (Forrest and Murie, 1998). This raised considerable challenges for local and state funding of council housing: the residualized stock was more expensive to maintain, but the formula used for the disbursement of central grant aid to local authorities was not sensitive to this. Further retrenchment of the role of local authorities in providing social housing has continued to occur. Investment in existing council stock has been increasingly conditional on its transfer out of council ownership to housing associations or other not-for-profit organizations.

These dramatic changes in tenure have occurred within the context of a continued rise in the overall number of households. According to the census, in 2001 there were approximately 20.5 million households in England, an increase of just over 6 percent since 1991. An aging population and delays in household formation have meant that the growth in single-person households accounts for the absolute growth in the number of households. In 2001 there were 6.2 million single-person households in England, an increase of 1.2 million (24 percent) since 1991. Table 2.1 sets

Table 2.1 Number of households and market share by tenure in England, 1981, 1991 and 2001.

Tenure	**1981**		**1991**		**2001**	
	Number	**Market share[a] (%)**	**Number**	**Market share[a] (%)**	**Number**	**Market share[a] (%)**
Owner occupation[b]	9,629,944	57.8	12,626,345	67.6	14,054,122	71.0
Private rented	1,853,880	11.1	1,743,283	9.3	1,798,864	9.1
Local authority	4,811,177	28.9	3,712,925	19.9	2,702,482	13.7
Registered social landlord[c]	352,226	2.1	596,917	3.2	1,238,246	6.3
All tenures	**16,647,227**	**100.0**	**18,679,470**	**100.0**	**19,793,714**	**100.0**

[a] Percentages may not sum exactly to 100 due to rounding.
[b] Includes shared ownership in 2001.
[c] Includes housing associations: see footnote 4 on page 39.
Source: Census 1981, 1991, 2001.

Table 2.2 Change in number of households and market share by tenure in England, 1981–2001.

Tenure	1981–2001		1991–2001		1981–1991	
	Absolute change	**Increase in share**[a] (%)	**Absolute change**	**Increase in share**[a] (%)	**Absolute change**	**Increase in share**[a] (%)
Owner occupation[b]	4,424,178	22.7	1,427,777	5.0	2,996,401	16.9
Private rented	–55,016	–18.4	–55,581	–2.6	–110,597	–16.2
Local authority	–2,108,695	–52.8	–1,010,443	–31.3	–1,098,252	–31.2
Registered social landlord[c]	886,020	195.7	641,329	95.8	244,691	51.0

[a] Percentages may not sum exactly to 100 due to rounding.
[b] Includes shared ownership in 2001.
[c] Includes housing associations: see footnote 4 on page 39.
Source: Census 1981, 1991, 2001.

out the number of households and the market share of the main tenures for the three census years of 1981, 1991 and 2001, while Table 2.2 shows this in terms of absolute and relative change over the same period.

As we have already noted, there has been an increasing expectation by the UK government that housing associations are the main providers of social housing. Increases in housing associations' stock portfolios through 'large scale voluntary transfers'[5] and other mechanisms have contributed to significant and increasing growth of the sector since the early 1990s (Mullins and Pawson, 2005). Some 1.2 million households in England now rent from a housing association, representing around 6.3 percent of all households. These figures do not include shared ownership arrangements where housing associations provide management services or an equity stake, so the level of housing association involvement in tenure arrangements is likely to be a little higher[6].

The number of households renting from a housing association has increased almost fourfold in the 20 years to 2001. Households renting from housing associations increased by some 51 percent in the period 1981–1991 and by nearly 96 percent between 1991 and 2001. At the same time, the parallel retrenchment of the local authority sector can also be seen. Council housing's market share has more than halved since 1981. In 2001, just over 2.7 million households rented from their local authority.

[5] 'Large scale voluntary transfers' (LSVTs) are a mechanism by which entire estates or stockholdings of local authority housing are transferred to the ownership of registered social landlords (RSLs).
[6] In 2001, nearly 134,000 households had a shared ownership arrangement in England (2001 Census).

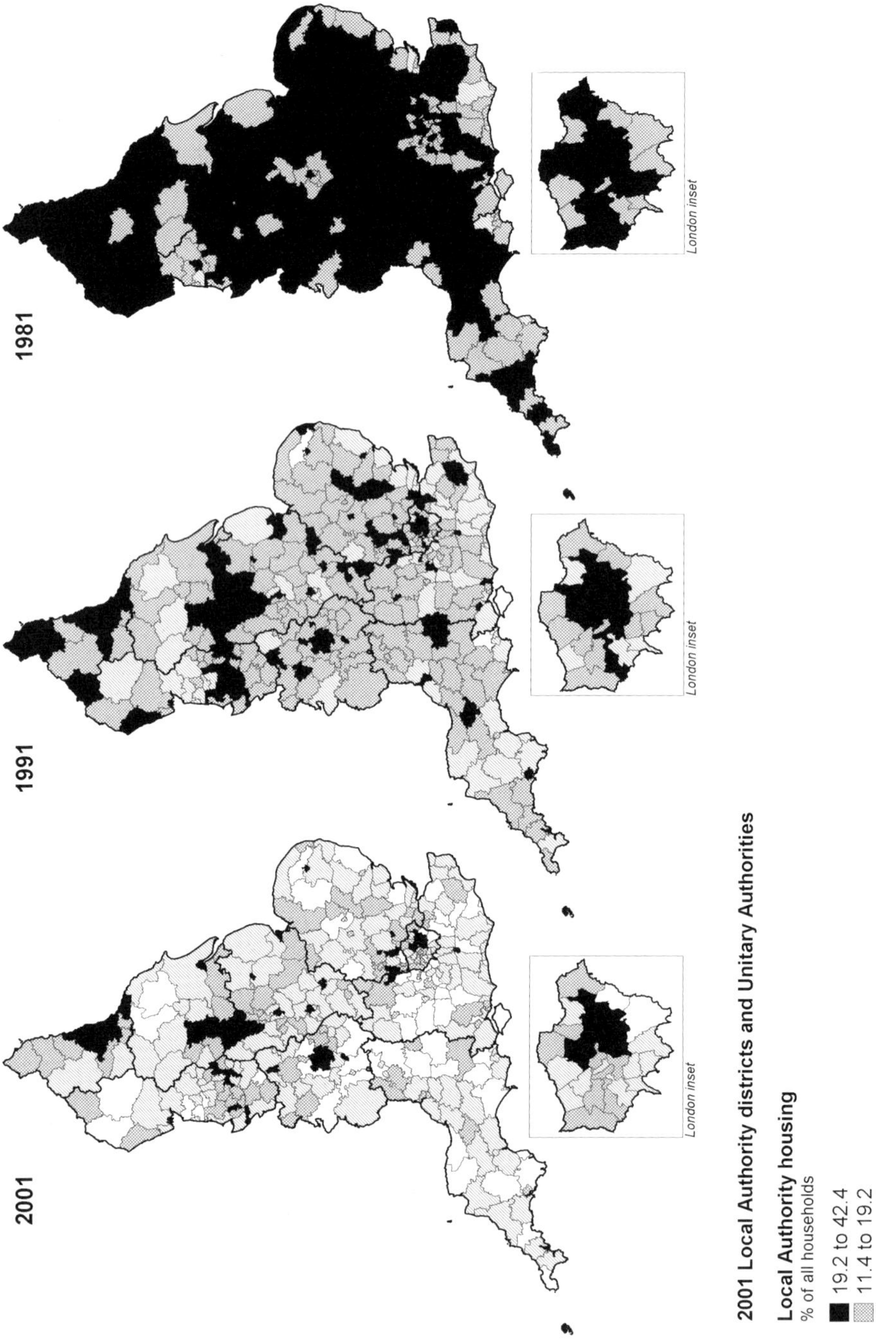

Source: 2001 Census, 2001 Census Output Area Boundaries, 1991 Census Small Area Statistics, 1981 Census Small Area Statistics.
© Crown copyright 2003. Crown copyright material is reproduced with the permission of the Controller of HMSO

Figure 2.2 Maps showing concentrations of local authority housing in England, 1981, 1991 and 2001.

In all, the local authority sector in England averaged a loss of over 100,000 households per year between 1981 and 2001. From the position in 1981 where 4.8 million households rented from their local council (a market share of about 29 percent), local authorities are now landlord to fewer than 14 percent of households. This equates to a 53 percent fall in market share.

Figure 2.2 shows that only a handful of (predominantly metropolitan) districts now have a local authority housing sector that exceeds around 19 percent. In 1981, the vast majority of districts in England had local authority sectors with over 19 percent market share. With the exception of London's East End, the remaining areas of significant local authority housing are in regions of England that have experienced low and changing demand, including the north-east of England (which has traditionally had a very high level of municipal housing but has also been subject to very high increases in levels of owner-occupation), South and West Yorkshire, Greater Manchester and the West Midlands conurbation.

Owner-occupation

More than ever, England is a nation of homeowners. The owner-occupied sector has continued to grow, with over 4.4 million more households owning their own home than in 1981. Over the same period, owner-occupation as a tenure increased its market share by around 23 percent, although the most spectacular rises were during the 1980s. Between 1981 and 1991, there were nearly 3 million new homeowners, an increase in share of about 17 percent. The 'Right to Buy' provisions of the 1980 Act may have accounted for as many as 43 percent of these new homeowners. A recent study for the Chartered Institute of Housing found that there were nearly 1.3 million sales of local authority and housing association homes to tenants in the years 1981–1990 inclusive (Murie and Ferrari, 2003). In net terms, nearly one million of the three million new homeowners have been gained from other tenures, with the balance – just over two million – attributable to growth in the number of households.

These trends have resulted in owner-occupation representing more than 70 percent of households in England. This equates to nearly 14 million homeowners either owning their home with a mortgage, owning it outright or, in a very small number of cases, having a shared ownership arrangement with a housing association. Alongside the housing association sector, home ownership is the only other main tenure to have increased its market share in both the periods 1981–1991 and 1991–2001.

The map in Figure 2.3 shows the areas of England that have most increased their share of owner-occupation since 1981. The two main agglomerations of districts that have significantly increased in terms of owner-occupation are in the ex-industrial north-east and in north-central London boroughs. Additionally, however, it is noticeable that a number of

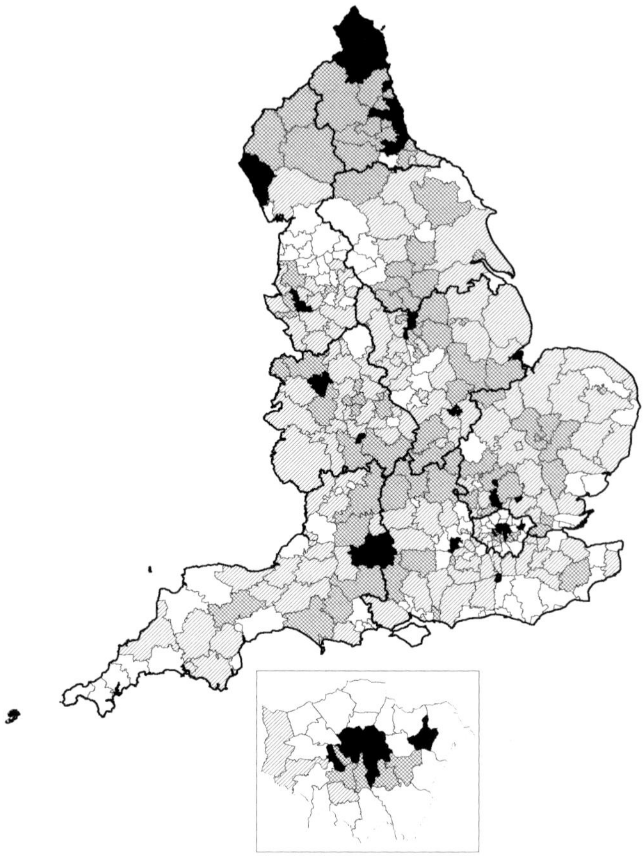

2001 Local Authority districts and Unitary Authorities

Owner occupation 1981-2001
% change in tenure share

- ■ Above 43.2% (above twice average) (31)
- 21.6 to 43.2% (above average) (83)
- 10.8 to 21.6% (below average) (133)
- □ Below 10.8% (below half average) (107)

Source: 2001 Census, 2001 Census Output Area Boundaries, 1991 Census Small Area Statistics, 1981 Census Small Area Statistics.

Figure 2.3 Map showing change in levels of owner-occupation in England, 1981–2001.

districts containing New Towns[7] have significantly grown their market share of owner-occupation. These include Telford, Redditch, Corby, Runcorn (part of Halton district), Peterlee (Easington) and Newton Aycliffe (Sedgefield) as well as several New Towns surrounding London (Bracknell, Crawley, Welwyn Garden City, Stevenage and Harlow). This growth is likely to have been driven by sales of municipal housing built by the New Town Development Corporations, which proved popular among residents exercising their right to buy.

Housing and estate archetypes in the New Towns have generally (although not without exception) represented relatively popular traditional property types in low-density settings. This is more the case in the first phase of New Towns than in the second phase New Towns which were characterized by more non-traditional housing types, an increased use of 'Radburn' estate design principles[8] and high-capacity, grade-separated highways. However, even the second phase New Towns (such as Runcorn and Telford) have demonstrated large increases in owner-occupation. This is potentially a result of their larger maximum planned populations and increased private housing provision than any significantly greater degree of Right to Buy uptake.

The role of housing as an investment commodity and as a hereditary asset continues to grow and is now a strong feature of the relationship between housing and the economy (the fourth phase illustrated in the first part of this chapter). As would be expected, there is an increasing number of outright owners, now accounting for nearly 43 percent of the sector (Table 2.3). Reform of pension rules in the UK indicates the ongoing expectation that housing will be an important component of personal wealth creation and supplementing the welfare state.

Renting

Whilst its market share has declined since 1991, private renting has nonetheless increased in absolute terms, partly the result of increased investment in the sector as an alternative to an unfavourable stock market.

[7] New Towns are purpose-built settlements built by New Town Development Corporations in England and Scotland under the provisions of the 1946 New Towns Act. The intention in most cases was that these settlements would be self-contained and would house an overspill population displaced from the major cities. Burton and Hartley (1997) have comprehensively documented the New Towns programme.

[8] Modelled after the principles of Radburn, New Jersey, such designs emphasize a separation of pedestrian routes from roads and are characterized by properties where the 'front door' faces open space while the 'back door' leads to communal garaging. Work on 'space syntax' has been used to demonstrate the crime and personal safety issues and the breakdown of defensible space that such layouts can promote. Consequently, Radburn layouts have tended to be relatively unpopular among residents.

Table 2.3 Type of owner-occupying households in England, 1991 and 2001.

Year	Owner-occupying households	Owned outright		Buying, including shared ownership	
		Number	%	Number	%
2001	13,920,429	5,969,670	42.9	7,950,759	57.1
1991	12,626,345	4,507,435	35.7	8,118,910	64.3

Source: Census 1991, 2001.

There were nearly 62,000 more households renting from a private landlord in 2001 than there were in 1991. But the sector's market share has declined slightly to 9.1 percent, a fall in percentage terms of some 2.4 percent. This was due to growth in the number of households disproportionately favouring other tenures, mainly owner-occupation. This drop in market share continues a general trend of decline (albeit at a reduced rate) in the private rented sector, which saw the sector decline from an 11.1 percent share in 1981.

It should be noted, however, that this is a national trend and the marginality of changes to the tenure at this level will mask more complex and significant variation at local level. The local economic structure of areas and in particular their level of employment in sectors that rely on a mobile or flexible workforce will contribute to this. Furthermore, in the last decade there has been an explosion in the popularity of city-centre flats and apartments in major urban areas. Gains to the private rented sector have been slightly offset by a significant owner-occupation market among city-centre developments; also institutional investment in such developments often means that the rate of vacancy among properties is high (see, for example, Barber and Blackaby, 2003).

Spatial differentials in tenure change

The national figures belie significant variation at a regional level. Just as tenure is a product of a number of forces (including supply constraints, cultural expectations and preferences, and housing needs shaped by employment and deprivation) so the varied socio-economic landscape of the country is reflected in its tenure composition. Table 2.4 provides a breakdown of changes in tenure by region.

The scale of change in some regions is telling. In the north-east, there was traditionally a very low level of owner-occupation. In 1981, it accounted for just 45 percent of households – a lower proportion than existed in London where private renting and housing association provision have traditionally been high. Conversely, social renting in the north-east was extremely prevalent, with over 40 percent of households in the north-east

Table 2.4 Percentage of households by tenure and region, 1981–2001.

Region	Owner-occupation (%)			Private renting (%)			Housing association (%)			Local authority (%)		
	1981	1991	2001	1981	1991	2001	1981	1991	2001	1981	1991	2001
East	60.79	71.01	74.58	9.96	8.93	7.83	1.57	2.33	5.08	27.69	17.73	11.99
East Midlands	59.85	70.66	73.86	10.15	8.02	7.51	1.34	2.13	3.78	28.65	19.19	14.28
London	48.61	57.21	57.24	16.56	13.84	14.77	4.09	5.60	9.36	30.74	23.35	17.64
North-east	45.30	58.95	64.78	8.95	6.72	6.44	2.58	3.81	5.40	43.17	30.52	22.95
North-west	59.57	68.12	70.80	8.92	7.41	7.90	2.18	3.68	6.71	29.33	20.79	13.99
South-east	64.61	73.83	75.68	12.18	10.01	9.06	1.61	2.92	6.83	21.60	13.24	7.61
South-west	63.77	72.83	75.31	12.55	10.77	9.98	1.61	2.08	6.02	22.07	14.31	8.05
West Midlands	57.56	67.67	71.37	8.14	6.89	6.60	1.96	2.80	6.56	32.34	22.64	14.78
Yorkshire and Humber	56.27	65.95	69.57	9.92	8.27	8.18	1.43	2.41	3.88	32.37	23.37	17.90
England	**57.85**	**67.62**	**70.33**	**11.14**	**9.31**	**9.09**	**2.12**	**3.20**	**6.26**	**28.90**	**19.88**	**13.65**

living in council housing. Against this backdrop, the north-east, the Midlands and Yorkshire and the Humber have all witnessed large increases in the number of owner-occupiers, although these increases have abated somewhat since 1991. The north-east, which started from an exceptionally low base, has dramatically increased its proportion of owner-occupation by 43 percent and now has a larger market share of owner-occupation than London.

All of the English regions are continuing to grow their base of owner-occupiers. However, in the southern regions (and the north-west, which has historically had a relatively high proportion of owner-occupied housing) the gains are somewhat smaller. There is progressively less potential for growth in market share for owner-occupation in these areas. Instead, the tenure pattern demonstrates a strong continued demand for affordable housing. The housing association sector in particular has grown extensively in these regions, although some of this growth has been as a result of transfers from the local authority sector.

In summary, the scale of changes in the tenure of housing provision in England over the last two decades, fuelled by the expansion and deregulation of the financial services industry in the 1980s and the introduction of the Right to Buy in 1980, has been remarkable. The growth of owner-occupation as the principal form of housing provision has continued unabated and is now far and away the dominant tenure in England. Even more significant, at least in relative terms, has been the growth of housing association provision. This has largely been at the expense of a shrinking local authority sector. In the south-east of England it has almost supplanted municipal housing as the main form of social housing provision. At the same time, the number of households renting privately has risen. Labour market mobility demands a healthy private rented sector, and the renaissance of city-centre living has also contributed to this. Yet, the sector has become more fragmented and has actually lost relative market share to other tenures. Fragmentation has occurred partly as a product of the expansion of owner-occupation. Some unpopular former council housing, which was sold under Right to Buy, has emerged in the private rental sector. Often this is very poor quality and is occupied by very low-income families. At the same time, new city centre apartments have expanded the top end of the sector.

This fragmentation of the private rental sector points to the uneven starting points in the historic differentials in tenure associated with regional economies, cultural traditions and planning policy. As a consequence, some of the most significant changes have occurred in regions and districts where there has traditionally been only a small owner-occupation sector. Large amounts of supply, often in green-belt areas, coupled with low price differentials in comparison with the south, led to a situation in the late 1990s where the gap between low-income market housing and social housing provision became marginal in the northern regions (Nevin et al., 2000). The objective of the Housing Market Renewal programme in many of these areas has been to re-balance the tenure profile so that it more closely matches aspirations and regional averages (Cole and Nevin, 2004). This will inevitably mean a net reduction in the number of social rented properties in areas where gains in owner-occupation have been greatest.

Spatial patterns of ethnic minority residence in England 1991–2001

Thus far, this chapter has concentrated on the role of long-term changes in the pattern of housing tenure in underlying housing market and neighbourhood change in the UK. The second factor we wish to discuss is just as significant. The rapid growth of a minority ethnic population in England since World War II has had profound implications for local housing markets and neighbourhoods. The relationship between the housing market and levels of deprivation is a complex one, and this is especially the case in areas with a significant ethnic minority population. Furthermore, housing market processes have been profoundly shaped by patterns of choice and constraint (Murie, 1974), again especially where a new, immigrant population was concerned.

Immigrants to the UK in the 1950s and 1960s were restricted in access to council housing (Rex and Moore, 1967) resulting in under-representation of sections of the black and minority ethnic (BME) population in certain tenures, especially the social rented sector (Lee and Murie, 1997). These historical features, mainly affecting the inner cities, have combined with cultural factors to produce a highly segmented and segregated structure of black and minority ethnic residential settlement patterns. This has important implications for housing market renewal, and 'urban renaissance' more generally, for three principal reasons:

- Changes in patterns of housing tenure reflect historical legacies of access rights and citizenship; hence these processes will be reflected in the continuing residential settlement patterns of minority ethnic groups. Changes in tenure will also reflect changing demand amongst BME communities, which will ultimately impact on housing renewal policies.
- Secondly, as British inner cities have become synonymous with high concentrations of minority populations and high levels of deprivation, there has been an implicit assumption on the part of policymakers that 'poor black' communities will always be amongst us without taking into account their changing housing aspirations. These trends imply a changing role for housing in areas of established BME residence within the inner cities.
- Finally, whilst the process of de-industrialization has resulted in significant population loss in the metropolitan urban areas of the north and Midlands, a comparatively more youthful minority ethnic population profile offers the potential for growth in the context of decline and outward migration of the white population.

On this final point, whereas the non-white population in England increased by 55 percent from 2.88 million to 4.46 million between 1991 and 2001 (Table 2.5), over the same period the white population of England increased by less than 2 percent (Table 2.6). Moreover, the total increase in the white population of roughly 733,000 was less than half the 1.58 million

Table 2.5 Black and minority ethnic population by region, 1991–2001.

Region	**1991**		**2001**		**Change 1991–2001**		**Share of 2001 BME population**
	No.	**% of total population**	**No.**	**% of total population**	**No.**	**%**	
East	160,149	3.2	263,137	4.9	+102,988	+64.3	5.9
East Midlands	186,677	4.7	271,792	6.5	+85,115	+45.6	6.1
London	1,337,728	20.1	2,068,888	28.8	+731,160	+54.7	46.4
North-east	35,688	1.4	60,027	2.4	+24,339	+68.2	1.3
North-west	244,819	3.7	374,271	5.6	+129,452	+52.9	8.4
South-east	222,845	3.0	391,656	4.9	+168,811	+75.8	8.8
South-west	61,498	1.3	113,123	2.3	+51,625	+83.9	2.5
West Midlands	422,102	8.2	593,008	11.3	+170,906	+40.5	13.3
Yorkshire and Humber	212,514	4.4	323,567	6.5	+111,053	+52.3	7.3
England	**2,884,020**	**6.2**	**4,459,469**	**9.1**	**+1,575,449**	**+54.6**	**100.0**

Source: 1991 Census Small Area Statistics (Enumeration Districts); 2001 Census Area Statistics.

Table 2.6 'White' population by region, 1991–2001.

Region	1991		2001		Change 1991–2001		Share of 2001 'white' population
	No.	% of total population	No.	% of total population	No.	%	
East	4,864,858	96.8	5,125,003	95.1	+260,145	+5.3	11.5
East Midlands	3,753,552	95.3	3,900,380	93.5	+146,828	+3.9	8.7
London	5,304,319	79.9	5,103,203	71.2	−201,116	−3.8	11.4
North-east	2,498,852	98.6	2,455,416	97.6	−43,436	−1.7	5.5
North-west	6,461,748	96.3	6,355,495	94.4	−106,253	−1.6	14.2
South-east	7,219,050	97.0	7,608,989	95.1	+389,939	+5.4	17.0
South-west	4,525,658	98.7	4,815,316	97.7	+289,658	+6.4	10.8
West Midlands	4,712,132	91.8	4,674,296	88.7	−37,836	−0.8	10.5
Yorkshire and Humber	4,605,995	95.6	4,641,263	93.5	+35,268	+0.8	10.4
England	**43,946,164**	**93.8**	**44,679,361**	**90.9**	**+733,197**	**+1.7**	**100.0**

Source: 1991 Census Small Area Statistics (Enumeration Districts); 2001 Census Area Statistics.

increase in the non-white population. The growth of the BME population over the decade to 2001 has therefore outpaced growth of the white population both in relative and absolute terms.

Across England, 110 local authority districts saw a decline in their white population whilst the remaining 244 districts increased their white population. Although some of this increase will have resulted from 'natural change' in birth and death rates, growth and decline of the white population in different parts of the country will be related to a considerable extent to differential patterns of migration. Population falls in one area will correspond to an increased population in another part of the region. The tendency for such movements to be away from metropolitan areas is well known, leading to its characterization as 'white flight'.

Despite significant growth in the BME population, none of the individual communities exceeded three percent of the population of England in 2001. However, the patterns of residence of non-white households reflect strong regional differences and have resulted in different levels of concentration for different communities. Table 2.7 and Table 2.8 show the regional population of the main ethnic minority groups for 1991 and 2001 respectively[9].

The following paragraphs summarize some of the key points from these tables.

[9] It should be noted that these data represent the largest ethnic groupings. The size of the 'other' category is some indication of the diversity of smaller minority ethnic population groups. For example, in England in 2001, around 786,000 people stated an ethnic origin that was not in one of the main groups used for aggregate analyses.

Table 2.7 Population by ethnic group and region, 1991.

Region	Population (thousands)							
	Black Caribbean	Black African	Black Other	Indian	Pakistani	Bangladeshi	Chinese	Other
East	21.6	5.5	13.6	39.2	24.6	11.1	12.1	33.2
East Midlands	24.4	3.2	10.6	98.5	17.3	4.3	7.5	21.3
London	289.6	161.4	80.4	345.4	87.0	85.0	55.8	231.8
North-east	1.0	1.3	1.5	7.4	9.1	3.3	4.6	7.6
North-west	21.5	9.4	16.1	55.7	77.1	14.8	17.4	32.9
South-east	23.1	8.2	13.4	64.0	35.7	8.2	17.9	52.1
South-west	12.4	2.8	6.4	11.0	4.1	2.3	6.4	16.5
West Midlands	77.5	5.1	19.0	158.1	98.6	19.5	9.8	35.4
Yorkshire and Humber	21.2	4.8	10.1	40.5	94.4	8.2	8.2	25.1
England	**491.7**	**201.4**	**168.6**	**819.5**	**449.6**	**159.2**	**140.5**	**454.3**

Source: 1991 Census Small Area Statistics (Enumeration Districts).

Table 2.8 Population by ethnic group and region, 2001.

Region	Population (thousands)							
	Black Caribbean	Black African	Black Other	Indian	Pakistani	Bangladeshi	Chinese	Other
East	46.3	23.2	5.4	51.2	38.8	18.3	20.5	59.8
East Midlands	47.1	12.5	3.8	122.2	28.0	7.1	12.9	38.4
London	414.5	413.1	60.2	436.8	142.7	154.2	80.3	367.2
North-east	3.8	4.3	0.5	10.1	14.1	6.3	6.0	15.1
North-west	42.4	25.6	5.4	72.0	117.1	26.2	26.9	58.5
South-east	51.2	34.4	4.8	89.6	58.4	15.2	32.8	105.6
South-west	25.6	9.9	2.5	16.3	6.9	4.9	12.8	34.5
West Midlands	122.2	15.8	10.0	178.6	154.3	31.6	16.3	64.8
Yorkshire and Humber	39.7	13.9	3.5	51.6	146.5	12.4	12.4	44.7
England	**791.1**	**550.4**	**93.4**	**1027.0**	**707.6**	**275.2**	**221.1**	**786.2**
England change 1991–2001(%)	**+60.9**	**+173.3**	**−44.6**	**+25.3**	**+57.4**	**+72.9**	**+57.4**	**+73.1**

Source: 1991 Census Small Area Statistics (Enumeration Districts).

- **Indian population**. The largest population describing itself as Indian resides in London (437,000), where it exceeded six percent of the total population in 2001. The West Midlands (179,000) and the East Midlands (122,000) had the second and third largest Indian communities. Whilst the Indian population only represented around three percent of the general population in these two regions, it was by far the largest BME community, representing 30 percent of the West Midlands minority ethnic population and 45 percent of the East Midlands minority ethnic population in 2001.

- **Pakistani population**. The Pakistani population represents the largest proportional share of population in the West Midlands and Yorkshire and the Humber (almost three percent in both regions) and, in absolute terms, continued to be more populous in the West Midlands (154,000) and Yorkshire and the Humber (146,000) than in London (143,000). There is also a sizeable Pakistani population in the north-west (117,000). Unlike the Indian population, which declined as a share of total non-white population, the Pakistani population increased its percentage share. In 2001, it represented just fewer than 16 percent of the non-white population, having grown by 58 percent over the decade to 2001, from 450,000 to 708,000. In proportional terms, the Pakistani population is the most dominant minority ethnic group in Yorkshire and the Humber (45 percent), the north-west (31 percent) and the West Midlands (26 percent).
- **Bangladeshi population**. The Bangladeshi population grew by 75 percent and was the second-fastest growing minority ethnic group over the decade, increasing from 159,000 to 275,000 (from 5.4 percent to 6.2 percent of the black and minority ethnic population). By 2001, Bangladeshis represented just over 0.5 percent of the total population of England and were mainly concentrated in London (154,000). London-based Bangladeshis represent 56 percent of the Bangladeshi population in England. After London, the regions of the West Midlands (31,000) and the north-west (26,000) have the largest Bangladeshi populations. These populations represent 11.4 percent and 9.4 percent respectively of the total Bangladeshi population in England. Whilst the Bangladeshi community is rapidly growing, it is not yet the largest ethnic community in any region.
- **Black population groups**. The black population is represented in the Census outputs by three categories: Black Caribbean, Black African and Black 'Other'. Together, these groups have increased by more than two-thirds between 1991 and 2001; this amounted to an increase in absolute terms of 576,000. The largest increase has been in London (356,000), which accounted for 62 percent of the total growth of the black population of England. London continues to be the region with the largest concentration of black residents. At the 2001 census, more than 12 percent (888,000) of London's population described themselves as from a black ethnic group. More than half (52 percent) of England's Black Caribbean population and almost three quarters of the Black African population live in London according to the 2001 census. Outside London, the second largest black population is in the West Midlands region (147,000), however this still represented less than 3 percent of the resident population.

- **Chinese population**. The Chinese population increased by around 57 percent between 1991 and 2001, from 140,000 to 221,000. The most significant increase was in the south-east region, although the largest increase in absolute terms (24,000) was in London. The Chinese community is not the majority ethnic group in any region, but continues to be heavily concentrated in three regions where it makes up almost two-thirds of the total Chinese population in England. These are London (80,000), the south-east (33,000) and the north-west (27,000).

Unsurprisingly, the most significant growth in minority ethnic population groups – in absolute terms – has been in London and metropolitan parts of the north and Midlands. Natural population growth among an already large population and the attraction of new immigrant groups to large existing communities contributes to this. Almost half (46.4 percent) of England's total black and minority ethnic population live in London. London continues to be the region in which the majority of the Black Caribbean, Black African, Black Other and Bangladeshi population live. It is also home to a significant proportion of the Indian and Chinese population as well as almost half of the minority ethnic population in groups other than the main ones. Outside London, the region with the largest share of the black and minority ethnic population is the West Midlands, where more than one in seven of England's black and minority ethnic residents live (13.3 percent).

Having said that, the greatest proportional increases in the black and minority ethnic population occurred in regions that had a relatively small such population in 1991. Close examination of the maps in Figure 2.4 reveals that areas of large relative increases in the non-white population between 1991 and 2001 have been in local authority districts that are *adjacent to* existing concentrations. For example, Birmingham, Leicester, Manchester and most of the London boroughs all had high proportions of minority ethnic populations in 1991, continue to do so in 2001, but have below average proportional increases in these population groups. On the other hand, many of the surrounding districts, including many rural areas, have above average growth rates.

The evidence from the British census indicates that whilst there has been some movement of the black and minority ethnic population, there are differences in this experience for different groups. Different starting points in migration as well as historical and cultural differences in the consumption of housing, such as those outlined by Cameron and Field (2000), begin to account for these differences. Clearly it is beyond the scope of this chapter to account for the complexity of spatial changes across ethnic minorities' residential patterns, or changes to their aspirations. What we do note, however, is that whilst there have been significant

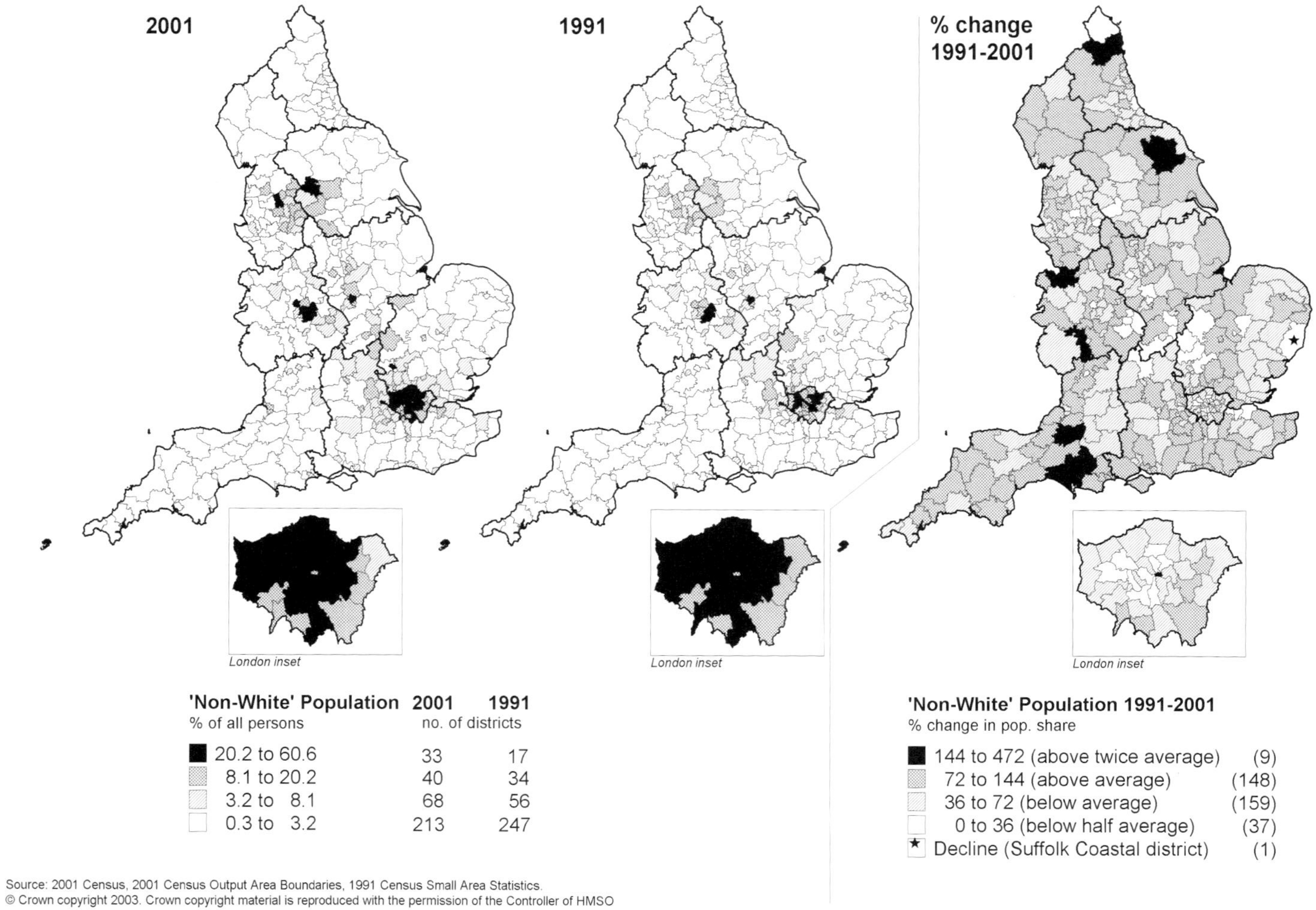

Figure 2.4 BME ('non-white') population by local authority district, 1991–2001.

Table 2.9 Index of dissimilarity by ethnic group, 1991 and 2001.

Population group[a]	**Index of dissimilarity, *D***		**Change**
	1991	**2001**	
Black African	70	66	−6.0%
Bangladeshi	64	65	+1.3%
Pakistani	63	63	+0.3%
Black Other	50	63	+24.7%
Indian	60	59	−2.4%
Black Caribbean	65	58	−11.1%
Chinese	35	35	+1.6%
Other	45	42	−7.1%
All black and minority ethnic population	**53**	**51**	**−4.9%**

[a] Table sorted by score on 2001 index.
Source: calculated from 1991 Census and 2001 Census using data standardized to 2001 local authority districts.

changes in consumption patterns of housing (measured in terms of housing tenure), these do not necessarily translate into significant changes in the underlying structure of British cities. Despite large relative gains outside the metropolitan cores, the general pattern has been large absolute growth in established and contiguous areas of BME settlement.

To aid this articulation more clearly, we summarize the evidence on BME population change for the period 1991–2001 using a measure of segregation. We have computed the index of dissimilarity, *D*, for each minority group with reference to the pattern of white population across England. *D* can be interpreted as the percentage of each minority group that would need to move residence in order that the total distribution of that group is equal to the distribution of the white population. Table 2.9 shows that for the BME population as a whole there has been only a slight decrease in segregation, with the *D* reducing from 53 to 51 over the decade. Whilst there has been a small reduction in the level of segregation across all BME groups, this should be seen in the context of a significant growth and expansion of the minority ethnic population, including some limited movement beyond the areas of existing concentration. Indeed, for some groups, notably the Bangladeshi and Pakistani population, *D* has increased. This would tend to reinforce the assertion that patterns of residence for minority ethnic communities are dependent on patterns of housing consumption (including tenure), choice and constraint, as well as cultural traditions or preferences.

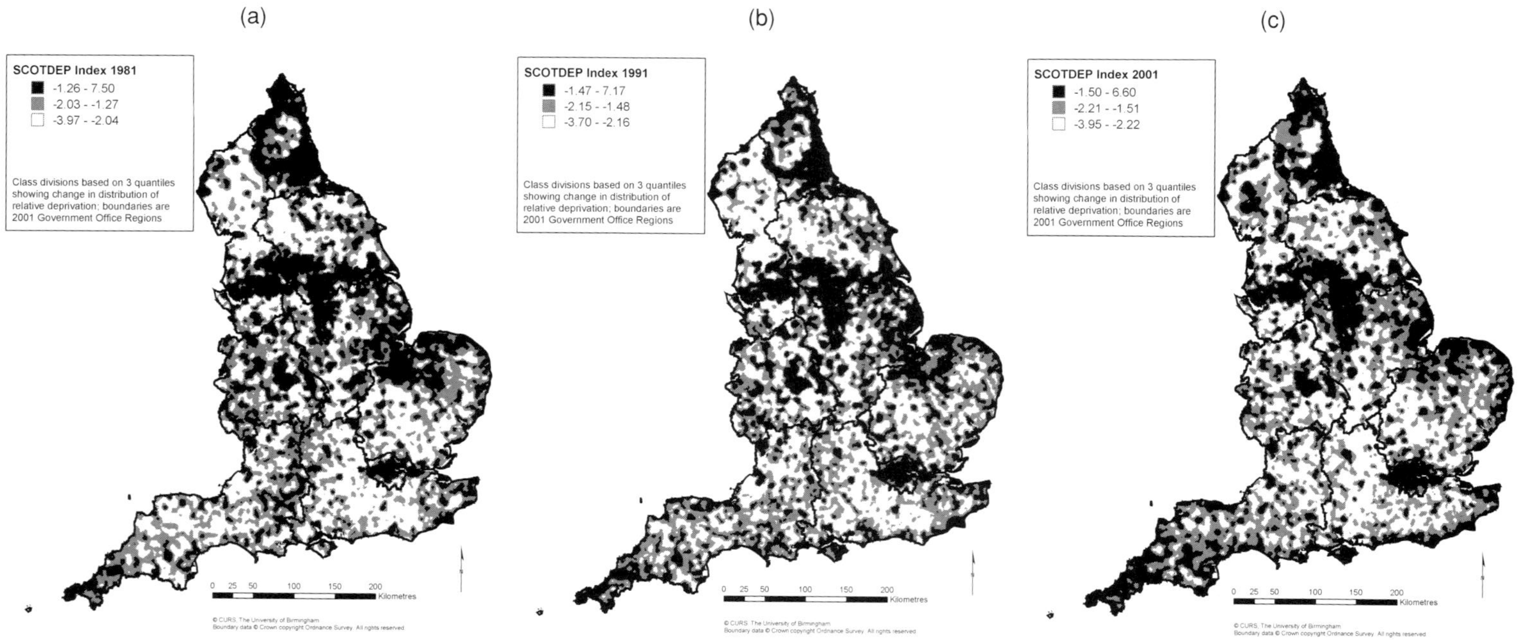

Figure 2.5 Relative change in the Scotdep index, 1981–2001 (based on relative distribution of scores in each census year).

Spatial patterns of deprivation in England 1981–2001

Patterns of choice and constraint and the historical pattern of migration and the role of black and minority ethnic communities in English cities have meant that patterns of ethnic segregation have become synonymous with concentrations of deprivation (Rex and Moore, 1967; Lee and Murie, 1997). Whilst existing patterns of deprivation are something of a blunt instrument in identifying areas at risk of low or changing demand for housing, it is clear that a high level of deprivation is nearly always implicated as a 'precondition' of an area experiencing problems of low demand (Lee and Nevin, 2003). Concentrations of deprivation reflect the historical legacy of housing and planning processes as well as the choices made by households in the housing market. Therefore, understanding the changing pattern of deprivation will partly act as a guide to forecasting future problems of low demand, especially when compared to tenure changes and changes in spatial patterns of ethnic segregation.

To conclude this chapter's brief résumé of some of the significant structural and spatial changes impacting on England's neighbourhoods and housing markets, we provide some evidence on the changing pattern of deprivation. Again, small-area data from multiple Censuses (1981, 1991 and 2001) have been standardized and combined to provide a comparative picture over two decades.

In the absence of a consistent measure for deprivation, especially since there have been changes in government definitions of deprivation and alterations to the geographical base over the past 20 years, we have recreated a standard measure of multiple deprivation (the 'Scotdep' index) developed originally by researchers at the University of Edinburgh (Carstairs and Morris, 1991). The Scotdep index combines four indicators into a single index. These are: male unemployment; households with no car; persons in a low social class (equivalent to social classes IV and V); and households in overcrowded accommodation (more than one person per room). Research has shown the Scotdep index to be the most valid and consistent measure of deprivation (Lee et al., 1995) and has the advantage that it can be recreated using historic Census data[10].

For the period 1981–2001, Figure 2.5 shows that arguably a more polarized pattern of deprivation has emerged. Evidence of increasing deprivation in coastal locations (such as parts of East Anglia and the north-west region)

[10] To account for the changing spatial geography, a GIS (geographical information system) was used to aggregate data up from the smallest statistical unit available in each census year and match to current local authority boundaries; GIS was also used in preparing thematic maps of change by assigning small area data to 100 m square cells.

and in the remote south-west of England highlights the fragility of these areas' economies, which are often seasonal. At the same time the hinterlands surrounding the major urban areas, the conurbations of the Midlands and the shire counties surrounding London have become much less deprived.

However, for the older industrial urban areas in the north and Midlands, as well as inner London, there has been little significant change in the pattern of deprivation between 1981 and 2001. At the district level, 52 local authorities were ranked in the 'most deprived' quintile (71 districts) in each of the census years from 1981 to 2001. A further 36 local authorities were ranked amongst the most deprived quintile in at least one, but not all of the census years. London boroughs dominate the list, but outside London it is the north-west, north-east and the West Midlands regions of England that feature most prominently.

The overall trend is one of polarization on two dimensions. First, between rural and urban areas and second, between well connected and more remote locations with fragile economies. Significantly, these trends apply equally to prosperous regions and those regions that still lag behind economically, but the overall pattern of change is highly consistent with changes in land use patterns and the housing market over the same period. This particularly relates to the continued shift towards owner-occupation and house price increases.

It is likely that continued restructuring of the economy and the pre-eminence of the service sector are driving this pattern of deprivation. This result, arguably, has been a spatial mismatch between the emergent service economies (located outside the industrialized metropolitan city-regions of the north and Midlands or in new, growing service corridors in the south of England) and deprived populations, sometimes with large minority ethnic populations. These trends are driven by investor preferences for greenfield sites and locations in the south-east, and are reinforced by the limited housing pathways and quality of the 'urban offer' in the older industrial urban core of the north and Midlands. Earlier in this chapter, we showed how the contiguous growth of minority ethnic communities has occurred in parallel with a loss of white population. The pattern amongst local authorities witnessing a white population increase coincides with those becoming less deprived, reflecting a spatial mismatch between minority ethnic population growth on the one hand and the restructuring of the economy on the other. 'Prospering' service sector towns (using the official government classification) in south Gloucestershire, Milton Keynes and East Riding are those that had the largest absolute increases in white population over the decade to 2001 (+23,000, +21,750 and +20,500 persons respectively).

Conclusion: housing market renewal in context

In this chapter we have described the context in which the process of housing market change, and the policy of market renewal, is taking place in England. The recognition of low and changing demand for housing in the late 1990s resulted in a wide-ranging academic and policy debate and the establishment of a housing market renewal programme to tackle the problem of low demand. This was part of a broader sustainable communities agenda designed to tackle social exclusion and deliver urban renaissance. The Sustainable Communities Plan set out a vision for the English regions with different levels and types of investment to tackle affordability pressures in the south and low demand in the north and Midlands.

One of the main concerns underpinning the current drive towards 'sustainable communities' is the widening gap between the top and bottom ends of the housing market. This naturally creates 'winners' and 'losers' at a variety of spatial scales. The literature on low and changing demand particularly emphasizes the role of regional economic differentials in explaining widening market inequalities at a number of scales from regional down to neighbourhood (Webster, 1998; Holmans and Simpson, 1999). We have also shown in this chapter how these processes coincide with, and are driven by, structural changes in housing tenure, ethnicity and deprivation.

Polarization in the housing market is a problem because it threatens to deepen and reproduce the cleavages in socio-economic space we have already outlined. A polarized housing market – with more low-value properties, more high-value properties and less in between – constrains 'vertical' mobility in the market. There becomes a lack of clear or viable housing pathways for different households thus making it much more difficult for households to use housing as a strategy for improving their own wealth and/or escaping processes of social exclusion. Furthermore, when polarization is evident between large areas, such as between regions, there can be an effect on spatial (horizontal) mobility of households. This dampens the ability for lower-income population groups to respond to labour market pressures. This can be seen in the south-east of England where poor housing affordability among lower-income 'key workers' results in longer commuting patterns or public service inefficiencies.

Given the progress made by the housing market renewal pathfinders in tackling the worst manifestations of low demand (through demolition, clearance and the improved management of social housing), recent rises in house prices have threatened the *raison d'être* of the market renewal programme (Ferrari and Leather, 2005). But whilst there have been remarkable changes in the profile of housing tenure across the English regions, the same level of change is not reflected in either levels of deprivation or in the residential settlement patterns of ethnic minority populations. Both these factors will continue to determine the success or failure of the Sustainable Communities Plan and urban renaissance agendas.

As we noted above, the historical development of cities has followed a pattern of housing investment that has been influenced by phases of economic development. Outward migration of an economically active and 'successful' white population has broadly mirrored an expansion and spatial consolidation of the ethnic minority population. This has continued the trend in which British inner cities have become synonymous with high concentrations of black and minority ethnic populations and high levels of deprivation and an implicit assumption on the part of policymakers that there will always be 'poor black' communities. As housing aspirations have changed and second and third generation minority groups have newfound affluence, so their aspirations have moved in line with this rising affluence. These trends imply a changing role for housing in areas of established minority communities within the inner cities. The aspirations and historical consumption patterns of housing will mean that different strategies will need to be adopted in different regions reflecting these patterns of residence amongst the different communities.

But looking at the broader picture, the evidence suggests that this is unlikely to affect deep-rooted trends or alter the fundamental structure or path dependency of the urban form. The housing market renewal programme will need to deliver a change agenda, which reconciles growth from the BME community and change in use of housing and neighbourhoods. All of these aspects need to be taken into account in developing strategies for housing renewal and require a broader perspective on the structuring of British cities beyond cyclical house price trends which point policymakers towards shorter-term solutions and fail to acknowledge the underlying function and trajectory of neighbourhoods. Without this perspective, housing renewal will fail in its longer-term objectives of delivering sustainable communities and urban renaissance.

References

Barber, A. and Blackaby, B. (2003) *Filling the Gap: the Need for Middle Market Housing in Birmingham City Centre.* Birmingham City Council, Birmingham.

Bramley, G. and Pawson, H. (2002) Low demand for housing: incidence, causes and UK national policy implications. *Housing Studies* **39** (3): 393–422.

Burton, A. and Hartley, J. (eds) (1997) *The New Towns Record, 1946–1996 (multimedia resource).* IDOX, Glasgow.

Cameron, S. and Field, A. (2000) Community, ethnicity and neighbourhood. *Housing Studies* **15** (6): 827–43.

Carstairs, V. and Morris, R. (1991) *Deprivation and Health in Scotland.* Aberdeen University Press, Aberdeen.
Champion, A. (2002) Flight from the cities? In: Bate, R., Best, R. and Holmans, A. (eds) *On the Move: the Housing Consequences of Migration.* Joseph Rowntree Foundation, York.
Cole, I. and Nevin, B. (2004) *The Road to Renewal: the Early Development of the Housing Market Renewal Pathfinder Programme in England.* Joseph Rowntree Foundation, York.
Crook, T., Currie, J., Jackson, A., Monk, A., Rowley, S., Smith, K. and Whitehead, C. (2002) *Planning Gain and Affordable Housing: Making it Count.* York Publishing Services, York.
Davoudi, S. and Healey, P. (1995) City Challenge: sustainable process or temporary gesture? *Environment and Planning C: Government and Policy* **13** (1): 79–95.
Department of the Environment, Transport and the Regions (DETR) (1999) *Report by the Unpopular Housing Action Team.* Report of the Social Exclusion Unit's Policy Action Team 7 (PAT 7). Department of the Environment, Transport and the Regions, London.
Ferrari, E. and Leather, P. (2005) Evaluating housing market renewal: national or neighbourhood policy? Paper presented at AESOP Congress, 13–17 July 2005, Vienna.
Forrest, R. and Murie, A. (1998) *Selling the Welfare State.* Routledge, London.
Galster, G. (1997) Comparing demand-side and supply-side housing policies: submarket and spatial perspectives. *Housing Studies* **33** (10): 561–77.
Holmans, A.E. and Simpson, M. (1999) *Low Demand: Separating Fact from Fiction.* Chartered Institute of Housing for the Joseph Rowntree Foundation, Coventry.
House of Commons Transport Local Government and the Regions Committee HC240-I (2001–02). *Empty Homes.* House of Commons Transport Local Government and the Regions Committee, sixth report of session 2001–02 (volume I).
Jones, C. and Murie, A. (2005) *The Right to Buy.* Blackwells, Oxford.
Lee, P. and Murie, A. (1997) *Poverty, Housing Tenure and Social Exclusion.* Policy Press, Bristol.
Lee, P. and Murie, A. (1999) Spatial and social divisions within British cities: beyond residualisation. *Housing Studies* **14** (5): 625–40.
Lee, P. and Murie, A. (2004) The role of housing in delivering a knowledge economy. *Built Environment* **30** (3): 235–45.
Lee, P. and Nevin, B. (2003) Changing demand for housing: restructuring markets and the public policy framework. *Housing Studies* **18** (1): 65–86.
Lee, P., Murie, A. and Gordon, D. (1995) *Area Measures of Deprivation: a Study of Current Methods and Best Practices in the Identification of Poor Areas in Great Britain.* Centre for Urban and Regional Studies, University of Birmingham, Birmingham.
Lowe, S., Spencer, S. and Keenan, P. (eds) (1998) *Housing Abandonment in Britain: Studies in the Causes and Effects of Low Demand Housing.* Centre for Housing Policy, University of York, York.
Lyons, M. (2003) Spatial segregation in seven cities: a longitudinal study of home ownership, 1971–1991. *Housing Studies* **18** (3): 305–26.
Mullins, D. and Pawson, H. (2005) The land that time forgot: reforming access to social housing in England. *Policy and Politics* **33** (2): 205–30.
Murie, A. (1974) *Household movement and household choice.* CURS occasional paper No. 28. Centre for Urban and Regional Studies, University of Birmingham, Birmingham.
Murie, A. and Ferrari, E. (2003) Reforming the right to buy in England. Paper commissioned for *The Right to Buy – A Symposium for Debate.* Chartered Institute of Housing, Coventry.

Murie, A., Nevin, B. and Leather, P. (1998) *Changing demand and unpopular housing*. Working paper No. 4. The Housing Corporation, London.
Nationwide Building Society (2005) *House Prices* (quarterly house price index bulletin). July 2005. Nationwide Building Society, Swindon.
Nevin, B., Lee, P., Goodson, L., Murie, A. and Phillimore, J. (2000) *Changing Housing Markets and Urban Regeneration in the M62 Corridor*. Centre for Urban and Regional Studies, University of Birmingham, Birmingham.
Office of the Deputy Prime Minister (ODPM) (2003a) *Sustainable Communities: Building for the Future*. ODPM, London.
Office of the Deputy Prime Minister (ODPM) (2003b) *Market Renewal Pathfinders: Learning Lessons – 2003 National Conference Report*. ODPM, London.
Office of the Deputy Prime Minister (ODPM) (2004) *Home Ownership Solutions for Low Demand Areas*. ODPM, London.
Payne, J. and Payne, G. (1977) Housing pathways and stratification: a study of life chances in the housing market. *Journal of Social Policy* **6** (2): 129–56.
Power, A. and Mumford, K. (1999) *The Slow Death of Great Cities? Urban Abandonment or Urban Renaissance*. Joseph Rowntree Foundation, York.
Rex, J. and Moore, R. (1967) *Race, Community and Conflict: a Study of Sparkbrook*. Oxford University Press, London.
Saunders, P. (1990) *A Nation of Home Owners*. Unwin Hyman, London.
Urban Task Force (UTF) (1999) *Towards an Urban Renaissance*. Final report of the Urban Task Force. UTF, London.
Webster, D. (1998) Employment change, housing abandonment and sustainable development: structural processes and structural issues. In: Lowe, S., Spencer, S. and Keenan, P. (eds) *Housing Abandonment in Britain: Studies in the Cause and Effects of Low Demand Housing*. The University of York, York.
Yates, J. and Whitehead, C. (1998) In defence of greater agnosticism: a response to Galster's 'Comparing demand-side and supply-side housing policies'. *Housing Studies* **13** (3): 415–23.

3

More Pluribus, Less Unum? The Changing Geography of Race and Opportunity*

Xavier de Souza Briggs

Location matters – for economic returns, quality of life and many other reasons. But the value of a given location as a place to live, work, invest, or go to school can shift profoundly over time as communities grow and their makeup changes. This chapter takes a critical look at evidence on the key forces that drive metropolitan growth patterns in America; the changing geography of race and opportunity associated with those growth patterns; and the social and economic consequences of that geography, focusing on access to good schools and jobs. The chapter also outlines important evidence about the geography of crime and insecurity, poor health and environmental hazards.

Growth and its discontents: America in comparative perspective

America's metropolitan areas are both very sprawling and very segregated by race and class, a pattern that is especially apparent when US cities and suburbs are compared with those of Europe or other wealthy regions around the world (Nivola, 1999; Friedrichs et al., 2003)[1]. As both liberal

* This chapter is reproduced from X. de Souza Briggs (2005) *Geography of Opportunity: Race and Housing Choice in Metropolitan America* with permission of the Brookings Institution Press.

[1] Like the US, a number of European nations are experiencing sharp increases in racial and ethnic diversity, thanks both to immigration and to comparatively low rates of natural increase among native ethnic groups. Sweden now has a larger foreign-born population than the US has, and particular cities in the European Union – London, Malmö,

and conservative observers have shown, there was nothing 'natural' or inevitable about these segregated outcomes. As much as individual housing preferences, deliberate policy decisions regarding transportation, education, housing finance, land use and taxation and expenditure across all levels of government, and even those regarding energy policy, determine where people live, how much they depend on cars to move around, how much of their income must be spent on housing and, to a striking degree, who their neighbors will be. Policy decisions also determine the benefits and costs that those housing and mobility patterns entail (Danielson, 1976; Massey and Denton, 1993; Altshuler et al., 1999; Nivola, 1999; Ellen, 2000; Wilson and Hammer, 2001; Savitch, 2002; Glaeser and Kahn, 2004). For its part, through mortgage insurance and other policies, the federal government directly encouraged segregation by race and class until roughly the middle of the twentieth century (Massey and Denton, 1993). Federal housing and community development programs have had segregating effects since then, in spite of affirmative policies meant to desegregate communities. Add direct and indirect effects through policies that shape land use, real estate practices and lending, and the local development game is hardly the textbook example of a free market. Rather, it is a competitive, unevenly regulated and subsidized regime.

The 1990s brought more of the century-old urban development pattern known as sprawl as well as increased decentralization and specialization in local government. During the 1990s land consumption in metropolitan areas proceeded at twice the rate of population growth, and even metropolitan areas that lost population, such as Buffalo and Pittsburgh, continued to develop at the fringe (US Department of Housing and Urban Development, 2000). Contrary to the popular perception that sprawl is king in western cities, the west is home to some of the densest metropolitan areas in the US, and many of them became denser over the 1980s and 1990s, while metro areas in the south sprawled rapidly to accommodate high population growth and, in the north-east and mid-west, sprawled at remarkably high rates in spite of their slow population growth rates (Fulton et al., 2001). Overall, population density per acre of land has steadily decreased for metropolitan areas, as households demand larger housing, rely increasingly on cars to meet their transportation needs and continue to migrate toward warmer and less dense areas of the country. The social costs and benefits of sprawl are widely debated, and divergent trends – having more cars on the

---

Marseilles – face the opportunities and challenges of this new diversity at much higher percentages. In Europe, increased diversity and some nativist backlash in the political arena contribute to interest in 'social exclusion', as Europe strengthens its regional identity and relaxes certain border controls. In general, segregation patterns in immigrant gateway cities in Europe parallel the segration patterns for Asians and Hispanics in the United States.

road but also stricter pollution emissions standards, for example – make for confusing net patterns over time[2].

From the standpoint of race and housing choice, sprawling new growth creates exit options from older, built-up areas – but more often for white households and middle- and upper-income minorities than for other groups. One way Americans separate themselves from urban problems is by leaving them behind and creating new local governments as suburban communities develop. Paralleling the trend in sprawl, the number of municipal governments increased by 15 percent between 1952 and 2002, and excluding school districts, the number of special-purpose governments – for managing water resources, public hospitals, cemeteries, fire services, metropolitan transportation and other functions – roughly tripled during the period (US Census Bureau, 2002). While the incorporation of new cities and towns within larger metropolitan areas means greater autonomy and choice for residents of those places, metropolitan areas with higher levels of government fragmentation (more jurisdictions) per capita also exhibit higher levels of economic inequality between cities and suburbs (Ellen, 1999).

Growth is inevitable, but the form it takes, inclusionary or exclusionary, is not. Consider these projections: over the next 25 years, the nation will lose about 15 million housing units and gain about 30 million households. This means that the country will need to create about 45 million housing units somewhere – or about 37 percent as many as the number of total units in the nation's housing stock at the end of the first quarter of 2004 – to accommodate a larger and more diverse population (Nelson, 2003; US Department of Housing and Urban Development, 2004)[3]. Patterns of new development, currently dominated by the sprawl model, are therefore enormously important.

Mobility matters too, of course. About half (49 percent) of the nation's population moved between 1995 and 2000, continuing a migration rate that has changed little in the past few decades (Schacter et al., 2003). These moves have affected investments in both newer and older communities, as well as community makeup. Whites and the middle class continued a net

---

[2] For an in-depth review of sprawl's causes and social and environmental consequences, as well as long-run scenarios for population growth in America under alternative approaches to managing growth, see Burchell et al. (2002). Detailed and widely read progressive commentaries include Moe and Wilkie (1997); Benfield et al. (1999); and Squires (2002). For conservative critique and response, most of it founded on public choice theory in economics, see Glaeser and Kahn (2004); Staley (1999, 2000); and discussion in Henig (2002).

[3] Based on analyses of the American Housing Survey, conducted by the Census Bureau and funded by the US Department of Housing and Urban Development, the nation loses about 0.6 percent of its housing stock annually. Meanwhile, projections for continued population growth and household size are based on trends in births, deaths, household formation and immigration (Joint Center for Housing Studies, 2003).

migration out of cities in the 1990s, citing two primary reasons for leaving: crime and poor school quality. White Americans strongly associate these problems with increased minority presence in a community, particularly with the presence of blacks.

It is increasingly clear that how the US handles increased racial and ethnic diversity will, for better or worse, be linked in the decades ahead to how it handles sprawl and other symptoms of uneven metropolitan development. On this score, there is a growing chorus of regionalists – those who advocate metropolitan responses to challenges that cross local government boundaries, including those associated with rapid sprawl: traffic congestion, concentrated poverty, overcrowded schools, 'zoning for dollars' and others[4].

Regionalism, which first appeared in the 1920s and enjoyed some prominence in the 1960s and 1970s, is not a single, coherent school of thought on urban policy but a set of hopes for wiser local development patterns and more rational governance to meet regionwide needs (Fishman, 2002; Henig, 2002). Regionalists emphasize, variously, that the benefits of planning, governing and investing resources across jurisdictional boundaries include enhanced fiscal efficiency, economic competitiveness, environmental sustainability and social equity (Rusk, 1999; Pastor et al., 2000; Henig, 2002; Orfield, 2002). But recent renewed hopes for more cooperation between cities and suburbs – in particular, for more regional planning and investment to address what David Rusk, the noted urban analyst and former mayor, terms the 'sprawl machine' – confront several powerful and closely linked barriers to change (Rusk, 1999). More optimistically, the barriers might be thought of as defining the key litmus test of the new regionalism. The first barrier is a very strong preference for local decision making (home rule) in American politics; in survey after survey, it seems, the more local, the better (Danielson, 1976; Fischel, 2001). Second is the dominant American lifestyle, which favors a car-bound and sprawling suburban model of consumption (instead of more compact, transit-oriented living), along with local fiscal autonomy, low taxes and a limited local obligation to meet the needs of the disadvantaged. And third – the focus here – is the geography of race and class.

As thoughtful observers of the new regionalism point out, social equity has been and will continue to be the most problematic plank in the

[4] Zoning for dollars, also known as fiscal zoning, is the establishment and use of local zoning regulations to favor real estate development that generates high property tax revenue and comparatively low municipal service costs (including public education costs). This practice often favors new retail stores, for example, over new housing development. Fiscal zoning is particularly effective at excluding modestly priced, multifamily housing, a key predictor of a community's economic and racial diversity over time (Fischel, 2001; Pendall et al., 2005).

platform (Jackson, 2000; Henig, 2002). For example, antisprawl, 'smart growth' advocacy across the nation has not, thus far, included strong or steady supporters of affordable and inclusionary housing (Downs, 2004). Yes, Americans of all races are working together more than ever before, as workplace statistics confirm (Estlund, 2003), but can they share communities? Can they live together, and educate their young people together, across persistent social divides? This question is a test of the new regionalism's political strength as well as its potential to improve lives and communities. The race and class segregation of American communities will help shape regionalism and determine its impact. Conversely, barriers to regional progress and smarter growth patterns will also be critical barriers to racial justice and social equity.

People and places: changing demography and settlement patterns

As newspaper headlines, Census Bureau reports and scholarly research document, the sharp growth in racial and ethnic diversity in recent years has been driven not only by immigration – the second great wave in the nation's history, which began with immigration reform in 1965, is also the first to be dominated by non-European immigrants – but also by a sharp decrease in the birth rate for non-Hispanic whites over the past generation. Eleven million immigrants came to this country in the 1990s, contributing to one-third of all population growth over that period, and more than 80 percent of those came from Africa, Asia, or Latin America and the Caribbean. Their growth fueled by immigration and higher-than-average birth rates, Hispanic Americans recently became the nation's largest minority (up from 22.4 million to 38.7 million between 1990 and 2002), overtaking blacks (36.6 million). Meanwhile, the 72 percent increase in Asians and Pacific Islanders (up from 7.4 million to nearly 12.7 million) was the largest in percentage terms (US Census Bureau, 2001, 2003; Suro and Singer, 2003)[5]. About 3.5 million persons identify themselves as

[5] The 2000 census brought a significant, and widely contested, change in the way data on racial-ethnic identity were collected, and these changes carry over for the population estimates through July 1, 2002 (reported in this chapter), that represent current indicators of growing racial and ethnic diversity. For the first time, respondents were allowed to check more than one race – an option chosen by more than 6.8 million Americans (2.4 percent of the population). But as before, the racial identity question was separated from an ethnic identity question about having 'Spanish, Hispanic, or Latino origin', which the federal government defines as 'a person of Cuban, Mexican, Puerto Rican, Central or South American, or other Spanish culture, regardless of race'. Together, the two questions generate no fewer than 126 possible identity combinations, meaning that racial-ethnic data from the year 2000 are not directly comparable with those of earlier censuses

Native Americans, a figure that also has increased in recent decades, because of both higher-than-average fertility rates and growth in ethnic pride among the nation's first peoples. But these conventional racial statistics obscure a changing, and some say blurring, color line: about 7 million persons in America now identify themselves as multiracial, and one in four marriages involving either Hispanics or Asians crosses group lines (Lindsay and Singer, 2003).

Over the 1990s, most central cities in America became 'majority minority' for the first time in history, and one-quarter of the central cities that saw growth – a key concern for every mayor, economic development director and business roundtable participant in the land – would actually have shrunk if not for growth in their Hispanic populations (Berube, 2003). While almost half of the nation's immigrants arrive in the New York, Chicago and Los Angeles metropolitan areas, the fastest rates of increase – the most rapidly felt local changes – are found in emerging gateways, such as Atlanta, Raleigh, Las Vegas and Nashville (Suro and Singer, 2003).

Ninety-five percent of the nation's foreign-born population lives in metropolitan areas, and while US history has cast central cities in the role of primary immigrant gateway, just over half of the foreign-born population is now living in the suburbs. The majority of both Asians and Hispanics now live in suburban areas. In some metropolitan areas, in fact, the suburbs are the first stop for immigrant workers and families, not a destination attained over generations (Frey, 2003).

Housing segregation and mobility patterns, by race and income

Analysts generally track residential segregation by race in two ways: by measuring how unevenly racial groups are distributed in a given metropolitan area and by measuring how much an 'average' member of each group is exposed to other groups at the neighborhood level (neighborhoods are typically defined as census tracts that contain, on average, a few thousand households). The two measures can lead to different impressions of how segregated a place is. Consider a hypothetical metropolitan area with only whites and blacks. The area may have a small black population that is very

---

(Katz and Lang, 2003). Population growth rates cited in this chapter, like most now reported by researchers and the media, now reflect 'race alone or in combination' responses for Hispanics and the four major non-Hispanic groups (Asian, black, Native American, white). For example, 'Asians' includes those who selected only that racial category and those who selected it in combination with another racial category, such as white, but not the small number who selected Asian plus Hispanic origin (as some Filipinos do). This means that although almost 98 percent of census 2000 respondents chose only one racial category, the total of the alone-or-in-combination responses exceeds (by 2.6 percent) the total population of the US.

evenly distributed across neighborhoods, but because there are so few members of the group to 'go around', the average white person will live in a neighborhood with very few blacks, while the reverse will be true for the average black: he or she will have mostly white neighbors. This hypothetical metropolitan area would have an encouragingly low 'unevenness' score but would also show very low average across-group 'exposure' for whites and very high exposure for blacks (Massey et al., 1996; Massey, 2001).

By these indexes, the latest census provides a very mixed picture of segregation levels and trends in the US at the turn of the twenty-first century. On one hand, as measured by unevenness of distribution, segregation rates in the most highly segregated metropolitan areas – the rates of black segregation from whites, most of all in older cities of the north-east and midwest – continued to fall. This continues a three-decade-long trend and represents a national drop in black–non-black segregation to its lowest point since 1920 (Glaeser and Vigdor, 2003). In general, neighborhoods that integrate black and white households became much more numerous, as well as more stable, in the 1980s than they were in the 1970s[6]. Over three-fourths (76.3 percent) of tracts integrated by 1980 remained so by 1990, while 16.6 percent of those tracts became majority black and 7 percent predominantly white. White losses in mixed neighborhoods were three times greater in the 1970s than in the 1980s. Where black–white neighborhood integration is concerned, economist Ingrid Gould Ellen finds that the primary correlates of racial stability are a neighborhood's history of past stability (stability is self-perpetuating, to some degree); its distance from a metropolitan area's main area of minority concentration (integrated neighborhoods that are more distant are more racially stable); its percentage of rental housing (among whites, renters are the most likely to integrate); a secure set of amenities (amenities help attract and retain a variety of households); and the black presence in the metropolitan area as a whole (holding other factors equal, racially integrated neighborhoods are more stable in the west, for example, where blacks constitute a smaller share of metropolitan populations) (Nyden et al., 1997; Ellen, 2000). Many integrated neighborhoods continued to be mixed in the 1990s, and the continued decline of black–white segregation during the 1990s appears to have come primarily

---

[6] Ellen (2000, pp. 12–19) defines a racially integrated neighborhood as a census tract in which blacks make up between 10 and 50 percent of the population. She then studies the non-Hispanic white shares of population of those tracts at the beginning and end of each decade. Analysts understand stability to include identifiable equilibrium points, at which the number of persons of a given racial group who are added to a neighborhood (either through birth or in-migration) is about equal to the number of persons of that group who are lost (either through death or out-migration). A perfectly stable racial mix would require a rare pattern indeed: complete equilibrium across all racial groups present in the neighborhood.

Table 3.1 Indexes of dissimilarity, 330 US metropolitan areas, 2000[a].

Index	White–black	White–Hispanic	White–Asian
Maximum index	84.7	75.4	58.8
Median index	52.3	38.2	35.2
Minimum index	20.2	11.6	14.5
(Standard deviation)	(13.7)	(12.1)	(7.9)
Change in median index, 1990–2000	−6.2	−4.6	−11.8

[a] Dissimilarity measures unevenness in two population distributions (whites versus black, say), on a 100-point scale. The index indicates the percentage of either group that would need to move in order to make the composition of each neighborhood (census tract) match that of the metropolitan area as a whole. Higher scores reflect greater unevenness (higher segregation). On measures of segregation, see Massey et al. (1996). The White House Office of Management and Budget designates census geography, with updates common in areas undergoing significant demographic transition. For 2000, 331 metropolitan areas were designated; segregation measures could not be calculated for all due to missing data.
Source: author's calculations using Lewis Mumford Center, University at Albany, census data file.

from the partial integration of neighborhoods that were once exclusively white (Rawlings et al., 2004).

At the same time, black–white segregation rates remain distressingly high in absolute terms, particularly in the rustbelt cities of the north-east and midwest. As sociologist John Logan described the decline, 'It's at a rate that my grandchildren, when they die, will still not be living in an integrated society' (National Public Radio, 'Segregation in the Cities', Talk of the Nation, May 10, 2001; Logan, 2003). Not only is the national median rate of segregation from whites higher for blacks than for other minority groups, but the black–white rates range upward to extremes that Hispanics and Asians do not experience in any metropolitan area (Table 3.1).

The rate of decline in black–white segregation has been especially slow in a number of large and highly segregated metropolitan areas, such as Detroit, Milwaukee, New York and Chicago. Logan notes that during the 1980s and 1990s these areas showed some of the highest rates of concentrated minority poverty in the nation. For blacks and for some Hispanic subgroups, extreme racial isolation is a marker for extreme social and economic distress. From the standpoint of integrating neighborhoods, in the 1990s both the lowest levels of residential segregation by race and the most encouraging improvements were in the west and south, in metropolitan areas than tended to be small or medium-sized and growing quickly and in which blacks constituted a modest share of the population (Glaeser and Vigdor, 2003).

In terms of exposure, the average white person in US cities and suburbs lives in a neighborhood that is overwhelmingly white (about 84 percent) and thus offers little exposure to other racial or ethnic groups. That typical neighborhood for a white person is just 7.1 percent black, 6.2 percent Asian and 3.2 percent Hispanic (Table 3.2). Notwithstanding the number of very

Table 3.2 Average neighborhood percentage racial composition, 331 US metropolitan areas, 2000[a].

Race	White	Black	Asian	Hispanic
Hispanic	3.2	9.4	11.7	42.1
Asian	6.2	2.6	19.3	4.2
Black	7.1	54.8	9.7	13.0
White	83.5	33.2	59.3	40.7

[a] Exposure (p^*) measures the composition of the average census tract of members of a given racial or ethnic group.
Source: author's calculations using Lewis Mumford Center, University at Albany, census data file.

racially isolated and poor ghettos and typically more mixed-income immigrant ethnic enclaves, Asians, Hispanics and blacks are, on average, much more exposed to other racial-ethnic groups than are whites. But over the 1980s and 1990s, high rates of immigration combined with concentrated settlement patterns to decrease the rate of Asian and Hispanic exposure to whites (Figure 3.1). In part, this reflects a clustering that is typical for newly arrived immigrants, especially those whose first language is not English. Immigrants arrive faster than they can 'diffuse' (spread out) across the housing market; moreover, ethnic enclaves offer assistance to new arrivals, from help in finding housing to securing a job and child care, obtaining familiar food, and celebrating religious and cultural traditions. A century ago, Italians, Irish and other groups showed a similar pattern of heavy immigration followed by residential and commercial clustering and therefore increased segregation. But immigrant ethnic enclaves can also be constraining, in that they limit contact with native English speakers, who

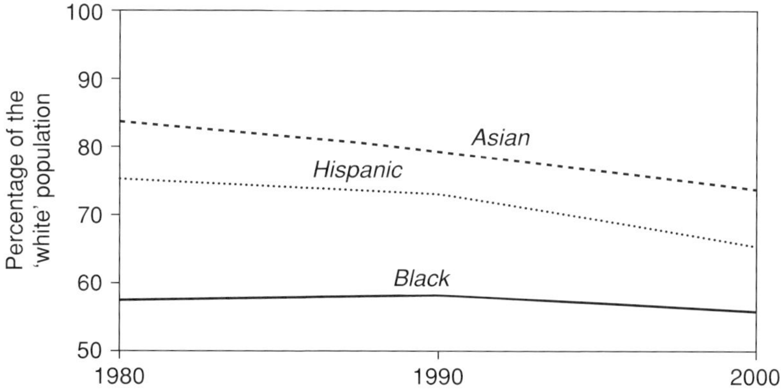

Figure 3.1 Minority-to-white neighborhood exposure in average tract, 325 US metropolitan areas, 1980, 1990 and 2000.[a]

Source: author's calculations using Lewis Mumford Center, university at Albany, census data file.
[a] On the exposure measure, see footnote for Table 3.2.

tend to have higher incomes, greater educational attainment and valuable social networks. The threshold question for any society committed to equal opportunity is whether wider housing choices exist for people who want to exercise them.

Over time, members of the two large, fast-growing immigrant groups, Asian and Hispanic American, move to the suburbs and take residence in more racially diverse neighborhoods. This is particularly true of the higher-skilled, middle-income members of those groups (Zhou, 2001; Alba and Nee, 2003). But even middle-income immigrants are relatively isolated from white neighborhoods – a pattern that is more pronounced for blacks – and some of the fastest growth in immigrant and black suburbanization is occurring in at-risk suburbs, which feature school failure, weak fiscal capacity and other problems long associated with vulnerable cities (Orfield, 2002). What is more, as suburbs became more racially and ethnically diverse in the 1990s, minority segregation in the suburbs increased. And in suburbs, just as in central cities, blacks are the group most segregated from whites.

It is a commonplace in the US to assume that, since housing discrimination has been illegal since 1968, segregation by race merely reflects differences in income among racial and ethnic groups: differences in where people live must simply reflect differences in what they can afford. Yet nationally the average black family earning more than $60,000 a year lives in a neighborhood with a higher poverty rate and lower educational attainment than the average white family earning less than $30,000 (Logan et al., 2003). In the housing market, we all make choices that are shaped in powerful ways by the choices of others. Their choices have a significant impact on the kinds of communities in which we can live. For example, blacks and other minorities with sufficient income can, barring discrimination, move into a racially integrated community, but they cannot ensure that the community will remain integrated. White avoidance of more racially diverse neighborhoods is a powerful factor in destabilizing integrated communities and prevents same-race communities from becoming stably integrated. Analyses focusing on neighborhood change over time and trajectories of individual households across neighborhoods over time both corroborate this pattern (Ellen, 2000; Quillian, 2002)[7].

On the other hand, geographers William Clark and Sarah Blue find that for blacks, Hispanics and Asians living in metropolitan areas with a high percentage of foreign-born residents (gateway communities), higher

---

[7] In his seminal analysis of segregative dynamics in housing markets, Schelling (1971) found that even small differences in the neighborhood preferences of various racial groups can lead to large and persistent rates of segregation. Avoidance, or 'self-steering' away from particular neighborhoods, captures a significant component of this dynamic.

education and, to a lesser extent, higher income are associated with higher rates of residential integration with whites (Zhou, 2001; Alba and Nee, 2003; Clark and Blue, 2004)[8]. These researchers stress that effects of class vary considerably by racial group and metropolitan area. Immigration and domestic migration patterns are obviously factors, because they shift the racial and economic profile of people who are in a given local (metropolitan) market for housing.

There are racial differences in income, of course, and so segregation by race does owe, albeit in modest measure, to differences in buying power. How segregated is the US, then, by those differences? Using census data from 1950 to 2000, sociologists Douglas Massey and Mary Fischer find that income segregation declined at the regional and state levels but increased within metropolitan areas – that is, at the level of neighborhoods and local jurisdictions – over the past half-century (Massey and Fischer, 2003). What this means is that income is more evenly distributed across states and major regions – a pattern, argue the researchers, that reflects growth in affluence in the south as well as increased poverty in the north-east and midwest over a long period of economic restructuring, immigration and significant internal migration across states. On the other hand, within any given metropolitan area, people of different incomes were less likely to share the same neighborhoods in the year 2000 than in 1970 or 1950. The trend is not limited to one racial group. For example, using the top and bottom fifths of the income distribution to describe rich and poor, respectively, Massey and Fisher find that segregation of the rich from the poor increased 34 percent among whites and 27 percent among blacks between 1970 and 2000.

Most of this increase appears to be associated with the economic dislocations of the 1970s and 1980s. In most regions of the country, as several studies have shown, the concentration of poverty in metropolitan areas actually declined somewhat in the 1990s (Jargowsky, 2003; Kingsley and Pettit, 2003). That is in part because some poor families gained income and in part because of migration patterns (higher-income households gentrifying urban neighborhoods; lower-income households moving out). Geographically concentrated poverty shifted away from in-demand urban neighborhoods and toward older suburbs, particularly in the nation's hundred largest metropolitan areas. There, between 1980 and 2000, the number of suburban census tracts with poverty rates of 30 percent or higher grew 89 percent, and the number of poor residents more than doubled.

---

[8] Neighborhood integration with whites, as well as 'suburban attainment', have long been used by analysts of race and immigration as indexes of minority group assimilation or progress over time.

Overall, Massey and Fischer indicate, income segregation has not decreased appreciably from its 1990 high. That is, even if the trend toward greater economic division did not get worse over the past decade, it did not reverse. One reason is that income inequality – which continued to increase in the 1990s, even though rising incomes at last lifted those on the bottom (Burtless and Jencks, 2003) – touches all racial groups in America. The sharp increase in residential segregation by income in America over the past 30 years thus reflects an expanded class divide within each racial group as well as a greater tendency for people with different financial resources, social contacts, political habits and educational access to live apart[9].

Whether in cities or suburbs, low-income, minority and female-headed families are more likely than other families to live in poor neighborhoods at a given point in time (Jargowsky, 1997). But as noted above, housing mobility is relatively high in America. One-third of all renter households move every year, and about half of the households in the country (renters and owners together) move every 5 years. In addition, neighborhoods can change dramatically 'around' someone who has not moved. So what kinds of neighborhoods are the various types of family exposed to over time? The limited empirical evidence on this question is sobering; when the long-run patterns of a representative national sample of black and white households are compared, race trumps both income and female-headed family structure as a predictor of the type of neighborhood to which families are exposed. Using the Panel Study of Income Dynamics for the 1979–90 period, sociologist Lincoln Quillian analyzed the probability that a family living in a poor neighborhood one year would be in a given type of neighborhood a year later (a non-poor neighborhood, the same poor neighborhood or a different poor neighborhood) (Quillian, 2003). He also analyzed moves into poor neighborhoods from non-poor ones and long-run patterns of 'recurrence' (falling back into poor neighborhoods in repeated cycles over time after a family once managed to live in a non-poor neighborhood). These shifts can occur because a family has moved, because the neighborhood has

[9] As Dickens (2003) and Levy (2003) note (in comments on the Massey–Fischer study), trend analyses that rely on summary measures of income segregation inevitably mask not only the details of social contact and their actual impact on human lives but also the local drivers of observed patterns. Poor people may have little contact with non-poor neighbors, those with contact may or may not benefit in ways we expect, and so on. Social researchers analyze many of these issues under the rubric of neighborhood effects and suggest a variety of ways that people at different income levels might affect each other, for better or worse, when they share neighborhoods (Jencks and Mayer, 1990; Leventhal and Brooks-Gunn, 2000; Ellen and Turner, 2003; de Souza Briggs, 2004). Moreover, we might want society to respond differently to choices by middle-income people of color – say, to leave poorer neighborhoods – as opposed to changes in income levels for people (of any racial background) residing in a given place. Though very different, both types of change can produce higher income segregation.

changed around the family, or for both reasons, so the probabilities capture effects of various structural advantages that some types of family enjoy.

Quillian finds that most blacks but only 10 percent of whites will live in a poor neighborhood at some point in a decade and that little of the difference is accounted for by racial differences in poverty rate or family structure. For example, when blacks in female-headed households with income below the poverty line were compared with whites in comparable households, 57 percent of blacks but only 27 percent of whites spent at least half of a 10-year span in a poor neighborhood. By this measure, even blacks in male-headed households with income above the poverty line face more risk (39 percent) than whites in female-headed, poor households – and far more than whites in comparable households (3 percent). Blacks leave poor neighborhoods often, but they fall back into such neighborhoods much more often than whites, leading Quillian to conclude, 'For African-Americans, the most difficult part of escape from a poor neighborhood is not moving out but staying out' (Quillian, 2003)[10]. It is not yet clear whether the sharp racial differences in neighborhood exposure changed in the 1990s or whether white–Hispanic and white–Asian differences are nearly as large as white–black ones.

School opportunity: the fall and rise of racial segregation

For many people of color and for some white people too, moving to the suburbs and even attaining middle-income status do not equate to 'making it' anymore – not if some middle- and upper-income families are increasingly able to seclude themselves from others. Perhaps the most important consequence is the geographic concentration of school failure. Housing integration, whether by race or class, is at best a proxy for access to opportunity or upward mobility in America. Access to good schooling, on the other hand, is a much more direct measure of an individual's prospects. Americans generally like to think of education as the great equalizer. Surely, even if wealth and income will always be somewhat unequal and the more affluent will always be able to afford somewhat better homes and neighborhoods, schools will reward anyone, of any generation, who strives. Few issues in public life, and certainly in domestic social policy, are more hotly debated than educational quality (Jencks and Phillips, 1998). But the

[10] In a related study (see note 9), I used simulation techniques to estimate potential effects of shifting transition probabilities (for example through housing policy and other tools), including escape (from poor areas to non-poor ones), moving over (from poor area to poor area) and falling back (from non-poor areas to poor ones). I find that even massive improvements in the riskiest transitions – doubling the odds of escape and halving those of moving over and falling back – would leave some households exposed to poor neighborhoods for significant periods over the long run.

trend in this debate, at least over the past 20 years, has been to pay much more attention to what resources schools have, particularly in terms of teacher quality, teacher expectations of students and physical facilities, than to the race and class makeup of the student body, which reflect segregated housing patterns. This lopsided focus is particularly worrisome in light of recent reforms stressing accountability and wider parental choice in education.

Why does enrollment composition matter? In 1966 the landmark federal study Equality of Educational Opportunity, led by sociologist James Coleman, indicated that differences in school resources explained only a modest portion of differences in school achievement (as measured by test scores), for both white and minority students. On the other hand, differences in enrollment – racial composition, parents' educational background, at-home resources – explained far more. Show me the neighborhoods where most of the kids have encyclopedias at home and plans to go to college, said Coleman – today one looks for internet access and plans for a professional career as well – and I will show you the high-scoring schools. Subsequent and more technically advanced studies yield similar results, as have studies of the long-run effects of careful desegregation experiments, which indicate particularly powerful effects on low-income minority students who attend desegregated schools with white peers (Coleman, 1966; Crain and Wells, 1994; Clotfelter, 2004). Compared with counterparts schooled in segregated minority schools, the desegregated students were not only more likely to attend college and have higher earnings, they had more racially diverse social networks as adults and were more comfortable navigating racially diverse environments throughout their lives.

This compelling and varied research evidence suggests that it is important not only to improve the contributions that schools and parents make to a child's achievement wherever the child happens to be enrolled but also to look for ways to teach disadvantaged children together with relatively advantaged peers. In public education, that has generally meant desegregating schools, since local school enrollment tracks the segregation of neighborhoods and municipalities by income and race. In an era of public school choice, it also means taking steps to further integration as school enrollments shift and new school options appear. In real estate, on the other hand, the effects of the segregated patterns that Coleman discovered two generations ago point to the marketing advantages of homogeneously affluent school districts, most of them racially homogeneous as well. Real estate agents are among the first to snap up reports of school test scores and college entry. Higher-scoring school districts fetch higher housing prices (other factors being equal), and economists find that changes in test scores register in housing prices too, underscoring the importance of school achievement to many homebuyers (Kane et al., 2003)[11]. In these high-performing schools districts, even buyers without children – buyers who, in effect, pay higher property taxes to support schools that they do not use – benefit financially from the favorable impact of good schools on housing prices.

As a policy objective, school desegregation since the landmark Brown v. Board of Education ruling, in 1954, has faced stiff competition from other cherished public priorities – preserving local decisionmaking in public education, respecting the primacy of parental choice, and specifically avoiding 'forced' busing, which often is unpopular with minority groups and whites alike – as well as from the private interests that those priorities protect, including property wealth (Clotfelter, 2004). For these and other reasons, school desegregation has received less and less support from the courts, which have released many local school districts from court-supervised desegregation efforts (Orfield and Eaton, 1996).

An earlier Supreme Court decision to block compulsory metropolitan desegregation programs, the 1974 Milliken decision, ensured the perpetuation of race and class barriers in education that are directly produced by segregated housing patterns. These barriers to equal opportunity at the elementary and secondary levels are all the more striking in the wake of the Supreme Court's 2003 Grutter rulings on undergraduate and law school admissions to the University of Michigan, which hold that diversity is a benefit to students of all races and, more controversially, that it is a compelling enough benefit to warrant a systematic approach to 'affirmative' admission of minority students to mostly white institutions of higher learning (Grutter v. Bollinger, 2003a, 2003b). For now, the courts have decided that institutions of higher learning can continue to swim upstream against the massive educational inequality by race to which segregation contributes in primary and secondary schools.

Today, almost 40 percent of public school students are minority, roughly double the share in the 1960s. This fact, along with the protection (through Milliken) of suburbs against metropolitanwide desegregation, has produced an increase in elementary and secondary school segregation by race as well as class. Hispanic students are the most segregated, by ethnicity,

---

11 Kane et al. (2003) add that housing values (in the Charlotte-Mecklenburg school district) did not respond quickly to changes in test scores and hypothesize that this may be because parents understand that scores fluctuate from year to year – are 'noisy', in the terms that analysts use – or because parents are more interested in the socioeconomic make-up of the schools than the test scores (alone) in a given year. Interestingly, given the lack of attention to school siting and neighborhood segregation in many public school choice debates, Kane et al. also find that parents place a high value on school proximity, holding other factors equal. For a critical review of the evidence on housing choice, housing prices, and school segregation, see Clotfelter (2004).

income and even language (Frankenberg et al., 2003). Asians are the most integrated group, and their college graduation rate, which is almost double the national average, is four times that of Latinos. After steady decline from the 1950s through the 1980s, black segregation in schools has increased to levels not seen in 30 years. A growing share of black and Hispanic students, particularly in the big-city school systems, attend schools that are virtually all non-white, characterized by high student poverty rates, limited school resources, less experienced and credentialed teachers, less educated parents, high student turnover, overcrowded and disorderly classrooms and a host of health and other problems.

These patterns are not limited to central cities. They reflect patterns of racial and economic change in many older suburbs and satellite cities as well, and those patterns in turn reflect avoidance by middle-class housing consumers of neighborhoods and school districts perceived to be shifting toward minority poverty (Ellen, 2000)[12]. Indeed, economist Charles Clotfelter finds that between-district segregation now explains a much larger share of overall segregation in metropolitan areas than was previously the case. In the most segregated metropolitan areas, between-district segregation represented less than 4 percent of all segregation in 1970, compared with an astonishing 84 percent in the year 2000. Between-district racial segregation is a greater problem in large metropolitan areas than in smaller ones, which offer fewer separate districts into which families may sort themselves. Such segregation – perhaps the most direct measure of the uneven geography of race and opportunity – is also much larger than segregation between public and private schools. Summarizing the long-run effects of countervailing court decisions – Brown and then Milliken – Clotfelter concludes, 'Whites who wanted to cushion or avoid the effects of (desegregating) actions had several means of doing so. The principal one was to seek out whiter school districts' (Clotfelter, 2004, p. 67).

What do persistent patterns of racial segregation imply in an age of school choice? When parents are dissatisfied with their assigned public schools, they can move into a new school district, send their children to private schools (religious or non-religious), teach their children at home, or – more and more, thanks to the federal No Child Left Behind Act of 2002 – make use of a growing number of public school options, including magnet

---

12 Orfield (2002) offers the particularly striking example of Matteson, Illinois, a suburb of Chicago that underwent significant economic and racial change in the 1980s and 1990s. Black families that moved into Matteson – pioneers, in that they were integrating a formerly all-white community – actually arrived with higher incomes and higher student test scores, on average, than those of their white neighbors. Yet perceptions that racial change spelled certain school decline and a threat to property values set in motion a self-fulfilling prophecy. Matteson has resegregated considerably, this time losing much of its white, middle-class population and income diversity as well.

and voucher programs and semiautonomous but publicly funded charter schools. The share of student enrollment in assigned public schools has decreased steadily, from 80 percent to 76 percent between 1993 and 1999 alone, a change explained almost entirely by the increase, from 11 percent to 14 percent, in the share of students attending 'chosen' public schools (Bielick and Chapman, 2003). These alternatives are increasingly important arbiters of segregation (or integration) in education.

This is not just a reflection of the interest in choice shown by government and reform advocates. Choice options are particularly popular with low-income and minority families, which have perhaps the greatest stake in the improvement of public education. But the evidence thus far indicates that without mechanisms to improve the information and transportation options available to all families and without the aim of racial integration, choice is compounding the problem of school segregation. According to a study by the Civil Rights Project at Harvard University, 10 years after the founding of the charter school movement in America, 70 percent of black charter school students attended highly segregated schools (with 90 percent or higher minority enrollments), compared to just 34 percent of black public school students, a significant difference, since fully one-third of all charter school students are black. Hispanic students make up just under 20 percent of both charter and public school enrollments, and their segregation rates are similar for both types of schools (42 percent of charter school students and 37 percent of public school students are in highly segregated schools). As the authors note, 'The justification for segregated schools as places of opportunity is basically a 'separate but equal' justification, an argument that there is something about the schools that can overcome the normal pattern of educational inequality that afflicts these schools' (Frankenberg and Lee, 2003, p. 53). But while studies show that parents in such chosen schools are generally more satisfied and more actively involved with their children's schools (Bielick and Chapman, 2003) – not a bad thing by any means – there is not yet any systematic research to show that charter schools actually perform better than assigned public schools. Moreover, many charter schools are founded by groups that wish to target specific populations, including racial groups. Political scientist Gary Orfield, a long-time analyst of segregation trends, reminds us that a variety of earlier choice experiments in racially integrated schools produced white flight. 'Those experiences,' says Orfield, 'were apparently unknown or overlooked by designers and supporters of many charter school policies' (Frankenberg et al., 2003, preface).

This assessment of how and where America is schooling its young people is not a defense of centralized decision making or of big, impersonal schools. There is evidence that certain kinds of flexibility in the classroom and in the organization of schools can improve learning, and studies of charter schools as a group inevitably lump together a wide array of approaches, making broad statements about performance unreliable. But segregation in public schools is steadily growing in ways that threaten the social and economic prospects of black and Hispanic students most of all, and charter schools and other choice options appear, if anything, to be exacerbating segregation. In schools as in housing, and in cities as well as in increasingly diverse suburbs, the objective of integration frequently conflicts with other objectives. As reports by the Civil Rights Project detail, the movement toward choice in education appears to present an opportunity – thus far untapped – to promote inclusive choice through expanded options and information, along with flexibility and innovation.

Geographic access to jobs

Much like the Coleman report, which stoked interest in the composition of public school enrollments, a classic economic analysis of the late 1960s has spurred considerable interest in the spatial mismatch between local job and housing locations. John Kain's influential study of racial differences in employment and metropolitan employment decentralization ('job sprawl') sounded an early warning of the massive restructuring of the US economy that would later devastate minority and less skilled workers, as well as the neighborhoods in which they were concentrated (Kain, 1968). Decentralization accompanied the erosion of an economy defined by heavily unionized, high-wage, blue-collar factory jobs concentrated in central cities (Bluestone and Harrison, 1982; Wilson, 1987). Deindustrialization (the emergence of a service economy driven by small to medium-sized firms rather than big factories) and sprawling physical development patterns (including the establishment of new job centers, or 'edge cities,' in suburban areas) both have contributed to the jobs–housing mismatch. By the mid-1990s, about 70 percent of all jobs in the manufacturing, retailing and wholesaling sectors, which tend to have many entry-level positions, were in the suburbs, and suburban job growth and business expansion continued to outstrip that in cities throughout the decade (Kasarda, 1995; US Department of Housing and Urban Development, 2000).

The jobs–housing mismatch disproportionately affects black and Hispanic households, as well as certain low-income Asian refugee groups, because these groups tend to live farther away from areas of job growth than other groups. The mismatch also affects job-hunting low-wage workers and welfare recipients regardless of race. To a significant degree, the link to job outcomes appears to be a function of unequal access to transportation by race: many suburban jobs are inaccessible by public transportation, and car ownership rates are much lower for blacks and Hispanics than for whites and for low-income people than for higher-income people.

But the mismatch also affects job outcomes through less direct factors, such as weaker social networks for transferring job information and endorsements and the stigma associated with living in a low-income, predominantly minority neighborhood (Ihlanfeldt and Sjoquist, 1998)[13]. Besides creating barriers to employment, the mismatch between housing and job locations means long commutes in many directions (for workers of all racial groups) and hellish traffic congestion in many metropolitan areas – both classic symptoms of sprawl (Putnam, 2000; Squires, 2002).

Underscoring the importance of housing choice in economic outcomes, economists Steven Raphael and Michael Stoll find that, during the 1990s, blacks' overall proximity to jobs improved somewhat (about 13 percent), largely because of black residential mobility within metropolitan areas; that is, because people, not jobs, moved. Still, blacks on average remain more physically isolated from jobs than members of any other racial group. Overall proximity to jobs did not change significantly for whites, Hispanics or Asians during the 1990s. Raphael and Stoll find that metropolitan areas with higher levels of residential segregation by race also show larger jobs–housing mismatches by race. In a related study, the researchers find that efforts to raise minority car-ownership rates to match those of whites, for example through ownership assistance (car voucher) programs that have a positive track record, could eliminate as much as 45 percent of the black–white differential and 17 percent of the Hispanic–white differential in geographic access to outlying jobs (Raphael and Stoll, 2000, 2002)[14].

Summary

The segregation of metropolitan areas by race and class shifted somewhat in the 1990s, bringing the risks long associated with inner cities into older suburbs as the latter became more racially and economically diverse. Poverty concentration declined in central cities, but the long-run trend, between 1970 and 2000, was one of wider class divide: segregation by income increased sharply for both blacks and whites. New evidence on household mobility and neighborhood change over time shows that blacks are much more likely than whites to get trapped in poor neighborhoods for long periods of time and to fall back into them repeatedly even after they manage to leave. While much more research is needed, particularly on

---

[13] Tilly et al. (2001) label the stigma 'space as a signal'. The pattern was first identified by Kirschenman and Neckerman (1991) in a survey of Chicago-area firms.

[14] A number of states have car-ownership programs targeting welfare recipients, to aid in the transition to work and help meet other life needs, such as access to childcare and family supports (Goldberg, 2001).

patterns of Asians and Hispanics, these tendencies lead to major racial differences in long-run exposure to particular types of neighborhood – those rich either in resources or in risk factors.

School segregation increased in the 1990s. Between-district segregation became the dominant component of racial segregation in education, underscoring the effect of unequal housing choice on educational inequality; and a growing share of the nation's children left assigned public schools for charter schools and other chosen alternatives. Minorities making those choices were more likely to attend segregated schools than were their counterparts in assigned public schools.

Access to the sprawling 'geography of jobs' likewise tracked housing choices. Minorities faced more spatial barriers, and blacks on average were the group most isolated from areas of job growth, though housing mobility reduced that isolation somewhat in the 1990s. Beyond the education and employment patterns specified here, a growing body of evidence suggests that housing segregation contributes to persistent racial disparities in exposure to crime and violence, physical and mental health status and health-related behaviors (disease, trauma, and other stressors, poor diet and exercise habits), and a variety of environmental health hazards, including pollution (Ellen et al., 2001; Pastor, 2001).

On the positive side, many minorities, including a growing number of immigrants, showed substantial housing mobility, and if performance improves in the schools with the lowest achievement scores, in time the movement for school accountability and choice could mitigate the high educational costs of living in segregated minority neighborhoods. But many minority families that moved to the suburbs in the 1990s, even if they became homeowners, did not escape the pattern that contains poverty, school failure and job isolation in particular geographic areas. The 'container' may no longer follow city limits, but it still contains (separates).

There is no reason to think that trends over the 1990s are anomalies or transient patterns that might reverse themselves. In the decade ahead, for example, the Census Bureau projects that fully two-thirds of all new household growth in America will come from racial and ethnic minority groups, both native and foreign born (Joint Center for Housing Studies, 2003). In this context, persistent inequalities by race – in the geography of residence, in access to quality schools and jobs and to safe and healthy neighborhoods, and in incomes – do not bode well for the society. True, none of these factors fully determines economic security, health, political influence, or other essentials of well-being, and the most important litmus test for any society is probably not equality of resources among groups at any point in time but fair and just access to opportunity and rewards over time – that is, whether groups are able to move up and whether they believe that they can do so. Still, these trends suggest reason for serious concern as the nation becomes more diverse and continues its attachment to sprawling, segregated growth at the local level. The uneven geography of opportunity – and in particular the limits on housing choice that are tied to race and income – are perhaps more important than ever before.

References

Alba, R. and Nee, V. (2003) *Remaking the American Mainstream*. Harvard University Press.

Altshuler, A. et al. (eds) (1999) *Governance and Opportunity in Metropolitan America*, pt. 1. National Academy Press, Washington.

Benfield, F., Raimi, M. and Chen, D. (1999) *Once there were Greenfields: How Urban Sprawl is Undermining America's Environment, Economy and Social Fabric*. Natural Resources Defense Council, Washington DC.

Berube, A. (2003) Racial and ethnic change in the nation's largest cities. In: Katz, B. and Lang, R.E. (eds) *Redefining Urban and Suburban America: Evidence from Census 2000, vol. 1*, pp. 137–54. Brookings, Washington DC.

Bielick, S. and Chapman, C. (2003) *Trends in the Use of School Choice, 1993–1999. Statistical Analysis Report 2003-031*. National Center for Education Statistics, US Department of Education.

Bluestone, B. and Harrison, B. (1982) *The De-Industrialization of America*. Basic Books, New York.

Burchell, R.W. et al. (2002) *Costs of Sprawl – 2000*. Transportation Research Board of the National Research Council. National Academy Press, Washington.

Burtless, G. and Jencks, C. (2003) American Inequality and Its Consequences. In: Aaron, H.J., Lindsay, J.M. and Nivola, P.S. (eds) *Agenda for the Nation*, pp. 61–108. Brookings, Washington DC.

Clark, W.A.V. and Blue, S.A. (2004) Race, class, and segregation patterns in US immigrant gateway cities. *Urban Affairs Review* **39** (6): 667–88.

Clotfelter, C.T. (2004) *After Brown: The Ruse and Retreat of School Desegregation*. Princeton University Press.

Coleman, J. (1966) *Equality of Educational Opportunity*. US Government Printing Office.

Crain, R.L. and Wells, A.S. (1994) Perpetuation theory and the long-term effects of school desegregation. *Review of Educational Research* **64** (4): 531–53.

Danielson, M. (1976) *The Politics of Exclusion*. Columbia University Press.

de Souza Briggs, X. (2004) *Traps and Stepping Stones: Neighborhood Dynamics and Family Well-Being*. KSG Working Paper No. RWP04-13. www.papers.ssrn.com

Dickens, W. (2003) Comment on Massey and Fischer. *Brookings-Wharton Papers on Urban Affairs* 2003: 1–30.

Downs, A. (2004) Growth management, smart growth, and affordable housing. In Downs, A. (ed.) *Growth Management and Affordable Housing: Do They Conflict?*, pp. 264–74. Brookings, Washington DC.

Ellen, I.G. (1999) Spatial stratification within US metropolitan areas. In: Altshuler, A. et al. (eds) *Governance and Opportunity in Metropolitan America*, pp. 192–212. National Academy Press, Washington.

Ellen, I.G. (2000) *Sharing America's Neighborhoods: Prospects for Stable Racial Integration*. Harvard University Press.

Ellen, I.G. and Turner, M.A. (2003) Do neighborhoods matter, and why? In: Goering, J. and Feins, J. (eds) *Choosing a Better Life? Evaluating the Moving to Opportunity Social Experiment*, pp. 313–38. Urban Institute, Washington.

Ellen, I.G., Mijanovich, T. and Dillman, K.-N. (2001) Neighborhood effects on health: exploring the links and assessing the evidence. *Journal of Urban Affairs* **23** (3–4): 391–408.

Estlund, C. (2003) *Working Together: How Workplace Bonds Strengthen a Diverse Democracy*. Oxford University Press.

Fischel, W.A. (2001) *The Homevoter Hypothesis: How Home Values Influence Local Government Taxation, School Finance, and Land Use Policies*. Harvard University Press.

Fishman, R. (2002) The Death and Life of American Regional Planning. In: Katz, B. (ed.) *Reflections on Regionalism*, pp. 107–23. Brookings, Washington DC.

Frankenberg, E. and Lee, C. (2003) *Charter Schools and Race: A Lost Opportunity for Integrated Education*. Civil Rights Project, Harvard University.

Frankenberg, E., Lee, C. and Orfield, G. (2003) *A Multiracial Society with Segregated Schools: Are We Losing the Dream?* Civil Rights Project, Harvard University.

Frey, W. (2003) Melting pot suburbs: a study of suburban diversity. In: Katz, B. and Lang, R.E. (eds) *Redefining Urban and Suburban America: Evidence from Census 2000, vol. 1*, pp. 155–80. Brookings, Washington DC.

Friedrichs, J., Galster, G. and Musterd, S. (2003) Neighborhood effects on social opportunities: the European and American research and policy context. *Housing Studies* **18** (6): 797–806.

Fulton, W. et al. (2001) *Who Sprawls Most? How Growth Patterns Differ Across the US*. Metropolitan Policy Program, Brookings, Washington DC.

Glaeser, E.L. and Kahn, M. (2004) Sprawl and urban growth. In: Henderson, J.V. and Thisse, J.F. (eds) *Handbook of Regional and Urban Economics: Cities and Geography, vol. 4*. North-Holland, New York.

Glaeser, E.L. and Vigdor, J. (2003) Racial segregation: promising news. In: Katz, B. and Lang, R.E. (eds) *Redefining Urban and Suburban America: Evidence from Census 2000, vol. 1*, pp. 211–34. Brookings, Washington DC.

Goldberg, H. (2001) *State and County Supported Car Ownership Programs Can Help Low-Income Families Secure and Keep Jobs*. Center on Budget and Policy Priorities, Washington.

Grutter v. Bollinger (2003a) 71 U.S.L.W. 4498.

Grutter v. Bollinger (2003b) 71 U.S.L.W. 4480.

Henig, J. (2002) Equity and the future politics of growth. In: Squires, G.D. (ed.) *Urban Sprawl: Causes, Consequences, and Policy Responses*, pp. 325–50. Urban Institute, Washington.

Ihlanfeldt, K.R. and Sjoquist, D. (1998) The spatial mismatch hypothesis: a review of recent studies and their implications for welfare reform. *Housing Policy Debate* **9** (4): 849–92.

Jackson, K. (2000) Gentleman's agreement: discrimination in metropolitan America. In: Katz, B. (ed.) *Reflections on Regionalism*, pp. 185–217. Brookings, Washington DC.

Jargowsky, P. (1997) *Poverty and Place: Ghettos, Barrios, and the American City*. Russell Sage, New York.

Jargowsky, P. (2003) *Stunning Progress, Hidden Problems: The Dramatic Decline of Concentrated Poverty in the 1990s.* Metropolitan Policy Program, Brookings, Washington DC.

Jencks, C. and Mayer, S.E. (1990) The social consequences of growing up in a poor neighborhood. In: Lynn, L.E. and McGeary, M.G.H. (eds) *Inner-City Poverty in the United States.* National Academy Press, Washington.

Jencks, C. and Phillips, M. (1998) The black-white test score gap: an introduction. In: Jencks, C. and Phillips, M. (eds) *The Black-White Test Score Gap*, pp. 1–53. Brookings, Washington DC.

Joint Center for Housing Studies (2003) *State of the Nation's Housing.* Harvard University.

Kain, J. (1968) Housing Desegregation, Negro Employment, and Metropolitan Decentralization. *Quarterly Journal of Economics* **32** (2): 175–97.

Kane, T.J., Staiger, D.O. and Samms, G. (2003) School Accountability Ratings and Housing Values. *Brookings-Wharton Papers on Urban Affairs* 2003: 83–138.

Kasarda, J. (1995) Industrial restructuring and the changing location of jobs. In: Farley, R. (ed.) *State of the Union: America in the 1990s, vol. 1, Economic Trends*, pp. 215–68. Russell Sage, New York.

Katz, B. and Lang, R.E. (2003) Introduction. In Katz, B. and Lang, R.E. (eds) *Redefining Urban and Suburban America: Evidence from Census 2000, vol. 1*, pp. 1–13. Brookings, Washington DC.

Kingsley, G.T. and Pettit, K. (2003) *Concentrated Poverty: A Change in Course.* National Neighborhood Indicators Project, Urban Institute.

Kirschenman, J. and Neckerman, K.M. (1991) 'We'd love to hire them but . . .': The meaning of race for employers. In: Jencks, C. and Peterson, P.E. (eds) *The Urban Underclass*, pp. 203–32. Brookings, Washington DC.

Leventhal, T. and Brooks-Gunn, J. (2000) The neighborhoods they live in: the effects of neighborhood residence on child and adolescent outcomes. *Psychological Bulletin* **126** (2): 309–37.

Levy, F. (2003) Comment on Massey and Fischer. *Brookings-Wharton Papers on Urban Affairs* 2003: 1–30.

Lindsay, J.M. and Singer, A. (2003) Changing faces: immigrants and diversity in the twenty-first century. In: Aaron, H.J., Lindsay, J.M. and Nivola, P.S. (eds) *Agenda for the Nation*, pp. 217–60. Brookings, Washington DC.

Logan, J. (2003) Ethnic diversity grows, neighborhood integration lags. In: Katz, B. and Lang, R.E. (eds) *Redefining Urban and Suburban America: Evidence from Census 2000, vol. 1*, pp. 235–56. Brookings, Washington DC.

Logan, J., Oakley, D. and Stowell, J. (2003) *Segregation in Neighborhoods and Schools: Impacts on Minority Children in the Boston Region.* Metro Boston Equity Initiative of the Civil Rights Project, Harvard University.

Massey, D.S. (2001) Residential segregation and neighborhood conditions. In: Smelser, N.J., Wilson, W.J. and Mitchell, F. (eds) *America Becoming: Racial Trends and Their Consequences, vol. 1*, pp. 391–434. National Academy Press, Washington.

Massey, D.S. and Denton, N.A. (1993) *American Apartheid: Segregation and the Making of the Underclass.* Harvard University Press.

Massey, D.S. and Fischer, M. (2003) The geography of inequality in the United States, 1950–2000. *Brookings-Wharton Papers on Urban Affairs* 2003: 1–30.

Massey, D.S., White, M.J. and Phua, V.C. (1996) The dimensions of segregation revisited. *Sociological Methods and Research* **25** (2): 172–206.

Moe, R. and Wilkie, C. (1997) *Changing Places: Rebuilding Community in the Age of Sprawl*. Henry Holt, New York.

Nelson, A. (2003) Top ten state and local strategies to increase affordable housing supply. Fannie Mae Foundation. *Housing Facts and Findings* **5** (1): 1, 4–7.

Nivola, P.S. (1999) *Laws of the Landscape: How Policies Shape Cities in Europe and America*. Brookings, Washington DC.

Nyden, P., Maly, M. and Lukehart, J. (1997) The emergence of stable, racially and ethnically diverse urban communities: a case study of nine US cities. *Housing Policy Debate* **8** (2): 491–534.

Orfield, G. and Eaton, S.E. (1996) *Dismantling Desegregation: The Quiet Reversal of Brown v. Board of Education*. New Press, New York.

Orfield, M. (2002) *American Metropolitics: The New Suburban Reality*. Brookings, Washington DC.

Pastor, M., Jr. (2001) Geography and opportunity. In: Smelser, N.J., Wilson, W.J. and Mitchell, F. (eds) *America Becoming: Racial Trends and Their Consequences, vol. 1*, pp. 435–67. National Academy Press, Washington.

Pastor, M., Jr. et al. (2000) *Regions that Work: How Cities and Suburbs Can Grow Together*. University of Minnesota Press.

Pendall, R., Nelson, C., Dawkins, C. and Knaap, G. (2005) Connecting smart growth, housing affordability, and racial equity. In: Briggs, X. (ed.) *The Geography of Opportunity: Race and Housing in Metropolitan America*. Brookings Institution Press, Washington DC.

Putnam, R. (2000). *Bowling Alone: The Collapse and Revival of Community in America*. Simon and Schuster, New York.

Quillian, L. (2002) Why is black-white residential segregation so persistent? Evidence on three theories from migration data. *Social Science Research* **31**: 197–229.

Quillian, L. (2003) How long are exposures to poor neighborhoods? The long-term dynamics of entry and exit from poor neighborhoods. *Population Research and Policy Review* **22**: 221–49.

Raphael, S. and Stoll, M.A. (2000) *Can Boosting Minority Car-Ownership Rates Narrow Inter-Racial Employment Gaps?* Working paper. Joint Center for Poverty Research, Northwestern University.

Raphael, S. and Stoll, M.A. (2002) *Modest Progress: The Narrowing Spatial Mismatch between Blacks and Jobs in the 1990s*. Metropolitan Policy Program, Brookings, December.

Rawlings, L., Harris, L. and Turner, M.A., with Padilla, S. (2004) *Race and Residence: Prospects for Stable Racial Integration*. Policy Brief 3. Metropolitan Housing and Communities Center, Urban Institute.

Rusk, D. (1999) *Inside Game/Outside Game: Winning Strategies for Saving Urban America*. Brookings.

Savitch, H.V. (2002) Encourage, then cope: Washington and the sprawl machine. In: Squires, G.D. (ed.) *Urban Sprawl: Causes, Consequences, and Policy Responses*, pp. 141–64. Urban Institute, Washington.

Schacter, J.P., Franklin, R.S. and Perry, M.J. (2003) *Migration and Geographic Mobility in Metropolitan and Nonmetropolitan America: 1995 to 2000*. CENSR-9. Census Bureau.

Schelling, T.C. (1971) Dynamic models of segregation. *Journal of Mathematical Sociology* **1**: 143–86.

Squires, G.D. (2002) Urban sprawl and the uneven development of metropolitan America. In: Squires, G.D. (ed.) *Urban Sprawl: Causes, Consequences, and Policy Responses*, pp. 1–22. Urban Institute, Washington.

Staley, S. (1999) *The Sprawling of America: In Defense of the Dynamic City*. Policy Study 236. Reason Public Policy Institute, Los Angeles.

Staley, S. (2000) *The Vanishing Farmland Myth and the Smart Growth Agenda*. Policy Brief 12. Reason Public Policy Institute, Los Angeles.

Suro, R. and Singer, A. (2003) Changing patterns of Latino growth in metropolitan America. In: Katz, B. and Lang, R.E. (eds) *Redefining Urban and Suburban America: Evidence from Census 2000, vol. 1*, pp. 181–210. Brookings, Washington DC.

Tilly, C. et al. (2001) Space as a signal: how employers perceive neighborhoods in four metropolitan labor markets. In: O'Connor, A., Tilly, C. and Bobo, L.D. (eds) *Urban Inequality: Evidence from Four Cities*, pp. 304–39. Russell Sage, New York.

US Census Bureau (2001) *Overview of Race and Hispanic Origin*. Census 2000 brief. US Government Printing Office.

US Census Bureau (2002) *2002 Census of Governments, vol. 1, no. 1, Government Organization*. GC02(1)-1. US Government Printing Office.

US Census Bureau (2003) *National Population Estimates to July 1, 2002 (Tables)*. US Government Printing Office.

US Department of Housing and Urban Development (2000) *The State of the Cities 2000*. US Government Printing Office.

US Department of Housing and Urban Development (2004) *US Housing Market Conditions: First Quarter 2004*. US Government Printing Office.

Wilson, F.D. and Hammer, R.B. (2001) Ethnic residential segregation and its consequences. In: O'Connor, A., Tilly, C. and Bobo, L.D. (eds) *Urban Inequality: Evidence from Four Cities*, pp. 272–303. Russell Sage, New York.

Wilson, W.J. (1987) *The Truly Disadvantaged: The Inner City, the Underclass, and Public Policy*. University of Chicago Press, Chicago.

Zhou, M. (2001) Contemporary immigration and the dynamics of race and ethnicity. In: Smelser, N.J., Wilson, W.J. and Mitchell, F. (eds) *America Becoming: Racial Trends and Their Consequences, vol. 1*, pp. 200–40. National Academy Press, Washington.

Changing Cities and Neighbourhoods

4

Reflecting on New York City's Housing Policy: 1987 to 2004

Jerilyn Perine and Michael H. Schill

Introduction

New York City has always been at the forefront of American housing policy. The first publicly owned housing developments in the nation – First Houses – was built in the City's Lower East Side in 1936, 1 year before the nation's public housing program was enacted. New York pioneered the use of below market interest rate financing of subsidized housing in its Mitchell-Lama[1] rental housing programs, a model that was to be used by the national government several years later. The city was also the first in the nation to outlaw racial discrimination in the privately owned housing market.

Since 1987, New York City has continued to lead the nation in innovative housing policy. Faced with widespread housing deterioration, arson and abandonment in the mid-1980s, the city began what has come to be known as the Ten Year Plan, a program that ultimately continued for over 15 years and which systematically built or rehabilitated over 200,000 units of housing, more than the total housing stock of many sizable cities. This unprecedented investment of over 6 billion dollars in housing construction re-wrote the textbook in terms of what a city could do to improve the housing and neighborhood conditions of its citizens.

In the first part of this chapter, we briefly summarize the demographic patterns that shape New York City's housing market. We also provide

[1] Created by New York State law, the program provided below market construction and permanent financing which required that the housing created would remain affordable to middle-income households for the term of the mortgage loan (typically 30 years).

a description of current housing conditions and problems. In the second section, we describe and assess the reasons why the city launched the Ten Year Plan, how the plan was structured and some of its impacts. In the third part, we show how the successes of the Ten Year Plan sparked a new set of challenges reflected in the New Housing Marketplace Program adopted by Mayor Michael Bloomberg in 2002. Finally, in the concluding section we reflect on some of the lessons of the Ten Year Plan, lessons that can help inform housing policy for cities throughout the nation.

Population and housing in New York City

New York City is, by far, the largest city in the US. After losing population for decades, the city began to grow in the 1980s reaching an all-time high of 8,008,278 residents in 2000 (Previti and Schill, 2003). True to its history as the major gateway into the US, New York's striking increase in population was largely attributable to unprecedented waves of immigration. In 2000, the city was home to 2,871,032 people who were born outside the US – roughly 36 percent of the total population. The largest groups of foreign-born residents include those from Latin America (655,190), Europe (557,492), the Caribbean (487,043), the Dominican Republic (369,186) and China (261,551).

The diversity of New York's population extends beyond the nativity of its residents. In terms of race, 47.5 percent of all residents are white, 28.4 percent are black and 10.9 percent are Asian, based on the 2000 US Census of Population. People of Hispanic origin (who can be either white or black in terms of race) constitute 27 percent of the total population and are rapidly growing both in absolute and relative terms. Racial and ethnic patterns are not uniform across the city's five boroughs. Blacks comprise approximately 38 percent of the population of the Bronx and Brooklyn, yet only 10.5 percent and 18.9 percent of the populations of Staten Island and Manhattan, respectively. Similarly persons of Hispanic background comprise 48.7 percent of the population of the Bronx, but only 12.1 percent of the population of Staten Island. Racial and ethnic minorities tend to earn lower incomes than white households. For example, according to the 2002 Census, in Brooklyn, the most populous of the five boroughs, the median income of white households was $44,404 compared to $33,184 among blacks and $26,661 among Hispanics.

In terms of the city's physical structure, density varies across the five boroughs. Manhattan is the most densely settled area in the US with over 66,940 persons per square mile of land area. On the other hand, Staten Island looks very much like an older suburb with a density of only 7587.9 persons per square mile.

With respect to housing, New York is primarily a city of renters. According to the 2000 Census, only 30.2 percent of all households own their homes in the city compared to a nationwide average of 66.2 percent. Again, these overall statistics mask a significant amount of heterogeneity in the city. In the borough of Queens, for example, the home ownership rate is 42.8 percent; in Staten Island it is 63.8 percent.

The housing stock of New York is among the oldest in the nation. According to the 2002 Housing and Vacancy Survey, 41.4 percent of all of the housing units in the city were built before 1930. The quality of the housing, however, is, on the whole, quite good and getting better. One indicator of housing quality is whether a housing unit has five or more 'maintenance deficiencies', which include holes in the ceiling, rodent infestation and broken plumbing. In 2002, only 2.8 percent of all housing units in New York had five or more of these problems.

Another indicator of housing quality is crowding. Typically in the US, an overcrowded housing unit is defined as one with over one person per room; an extremely overcrowded unit has more than 1.5 persons per room. In 2002, only 3.8 percent of all housing units in the city were extremely overcrowded.

Although the quality of housing in New York City is generally good, much of it strains the wallets of its residents. The 'rule of thumb' in the US is that rent should not consume more than 30 percent of one's income. By this indicator, New York's housing is extremely unaffordable – roughly 42.2 percent of all households pay more than 30 percent of their incomes for rent. Indeed, 459,335 households or 22.7 percent of all renters pay more than half their incomes in rent, a level of rent burden that can only be characterized as severe.

Interestingly, rent burdens in New York City are not extraordinarily higher than those of other large American cities. One of the reasons for this is the relatively high proportion of housing that is either rent regulated or subsidized. In 2002, 68.1 percent of the city's entire rental housing was rent regulated. When added to the housing that is subsidized through federal or city programs, nearly three-quarters of all housing rental housing units are not charging market rents.

The Ten Year Plan for housing

The antecedents

What came to be known as the Ten Year Plan had its roots in the demographic shifts that buffeted New York City in the years following World

Table 4.1 Population change in key US cities, 1950–2000.

City	Population						% change from 1950 to 1990
	1950	1960	1970	1980	1990	2000	
New York	7,891,957	7,781,984	7,894,862	7,071,639	7,322,564	8,008,278	+1.47%
Chicago	3,620,962	3,550,404	3,366,957	3,005,639	2,783,726	2,896,016	−20%
Philadelphia	2,071,605	2,002,512	1,948,609	1,688,210	1,585,577	1,517,550	−26.7%
Detroit	1,849,568	1,670,144	1,511,482	1,203,339	1,027,974	951,270	−48.5%
Baltimore	949,708	939,024	905,759	786,775	736,014	651,154	−31.4%
Cleveland	914,808	876,050	750,903	573,822	505,616	478,403	−47.7%
Milwaukee	637,392	741,324	717,099	636,212	628,088	596,974	−6.3%
St. Louis	856,796	750,026	622,236	453,085	396,685	348,189	−59.3%
Pittsburgh	676,806	604,332	520,117	423,938	369,879	334,563	−50.7%

Source: US Census Bureau Data.

War II[2]. In the 1970s and early 1980s, America's cities were swept with a wave of population loss, employment decline (which was particularly acute in the older industrialized cities that had already suffered a steady loss of manufacturing jobs and population in the 1950s and 1960s) and rampant housing abandonment. For some cities, the loss of jobs, people and housing was so acute that they never regained their economic or social strength – cities like Detroit with a population high of 1,849,568 in 1950, now has only 951,270 residents according to the 2000 census. Some other major cities that lost population in the last three decades are shown in Table 4.1.

New York's population has reached an all-time high of over 8 million people according to the 2000 census (which most analysts believe is an undercount). Its economy has been able to diversify from its reliance on small manufacturing based on its historic port activities, with its cultural life attracting tourism at an unprecedented scale, attracting new investment and generating significant employment. Table 4.2 highlights changes in key employment areas from the 1950s to the present.

There were many reasons for the decline in the decades following World War II of America's largest and oldest cities. A postwar shortage of housing for returning veterans was met by federal policies that subsidized the construction and purchase of small homes in outlaying areas of the cities. In addition, road systems needed to bring people, particularly middle-class

[2] For a more detailed analysis of the antecedents of the Ten Year Plan see Jerilyn Perine, 'Looking Back at the Ten Year Plan' (unpublished manuscript, 2004).

Table 4.2 New York City employment patterns, 1950–2001.

1950			2001		
Rank	**Industry**	**Employment**	**Rank**	**Industry**	**Employment**
1	Manufacturing	1038	1	Services	1465
2	Trade	754	2	Trade	619
3	Services[a]	507	3	Government	563
4	All government	374	4	FIRE	486
5	Finance, insurance, real estate (FIRE)	336	5	Manufacturing	230
6	Transportation, public utilities	331	6	Transportation	213
7	Construction	123	7	Construction	121

[a] Hotel, lodging, amusement, health, legal, professional, educational.
Source: US Department of Labor, Bureau of Labor Statistics.

families, out of America's cities and into its newly created suburbs. This process was heavily subsidized by the federal government[3].

These policies contributed not just to a population loss in America's older cities, but a specific one – the loss of middle-income families, who were predominately white[4]. For example, during the 1950s, New York's suburban population surpassed the urban population for the first time (Starr cited in Plunz, 1990, p. 274). The impact on the tax base of American cities and on their character was dramatic. Without the manufacturing jobs that provided a source of employment for workers without high educational levels, the population left behind, now more likely to be composed of lower income and minority households, would find far fewer jobs available to them than their predecessors. In 1966, there were just over one million manufacturing jobs in New York City. By 2001, this number had decreased to 230,000 jobs, a loss of over 75 percent of all manufacturing jobs in the city.

The housing stock left behind by the 'white flight' (Jackson, 1985) of the 1950s and 1960s began a slow decline that would culminate in the 1970s and 1980s with widespread disinvestment and abandonment, and

[3] One of the 'most important inducements to single-family residences' is found in The US Internal Revenue Code, which allows for the deduction of mortgage interest and property taxes from total taxable income (Jackson, 1985, p. 293). The value of these deductions is $63 billion, which is twice the amount allocated to all other federal housing programs in the US (Collins et al., 1998).

[4] White flight from inner cities was accelerated by a combination of postwar migration of black households from the south, racism and blockbusting activities by real estate brokers (Massey and Denton, 1993).

ultimately the rise of homelessness on a level not seen since the Great Depression of the 1930s[5].

The high inflation in the 1970s further weakened New York's real estate market as the costs of operating and maintaining buildings rose faster than tenant incomes. In addition, extensive and often misguided urban renewal policies of the 1960s and 1970s demolished entire communities, replacing them with highways and densely constructed high-rise buildings which were poor substitutes for the cohesive neighborhoods that had fostered stability and a sense of community for generations of low-, moderate- and middle-income New Yorkers[6].

In some cities like Detroit and New York City, arson, often paid for by the landlord, would further decimate communities and threaten the lives of those struggling to remain. Into this void, rampant crime would ravage those who were left behind.

As cities such as New York declined, financial institutions took actions that would accelerate the process. In many cities, banks, often aided and abetted by federal mortgage insurance underwriting policies, refused to underwrite mortgage loans in inner-city neighborhoods, particularly those composed of racial and ethnic minorities. This process, often called 'redlining', cut these communities off from credit, depressed sales prices and further reduced incentives to maintain properties.

With rents plummeting and middle-class families fleeing, many landlords simply began to collect whatever rent they could, stopped paying property taxes and carried out only the most basic maintenance and services. While the owners 'milked' their buildings in desperate and cynical attempts to recoup some of their investment, those tenants unable to flee

---

[5] In September of 1979, there were 1136 abandoned buildings in the Bronx, New York City, certified for demolition and 2121 unsafe buildings in the legal process, with most added to the demolition list (Dallas, 1979). 'About 140,000 people lived in Bushwick (Brooklyn) in 1970. By 1977, after rows of wood-frame houses had been destroyed by arson, the population had fallen by a third' (Roberts, 1987). Prior to 1982, the city did not classify displaced, burned out families as homeless. The city was not compelled by law to place these families, but could do so because there was an availability of affordable housing. In 1982, the city counted 7584 individuals as homeless. By 1988, the number had increased to 27,646, representing an increase of nearly 265 percent (New York City Department of Homeless Services Average Daily Census).

[6] 'During the 1960s the urban renewal and public housing programs that were supposed to revitalize declining US cities came under increasing criticism . . . Large urban public housing projects, beset by the rising number of crimes and falling revenues, began to deteriorate. The demolition in 1973 of Pruitt-Igoe, a massive high-rise housing project in St. Louis, symbolized the despair that surrounded the public housing program' (Hoffman, 2001). This is similar to Hulme, Manchester, which was constructed during the 1960s and 1970s in remotely located residential neighborhoods for its poor populations, tearing down gridded urban neighborhoods in the process. The project was 'a guaranteed failure' and by the 1990s was a wasteland, causing housing professionals to conclude that they had to start again (Boylan, 2000).

to greener pastures were left to fend for themselves, often without heat, hot water or basic services. It would not be long before some owners made the leap from irresponsibility to criminality, employing arsonists to collect building insurance.

Between 1970 and 1981 New York City lost an astounding 321,000 housing units through this downward spiral – nearly the number of units that would house the city of Baltimore today. This loss, most acute in the 3-year period between 1975 and 1978, was overwhelmingly concentrated in the city's lowest income communities located in the South Bronx (e.g. Mott Haven, Melrose), central Brooklyn (e.g. Bushwick, East New York, Bedford-Stuyvesant, Ocean Hill/Brownsville) and northern Manhattan (e.g. Harlem).

In response to this devastation, academics and public policy experts began to discuss 'planned shrinkage' (Plunz, 1990, p. 323)[7], a strategy to acknowledge that some communities simply may have lost too many people, too many buildings to warrant continued municipal services. They would simply need to be cut off and eliminated. This was met in many circles with horror, not least by the people still living in these communities. Tenant advocates soon became much more focused and organized and many grassroots organizations, created by residents themselves, began to develop. Many of these fledgling organizations, started in response to a direct assault on their communities, would form the core of the community and housing development organizations that would play an integral part in the reconstruction of these neighborhoods beginning in the mid-1980s and continuing through today[8].

The massive abandonment of New York City's housing stock caught all levels of government off guard. On 13 July 1977, a major blackout left New York City in the dark, and looting and arson occurred in a number of neighborhoods. By 1980, Bushwick in Brooklyn had lost 20 percent of its housing, one-third of its population and nearly half its businesses (Barnes, 2000; Mahler, 2005).

At the federal level, housing programs and funds were designed to spur construction and to remove blight through slum clearance (e.g. the Urban Renewal Program). These programs typically took years to implement and were not designed to respond to an immediate crisis that resulted in 10,000 abandoned buildings with 100,000 deteriorated units, half of which were inhabited by tenants living in deplorable conditions. In addition the federal

---

[7] Roger Starr, 'Making New York Smaller;' Starr, 'Letters,' p. 16 as cited in Plunz, 1990, p. 323.

[8] By 1994, more than 400,000 units of housing had been produced by CDCs, which account for 40 percent of all non-profit housing produced in the United States National Congress for Community Economic Development, 1995 (O'Regan and Quigley, 2000).

government had largely withdrawn from funding housing production as a result of President Richard Nixon's 'moratorium' on housing spending in 1973 and the end of the Section 8 New Construction and Substantial Rehabilitation programs in 1983 (Dolbeare et al., 2004)[9].

New York State's greatest involvement during this period was in the funding of new middle-income housing construction. Through the state's Mitchell-Lama Program, 269 projects with 105,000 housing units were created from the Program's inception in 1955 through the mid-1970s[10] (the City of New York simultaneously developed 60,000 units). This type of housing largely depended on a strong housing market and development sites often created through urban renewal. It did nothing to insulate the hardest hit communities from abandonment and population loss. Indeed it is clear that the development of the 15,000 unit Co-op City project in the Bronx removed thousands of stable households from the elegant rental housing of the Grand Concourse, further pushing the north-west Bronx into decline.

Aside from its involvement in the Mitchell-Lama Program, in the 1960s and 1970s New York City played a supporting role to federally funded housing programs, clearing land through Urban Renewal for the new construction of federal or state funded programs. Neighborhood planning was relegated to the city's housing department and became synonymous with Urban Renewal, relocation and demolition. While much needed new additions to the overall housing stock were created through Urban Renewal's high-rise solutions (which were mostly successful in New York City), the preservation of the existing housing stock was not a concern. Housing 'solutions' became identified with demolition and new construction – building 'new communities' not saving old ones. The decline of New York City's neighborhoods during this period highlighted the vast gap that existed between the needs of neighborhoods and the ability of government to craft appropriate responses.

In addition, by the mid-1970s the city's financial situation had deteriorated almost to the point of bankruptcy. Turning to the federal government during the Ford administration for help out of its overall fiscal crisis, the city was greeted with a resounding no, memorialized in a famous New York tabloid's headline – FORD TO CITY: DROP DEAD[11].

Housing advocates began to look to a little known law that allowed the city to foreclose on tax delinquent property and assume ownership of buildings. This procedure, formally known as an *in rem* proceeding, was

---

[9] Jacobs B.G. et al., 'Guide to Federal Housing Programs, 1986' as cited in Dolbeare et al., 2004.

[10] New York State Department of Housing and Community Renewal.

[11] *The New York Daily News*, 30 October 1975.

named after the Latin term for 'against the property'. Local government saw such foreclosures as a tool to increase tax revenues by pressing tax collection enforcement at a time when the city itself was on the brink of bankruptcy. Extensive lobbying and public pressure finally got the attention of city officials. The law was changed to permit the 'taking' of such properties through tax foreclosure annually, rather than after 3 years, in the hope that buildings could be taken into city ownership before their inevitable cycle of decline into complete abandonment came to pass.

Unfortunately for the city, few owners redeemed their properties and paid their taxes. Instead, the city became the owner of a massive inventory of the most dilapidated stock of housing; a stock the city was not prepared to manage. Rather than increasing tax revenues, this inventory would cost the city billions of dollars to maintain, ultimately renovate and then return to private ownership. The city's housing agency was transformed from the Housing Development Administration to the Department of Housing Preservation and Development (HPD) and virtually overnight became the owner and manager of nearly 100,000 units in over 10,000 buildings of the worst housing stock in the city.

Interim management of this housing stock required neighborhood-based property management services that were established directly in the communities involved. This work was difficult and often dangerous, as these properties were not just the center of blight and deterioration, but also became havens for drug and other illegal activities. More than $300 million would be spent annually on just the basic management of this housing stock (NYC Department of Housing Preservation and Development, 1987; Arthur Andersen Consulting, 1995). Many buildings would be lost altogether, being too dilapidated for renovation, with demolition the only option. In the late 1970s and early 1980s, HPD was demolishing an average of 1750 buildings per year[12].

The Ten Year Capital Plan for Housing

It would be nearly a decade before New York City's finances improved enough to allow it to invest significant capital in these buildings. With an ability to re-enter the municipal bond market, the city's leaders developed a plan to use the sale of general obligation bonds to raise capital to fund a long-term housing reconstruction plan. Known at the time as the Ten Year Plan, a reference to the longest budget planning cycle that the city undertakes, it laid out a broad-based strategy to rebuild the abandoned housing stock that had come into city ownership.

---

[12] Mayor Edward I. Koch, 'Of Decals and Priorities for the South Bronx', *The New York Times*, 19 November 1983, Section 1, p. 24.

With a long-term financial commitment of $5 billion, HPD's staff created an array of program models that provided the planning, funding and technical assistance to carry out large-scale new construction where buildings could not be saved and rehabilitation where they could be preserved. Never before had a municipal government stepped in to reconstruct entire communities, on a scale not seen since the US government set about the reconstruction of Germany through the Marshall Plan following World War II.

While funding was of great importance, reliance on the grassroots organizations originally created to fight for tenants' rights would also be integral to the plan's success. With names like Mid Bronx Desperados, Los Sures and the Northwest Bronx Clergy Coalition, they would form a community development sector that would grow to more than 80 organizations throughout the city. Tenants themselves would be trained and organized to form low-income cooperatives and take over ownership and management of their buildings. In addition, private owners who remained committed to the areas that they had worked in would also be a resource along with contractors who were willing to step into the role of affordable housing developer.

Program models included direct construction by HPD, funded and carried out exclusively with agency staff. More typical, however, were programs that provided non-profit groups and private, profit-motivated developers with grants or low-cost loans (many of which were written off over time) and those that blended city grant or low-cost loan funds with conventional bank financing. This 'sliding scale' of public subsidies would become an important element in the city's programs. As local markets began to recover, public subsidies could be decreased and ultimately withdrawn, leveraging more and more private funding.

Maintaining affordability for existing tenants was and continues to be a key element in all of the housing programs that are designed and implemented by HPD. This approach requires not just capital subsidies, but also frequently operating subsidies and direct tenant subsidies funded through the federal Section 8 Program. All of this work has since 1986 resulted in the expenditure of more than $6 billion, which, in turn, has assisted the new construction or renovation of more than 200,000 units of affordable housing. This housing has overwhelmingly been concentrated in the same communities that were so devastated by housing abandonment in the 1970s and 1980s. Overall approximately 60 percent of these units were preserved or created for low-income households, 30 percent for moderate-income households and 10 percent for the homeless (New York City Department of Housing Preservation and Development, 2004).

By the mid-1990s it became clear, however, that the considerable work that was being focused on the renovation and return to new ownership of the vacant city-owned *in rem* stock (and to a lesser extent the occupied *in rem* buildings), was not stemming the intake of new tax-foreclosed buildings. The cycle of abandonment was continuing (although at a slower rate) and the cost of *in rem* management had soared to more than $90 million per year. The Dinkins administration had imposed a moratorium on *in rem* vestings in 1992, but no system had been put in place to address newly abandoned buildings, other than preventing them from becoming city-owned.

At the beginning of the Giuliani administration the city engaged the consultancy firm Arthur Anderson and Co. to conduct a pro bono analysis of the actual costs of the interim management of the *in rem* stock. Their study concluded that buildings came into city ownership with average tax arrears of $35,000 and remained in city ownership on average for 19 years at a cost of maintenance of over $3 million per building. This was, of course, exclusive of the subsequent cost of rehabilitation that these buildings typically required, which in 1995 averaged between $60,000 and $80,000 per unit[13].

By 1995 the city embarked on a review of its policies related to tax-foreclosed properties and, as a result, changed its foreclosure law to allow the most distressed properties that were also in tax arrears to be transferred directly to a new qualified owner rather than be taken into city ownership. This Third Party Transfer Program would allow for a more direct link to new ownership and ensure that construction financing would be available[14]. In addition the city adopted a policy of residential and commercial tax lien sales. Residential liens, however, could only be sold if it was determined that they were not placed against distressed properties[15]. These dual policies would permit greater enforcement against tax delinquencies while, at the same time, ensuring that the most distressed stock would receive adequate subsidies for rehabilitation and new competent ownership.

Since the inventory of vacant *in rem* buildings had been greatly reduced, the city was left increasingly with the abandoned occupied *in rem* stock. These severely deteriorated occupied buildings, typically requiring the replacement of more than 80 percent of their structural beams and joists, needed a new focus and programmatic response if they were to be properly

[13] Costs from the New York City Department of Housing Preservation and Development. Please note that a typical scope of work for the rehabilitation of occupied *in rem* buildings includes complete replacement of building-wide systems – plumbing, electrics, heating plant, roof and windows as well as replacement of on average 80 percent of the structural beams and joists.

[14] New York City Council, Local Law 37 of 1996.

[15] The local law provided for a specific formula related to the lien to value of the building as well as the number of serious code violations per unit that would trigger an automatic finding of distress. In addition the Housing Department had the discretion to 'add' to the list of distressed properties based on other factors, such as direct inspection.

renovated. Prior to this time the scope of work carried out in occupied buildings was closely regulated and restricted, with the greatest amount of work (and therefore money) reserved for the vacant building stock which was seen as the first priority. Occupied buildings with physical needs that exceeded the programmatic cost limits imposed by the HPD were left to languish in the city management. If there were to be a serious attempt to return the entire *in rem* stock to private ownership with adequate rehabilitation, this approach would have to be dramatically changed.

New programs were designed that were based on the traditional program models. This provided for full financing of the needed renovation, an adequate scope of work, funds for interim relocation and restructuring of rents. Relocation and rent restructuring were particularly controversial – occupied buildings had in the past typically only undergone 'moderate' rehabilitation which could be carried out with tenants in place. Now the extensive scope of work, combined with concerns regarding lead paint, would make in-place rehabilitation a thing of the past. In addition, since a significant number of *in rem* tenants were formerly homeless and very low-income households, the specter of increasing rents, at a time when Section 8 rental subsidies were scarce, created a great deal of concern and suspicion among tenant advocates.

Ultimately, the new programs were designed with the following principles in mind:

- The scope of work would be determined based on the buildings' needs so that full rehabilitation would be completed.
- The scope of work would be equal for all buildings regardless of the program that they would ultimately be sold through. As a practical matter this meant that all buildings now had a fair chance to become tenant cooperatives[16].
- Low Income Housing Tax Credits would be used to provide significant operating reserves to maintain affordability in the event that Section 8 subsidies were not available.
- Both community-based non-profit organizations and for-profit companies (most notably small local building managers) would be encouraged to participate in the rehabilitation programs.

Between 1995 and 2003 more than 40,000 units were sold and preserved through this strategy, reducing the occupied city-owned stock still in city management to fewer than 6000 units in 2003 (City of New York Mayor's Management Report, 2003).

[16] Prior to this change buildings requiring extensive renovation work were prevented from entering the tenant cooperative program.

Assessing the Ten Year Plan

New York City's Ten Year Plan for housing represents the most significant intervention in the housing market that any American city has ever undertaken. Over 200,000 units of housing were built or rehabilitated. Entire neighborhoods that were once devastated are today thriving communities. The city today owns a fraction of the housing that it once did and most of these are already designated for rehabilitation under a variety of programs. While it is highly unlikely that any other city in the US or elsewhere would ever mount a housing program on the same scale as New York, the experiences of the city in implementing the Ten Year Plan would certainly be instructive to others designing smaller scale initiatives. In this section we briefly outline some of these innovations (for a more detailed analysis, see Roberts et al., 2003).

Flexibility without sacrificing replicability

Most housing initiatives, both at the federal and local levels, typically involve one program that is replicated over and over again in all types of neighborhoods. New York City's Ten Year Plan encompassed dozens of different programs that existed simultaneously. The wide variety of programs enabled the city to be flexible in creating strategies for particular buildings and neighborhoods. At the same time, many of the programs utilized similar tools and principles that resulted in economies of scale and replicability (Wylde, 1999).

Unlike most federal programs, New York City programs typically used a gap finance approach to subsidize the capital, and, indirectly, the operating costs of the housing. Typically, the city would determine the cost of a project, the amount of equity the owner could contribute and the amount of private market debt the owner could service (after operating costs were subtracted from rental income), based upon prevailing and expected market conditions and, in the case of occupied, rental buildings, the rent-paying ability of current tenants. The difference between the cost of rehabilitation or new construction, on the one hand, and the sum of equity and private market debt, on the other, would be filled with city financing, typically in the form of nominal interest rate mortgage loans.

New York's flexibility also extended to the ultimate owners of the properties. Rather than a 'one-size-fits-all' approach of public or private ownership, the city was able to create program variety to utilize for-profit real estate entrepreneurs as well as non-profit organizations (city-wide organizations and community-based developers). For example, local entrepreneurs were enlisted in the Neighborhood Entrepreneurs Program (NEP) to take title to abandoned buildings, rehabilitate them with city and federal dollars and operate them. Similarly, non-profit, community-based organizations participated in the Neighborhood Redevelopment Program (NRP) and its predecessor, the Community Management Program, taking over and rehabilitating over 5500 housing units. Indeed, New York City was even able to promote home ownership among thousands of former renters in its Tenant Interim Lease Program (TIL). Under TIL, the city conveyed rehabilitated multifamily properties to very low-income tenants who had proved their ability to manage their buildings. This approach allowed the city to tap the capacity of the full spectrum of qualified owners and developers.

Leveraging private resources and capacity

During the Ten Year Plan, New York reached out to leading financial institutions and made them active partners. Banks made the construction and permanent loans needed for the city's gap financing method of subsidizing the Ten Year Plan. In addition to leveraging private capital, the city's use of banks allowed it to leverage their capacity and expertise. One of the major achievements of the Ten Year Plan was the extremely low level of corruption or fraud that accompanied the expenditure of huge sums of public money. It is doubtful that HPD would have had the staff and expertise to effectively monitor the use of funds itself. Instead, it relied upon banks to monitor developers as they disbursed construction loans.

New York City also made use of a variety of intermediary organizations. For example, home ownership initiatives were typically overseen by the New York City Housing Partnership, an organization created by the city's business elite, and by two church coalitions in Brooklyn and the South Bronx. Community-based non-profit housing developers also worked with intermediary organizations – the Local Initiatives Support Corporation (LISC) and the Enterprise Foundation – in rehabilitating city-owned rental buildings. These groups created programs, provided technical assistance and, perhaps most importantly, through their joint New York Equity Fund, raised and invested equity through the syndication of the Low Income Housing Tax Credit to private investors, including many Fortune 500 companies. Because the two intermediaries were fiduciaries to the entities that invested in the Equity Fund, they provided cover and gave credibility to the work undertaken by the hundreds of community-based developers.

Perhaps nowhere has the New York's privatization of functions been more apparent than in its efforts to stem the abandonment of privately owned housing. The sale of tax liens inaugurated by the Giuliani administration in the mid-1990s and the transfer of tax-delinquent housing directly to developers as part of the Third Party Transfer Program were both efforts to leverage private resources, initiative and efficiency in service of the public interest.

Achieving a mix of incomes

One of the most contentious issues during the early years of the Ten Year Plan was the appropriate mix of incomes of the households that would live in the new or rehabilitated buildings. From the beginning, tenant advocates pressed the city to increase the proportion of units that would be allocated to poor families. In addition, New York, under pressure from consent decrees guaranteeing each homeless family housing, sometimes sought to house high proportions of homeless families in the buildings. Countervailing pressure came from the neighborhoods in which the housing was located. Many community residents were seeking to rebuild and stabilize their neighborhoods and felt that these efforts would be jeopardized by large numbers of households who might be prone to crime, substance abuse or mental illness. In addition to community pressure, evidence from the academic community was beginning to accumulate in the late 1980s and 1990s, that housing programs that concentrated poor people together sometimes caused more problems than they solved.

In the end, the city, for the most part, made significant efforts to promote a mix of incomes in the buildings that were constructed or rehabilitated under the Ten Year Plan. With the exception of supportive housing projects that sought to provide needed services to a formerly homeless population with special needs, most of the rental programs limited the proportion of homeless families to around 15 percent of all tenants. In addition, most of the Ten Year Plan programs sought to house a mixture of low- and moderate-income tenants. This income mixing served two purposes. First, it reduced the concentration of extremely needy people, providing children with role models of working families and reducing the prevalence of social problems. Second, it provided an opportunity for cross-subsidy. This was made possible, in part, because of New York City's ability to restructure rents as a result of the *in rem* vesting and use of city financing. Apartments, which were typically subject to stringent rent regulation, had their rents increased to achieve economic feasibility. Existing low-income tenants were spared the impact of the new higher rents. But relatively higher income and new tenants who could pay higher rents did so which, in turn, helped offset some of the extremely low rents paid by the poor.

Middle-income households were also well served by several of the Ten Year Plan's home ownership programs. Efforts to include this group had the salutary effect of increasing political support for the massive expenditure of city capital funds on housing and helped to stabilize neighborhoods and reduce concentrated poverty.

Promoting neighborhood revitalization

One of the greatest innovations of the Ten Year Plan was that it combined the twin objectives of housing production and neighborhood revitalization.

Most housing programs in New York City, and elsewhere in the nation that preceded the Ten Year Plan, either ignored the neighborhood impact of housing or had a negative impact. From the beginning of the Ten Year Plan, HPD sought to use the housing that would be produced to catalyze other forms of economic and community development.

Visual inspections and anecdotal accounts of once devastated neighborhoods, such as the South Bronx, demonstrate how housing has literally rebuilt communities from the ground up. Recent academic research by NYU's Furman Center For Real Estate and Urban Policy has also confirmed the veracity of these accounts. These studies have found that upon the completion of the development or rehabilitation of housing under the Ten Year Plan, the sales prices of properties within 500 feet of a rental project increased by between 1.4 and 6.7 percentage points relative to the prices of homes located in the same census tract but more than 500 feet from the project (Schill et al., 2002). The relative increase for sales proximate to ownership projects ranged from 2.3 to 7.0 percentage points. This provides empirical evidence that the housing built under the Ten Year Plan generated spillover benefits that were capitalized in nearby property values. These benefits might have been attributable to the elimination of blighted and dangerous conditions, the creation of good quality buildings or the in-migration of additional population.

New challenges for New York City: the new housing marketplace

With the massive reduction of the *in rem* stock that took place as a result of the Ten Year Plan, New York City faced the prospect of requiring a new paradigm for housing assistance as the twenty-first century began. For nearly 15 years the city's housing programs were the basis for the creation of not just affordable housing, but an entire affordable housing industry. Non-profit organizations, developers, contractors, property managers and banks seeking lending opportunities that met the requirements of the federal Community Reinvestment Act[17] (CRA) flocked to the city's housing programs and eagerly awaited new offerings of *in rem* property. For the city, a significant amount of financial resources was allocated each year for the exclusive purpose of renovating the *in rem* residential stock including the majority of the city's overall federal allocation of Community Development Block Grant Funds, the sub-allocation from the State of between $8

[17] The Act required that banks carried out community investment activities in low-income communities where they conducted other business, such as mortgage lending in underserved markets, and investing in economic development projects.

and $14 million per year in Low Income Housing Tax Credits as well as direct City Capital funds of between $300 and $500 million per year. With the inventory of *in rem* property depleted, the future of the city's housing program and its neighborhood-based focus was now in question.

In addition, and perhaps most importantly, the city had changed significantly since so many of its neighborhoods began their decline in the early 1970s. Its economy, decimated by the loss of 500,000 manufacturing jobs in the 1960s and 1970s, was now rebounding thanks in part to the dramatic drop in crime in the late 1990s that triggered growth in tourism and jobs in the service sectors. This drop in crime was felt most keenly in New York's poorest communities, which had also benefited from significant housing investment in the preceding decades. The 2000 census would reveal the fruits of all of this labor: the city's population was growing.

With this growth came new challenges for the city's housing market. In contrast to the soft housing market of the 1970s and mid-1980s, excess demand significantly outpaced supply in just about every income cohort and neighborhood, leading to large increases in both overcrowding and affordability. And yet the vast majority of the city's resources for housing were still directed at the *in rem* stock. In short, the city was still spending its time and money solving the problems of the 1970s and 1980s rather than the challenges of the 1990s and beyond. Little was being done to address either the problems that the city was facing now or the challenges that it would face in the future.

Once again it was time to re-evaluate the nature, purpose and direction of the city's housing policy and once again a new administration would provide the opportunity to undertake a shift in direction. Significant private capital could reasonably be expected to be leveraged by public investment with the improvement in the local housing market. In addition, the large and sophisticated affordable housing community, now squarely in place, could be relied upon to participate in the various targeting efforts.

Finally, New York City had the great advantage of the experience gained over the preceding two decades. While new production would be important to any housing initiative, neighborhood preservation, most specifically the preservation of the significant investments in low- and moderate-income communities made by the city under the Ten Year Plan, would assume major importance. That meant not only a focus on affordability and geographic targeting, but on code enforcement and ensuring that the existing low-income stock was monitored for signs of decline and appropriate intervention.

It is important to understand the underlying causes of the New York City's current housing shortage. Conventional economic theory would presume that with a growth in demand, the market would respond appropriately with an increase in supply. In fact at the highest end of the housing marketplace this is the case, evidenced by healthy vacancy rates (10 percent) for apartments renting for more than $2000 per month (Lee, 2003). However market-rate housing is satisfying no other segment of housing demand. Even people who can pay significant amounts for housing often cannot find appropriate housing in the city.

This lack of market response is a function of a number of factors including rather significant barriers to entry. The cost of building housing in New York is extraordinarily high – in fact the highest in the nation (Salama et al., 1999). When compared to other large US cities, such as Los Angeles, Chicago and Dallas, the 'hard' costs of construction in New York are between 21 and 55 percent higher. While many factors contribute to these high costs, labor and an overly complex and time-consuming regulatory environment are primarily to blame, along with the high cost of land acquisition that is discussed below. The impact of the burdensome regulatory environment cannot be overstated here, since it not only adds directly to the cost of housing, but it further distorts the marketplace by severely discouraging new players from entering the market.

A second significant factor adding to the extraordinary costs and insufficient market response to demand is the difficulty and expense associated with the acquisition of sites suitable for the construction of new housing. New York City is an old city by US standards, and extensively built out. Unlike other older US cities, such as Boston, Philadelphia or Chicago, culturally New Yorkers were always willing to demolish the old to make way for the new. The needs of the marketplace, and the need to constantly reinvent itself, overshadowed historic sentimentality. And while this bias sometimes had disastrous consequences, evidenced by the demolition of the iconic Penn Station in the 1960s, it also served New York City well as the city was willing to clear away old buildings to make way for the new needs of an ever-changing metropolis.

Yet the new millennium found New York City so overbuilt and with so little available land that even its notorious penchant for the 'new' could not be fulfilled. As sites became scarcer, prices invariably began to rise, often beyond the reach of housing for anyone other than the very rich, or ironically the very poor, for which government subsidies would fill the gap.

And yet derelict land abounded in various locations throughout the city, particularly along its waterfront, the vestiges of a long-dormant industrial past. In addition, many residential areas around the city were under-built, restricted by zoning categories that were laid down to reflect the original scale of the neighborhoods which often went back to the nineteenth century. Most frustrating of all, a significant number of housing units (perhaps as many as 20,000) were held vacant and off the market purposely. Sometimes this was the result of long-term property ownership disputes. In other cases, owners had small properties and feared the uncertainty and risk of renting to 'bad' tenants, particularly in a legal context that made eviction for non-payment of rent difficult. In yet other instances, owners of mid-size buildings (8–20 units) with ground floor commercial uses chose to avoid the difficult job of managing housing because they derived sufficient income from their commercial tenants (New York City Department of Housing Preservation and Development, 2002).

So once again the city found itself in a position of needing to re-think its housing policies and change direction in response to a significantly altered landscape. In December 2002, just one year after he took office, Mayor Michael Bloomberg unveiled his administration's housing plan that would allocate over $3 billion towards a 5-year effort to build and preserve over 65,000 units of housing. The *New Housing Marketplace Plan* recognized that most housing has been and will continue to be built by the private sector and that government's role should be:

- 'Regulating private market activities, both through land use controls as well as standards which govern construction and maintenance; and
- Providing financial incentives to encourage housing activities that will build or strengthen a private marketplace such as subsidizing home ownership through federal tax benefits, local tax incentives to encourage rehabilitation and new construction and low cost financing for construction and maintenance of housing for targeted income groups' (New York City Department of Housing Preservation and Development, 2002, p. 6).

Essentially the new plan was being constructed on the lessons learned from the original Ten Year Plan – the private sector would be encouraged through incentives to create or preserve affordable housing and the government would craft the appropriate financial incentives.

The plan sought to address the current housing shortage by making below-market funds[18] available not just for the new construction of affordable housing but also for the acquisition and clean-up of brownfield sites. The plan would also place a renewed emphasis on the preservation of the existing housing stock and focus resources in targeted neighborhoods. Of the 65,000 units to be constructed or rehabilitated, approximately 41,000 represented existing units that would be renovated and preserved or renovated and brought into active use.

---

[18] Generally subsidies are provided as construction and permanent financing at reduced interest rates. In some programs the government financing remains in the deal as a soft second mortgage, payable only if the first mortgage is satisfied with funds to spare. In this way the funds function as grants when needed, but as loans if there is sufficient profit in the project over time to repay the second mortgage.

Finally, the plan would address one of the shortcomings of the Ten Year Plan. In spending close to $6 billion on housing over more than 15 years, New York City never attempted to address some of the root causes of why the private housing market could not even satisfy the needs of middle-income residents. The New Housing Marketplace Plan begins to take on some of these fundamental issues. The Plan includes an initiative to adopt a model building code that would simplify housing construction. A second proposal would re-zone underutilized manufacturing areas for residential uses.

Despite some criticisms over specific elements[19], the New Housing Marketplace Plan was greeted with praise by much of New York City's housing community. Advocates and industry participants, alike, recognized that despite budgetary difficulties, the city was making a significant public commitment of funds toward housing development and, at the same time, seeking to address some of the underlying problems in the marketplace, such as the need for more targeted homeowner assistance in lower-income communities and the redevelopment of brownfield sites. In addition the administration was commended for the creation of a transparent process that included the creation of the Neighborhood Investment Advisory Panel that established working committees to help design, not just the specific components of the plan's programs, but a system to collect information and monitor the plan's progress and impact over time.

Conclusion

Over the past two decades New York City has been the center of housing innovation in the US. At a time when the federal government was reducing its commitment to housing, the city undertook a program that poured enormous resources and creativity into an effort to build housing and improve communities. Many of the lessons learned from New York's experiment in housing policy are relevant not just for other US cities, but for cities throughout the world.

At the same time, it is worth considering at the conclusion of this chapter some pitfalls that New York City encountered. These shortcomings

---

[19] Although the plan was greeted with enthusiasm by the city's affordable housing industry, three elements of the plan remain controversial: (1) whether sufficient funds were reserved for housing low income as opposed to moderate income families, (2) whether inclusionary zoning requirements should be part of the plan which would ensure that low income housing be built in newly rezoned areas such as the Brooklyn waterfront and (3) whether the commitment of $3 billion dollars was sufficient given the significant need for housing in the city.

might well affect any other city that seeks to create an ambitious housing program. New York City's Housing Department was the driving force behind the conception, design and implementation of both the Ten Year Plan and the New Housing Marketplace Plan. Whilst this ensured that the Department would develop the deep expertise needed to carry out the programs, it also invariably led to an over-identification of the plan with the Housing Department exclusively. While inter-agency co-operation certainly takes place[20], this falls short of the level required to ensure that all issues are resolved quickly.

Various mechanisms have been employed over the years to create greater coordination among the municipal agencies that could affect the success or failure of the city's housing programs. For example, during the administration of Mayor Edward Koch, delays in construction approvals resulted in the creation within City Hall of the Mayor's Office of Construction Coordination. Mayor David Dinkins established the Mayor's Office of Homelessness to try to coordinate housing and an array of services focused on this population. The Giuliani administration established the Mayor's Office of Environmental Coordination to oversee and 'streamline' the environmental approvals required to carry out development, and for a short time a Deputy Mayor was made responsible for neighborhood redevelopment[21]. While all of these efforts had some success, the absence of a single high-level commitment to an integrated neighborhood redevelopment strategy, led by the reconstruction of the neighborhood's housing stock, led to missed opportunities and often engendered mistrust and confusion in the communities themselves. So while planning was under way for a massive rebuilding of a long-vacant housing stock, vacant school buildings were being demolished or auctioned to the highest bidder. Today the city faces an extreme shortage of school seats, most acute in many of the same neighborhoods that underwent significant redevelopment.

A second shortfall of the city's implementation of housing policy was the lack of emphasis placed on public communication and the 'marketing' of its ideas, objectives and results. Public discourse tended to be relegated to responding to various 'attacks' and there were few public forums available for discussion rather than argument. While today the Koch administration's Ten Year Plan is widely hailed for its vision, scope and the quality of its execution, at the time it was implemented it was continually

---

[20] Note for example the high level of coordination in the New Marketplace Plan with the Department of Buildings which must undertake a redesign of the city's building code and with the Department of City Planning's work to rezone former manufacturing areas for residential use upon which at least one of the plan's program depends.

[21] Fran Reiter who served as Deputy Mayor during the Giuliani Administration.

criticized for not going far enough or fast enough, for including too many for-profit developers and even for carrying out a plot to destabilize certain neighborhoods by creating an excessive amount of housing for the homeless. The constant barrage of criticism led HPD to be suspicious of sharing information with the public and to be wary of seeking out public opportunities to persuade rather than impose new ideas and programs.

Third, the lack of an overall neighborhood-based approach, one that would incorporate other government functions (e.g. schools, police, fire, sanitation, economic development), made the collection of good neighborhood-based data difficult. For example, school districts, police districts and local community boards all have different boundaries, making data comparisons difficult. This weakness not only affected the neighborhoods themselves as previously noted, but it has hampered research and analysis. Perhaps even worse it contributed to the inaccurate view that what goes on in New York City is irrelevant to other cities because New York City is 'different'. Certainly, New York City is *different* from other cities. The sheer size of its housing stock – over three million housing units – distinguishes it from most other large cities in the US and Europe.

Nevertheless, once neighborhoods are looked at individually, more commonalities arise. East New York in Brooklyn with its low- and moderate-income, largely minority population living in small one- and two-family homes is not so different from North Philadelphia or East Baltimore. The problems facing the preservation of Harlem's historic brownstone stock are not so different from those of limestone townhouses in Liverpool or the elegant vacant rental buildings along Leipzig's main commercial streets.

Increased data collection at the neighborhood level and research on the impacts of housing programs[22] are the first steps toward sharing with the world information about New York's experiment in housing and neighborhood revitalization. At a time when so many of the world's older industrial cities are either undergoing major population losses and housing abandonment on the one hand, or excessive population growth and housing shortages on the other, New York City, which has moved aggressively to address both of these cycles, should be a significant resource for research, analysis and the exchange of ideas and experiences among housing practitioners and policy makers.

[22] The New Marketplace Plan specifically provided for a group of housing researchers to establish a method of collecting and disseminating the data that will result from the plan's implementation. In addition, HPD and the NYU's Furman Center For Real Estate and Urban Policy have joined together to create an integrated web-based information system to make neighborhood-level data on housing, education, crime and population dynamics widely available. See www.nychanis.com.

References

Arthur Andersen Consulting (1995) *Breaking the Cycle: Developing an Effective Intervention Strategy for Dealing with New York City's In Rem Housing Problem.* Arthur Anderson Consulting, New York.

Barnes, J. (2000) The Two Faces of Bushwick. *The New York Times*, 27 February.

Boylan, E. (2000) *Housing The Next Generation.* A conference sponsored by the New York City Department of Housing Preservation and Development, October.

City of New York Mayor's Management Report (2003) City of New York. www.NYC.gov

Collins, J.M., Belsky, E. and Retsinas, N.P. (1998) *Towards A Targeted Homeownership Tax Credit.* Joint Center for Housing Studies, Harvard University.

Dallas, G. (1979) Housing Blight in Bronx is a Sad Sight. *New York Daily News*, 25 November.

Daniels, G. and Schill, M.H. (2002) State of New York City's housing and neighborhoods: an overview of recent trends. *Economic Policy Review* **9** (2): 5–17. http://www.ny.frb.org/research/epr/03v09n2/0306schi.pdf

Dolbeare, C., Saraf, I. and Crowley, S. (2004) *Changing Priorities: The Federal Budget and Housing Assistance 1976–2005.* National Low Income Housing Coalition, Washington DC.

Jackson, K. (1985) *Crabgrass Frontier: The Suburbanization of the United States.* Oxford University Press, New York City.

Lee, M.W. (2003) *Selected Findings of the 2002 New York City Housing and Vacancy Survey* (submitted to the New York City Council on 7 February 2003).

Mahler, J. (2005) *Ladies and Gentlemen, the Bronx is Burning.* Farrar, Straus and Giroux, New York City.

Massey, D.S. and Denton, N.A. (1993) *American Apartheid: Segregation and the Making of the Underclass.* Harvard University Press, Cambridge.

New York City Department of Housing Preservation and Development (1987) *Capital Commitments. City of New York Mayor's Management Report, 1987.* NYC Department of Housing Preservation and Development, New York City.

New York City Department of Housing Preservation and Development (2002) *The New Housing Marketplace.* City of New York Housing Preservation Department, New York City.

New York City Department of Housing Preservation and Development (2004) *HPD Starts, Completions, Sales and Capital Commitment History: fiscal years 1987 through fiscal year 2004.* NYC Department of Housing Preservation and Development, New York City.

O'Regan, K.M. and Quigley, J.M. (2000) Federal policy and the rise of nonprofit housing providers. *Journal of Housing Research* **11** (2): 297.

Plunz, R. (1990) *A History of Housing in New York City.* Columbia University Press, New York City.

Previti, D. and Schill, M. (2003) *State of New York City's Housing and Neighborhoods.* Furman Center for Real Estate and Urban Policy, New York City.

Roberts, R.T., Salama, J.J. and Schill, M.H. (2003) New York City's Ten Year Capital Plan for Housing. *The Wharton Real Estate Review* Fall.

Roberts, S. (1987) Blackout Starts Bushwick's Rise From The Ashes. *The New York Times*, 9 July.

Salama, J., Schill, M. and Stark, M. (1999) *Reducing the Cost of New Housing Construction in New York City*. NYU Furman Center For Real Estate and Urban Policy, New York City.

Schill, M., Ellen, I., Schwartz, A. and Voicu, I. (2002) Revitalizing inner-city neighborhoods: New York City's Ten-Year Plan. *Housing Policy Debate* **13** (3): 529–66.

van Hoffman, A. (2001) Evolution of Community Development Policy. Fannie Mae Foundation.

Wylde, K. (1999) The contribution of public–private partnerships to New York's assisted housing industry. In: Schill, M. (ed.) *Housing and Community Development in New York City: Facing the Future*. State University of New York Press, Albany.

5

Changing Housing Markets, Race and Community: Policy and Practice in Birmingham

Austin Barber, Rick Groves and Alan Murie

Introduction

Race and residence have been themes in the British research literature for nearly 40 years. In that period, some of the most influential studies have been carried out in Birmingham (Rex and Moore, 1967; Henderson and Karn, 1987). This provides us with the opportunity to set the changing geography of residence in the city against a background of changing policy and practice. This chapter discusses key elements in the account of race and residence in Birmingham. It does this, not by providing a micro account of changes in policy, the market or other factors, but rather providing a picture of policy and practice at two points in time. The chapter focuses on two different phases of development of the economy and policy of the city, associated with the post-war welfare paternalism of the 1950s, and the contrasting preoccupation with issues of economic competitiveness since the 1990s.

There is a wider literature that refers to the different phases in the development of the economy of cities and of public policy. The chapter contributes to this through an account of a particular policy area in a single city. In this way it shows how the policies and debates around race and residence have changed over time and have reflected the shift in emphasis occurring nationally, internationally and locally. The chapter also enables us to assess the impact of different policy regimes on race and residence within the city. In this respect, it might be expected that policy approaches which are more preoccupied with a market agenda and competitiveness, and which are not designed to achieve significant re-distribution of resources, would be less in the interests of black and minority ethnic (BME) households and more likely to generate increased levels of segregation. However, the emerging pattern is more complex, as this chapter will show. The paternalistic welfare state regime was associated with very unequal treatment of different ethnic groups and recognized the needs of some categories of the population above others. Its contribution to patterns of segregation was as significant as the policy approaches associated with later periods.

Although the overt discrimination associated with state bureaucratic management of the housing market has significantly modified since the 1980s, the legacy remains in patterns of residence in the city today. The less overt practices that determined residential outcomes (matching applicants and dwellings and letting property) are more ingrained in social and bureaucratic practice. At the same time, the maturation of the parts of the market most heavily used by the minority communities and changes in the economic circumstances of second and third generation households also have an impact on patterns of segregation. The significance of public policy for the residential situation of BME households is arguably less in the second phase simply because of this maturation effect.

The contemporary picture is further complicated by important developments in Birmingham's housing market since the mid-1990s, most notably the emergence of a dynamic new market in private sector city centre housing linked to the wider regeneration of the central districts. Although this housing is geared mainly towards affluent households, the emerging evidence suggests that the city centre population may be more diverse in terms of ethnicity than is often assumed. A second significant development concerns the substantial influx of asylum seekers and refugees into the city. These individuals and households are less likely to be able to connect with family and kinship networks that are already established in the city, which will impact upon the spatial patterns of settlement.

This chapter begins by setting out briefly the historical context of development and migration in Birmingham before examining how two distinct policy regimes in the city's recent history have shaped the spatial configuration of race and residence in the city. The first concerns the period of the 'welfare state city' when the role of the local state was associated with urban managerialism, the provision of public services and amenities for the city's population. The second focuses on the more recent era of the 'competitive city' reflected in a shift toward urban entrepreneurialism, the mobilization of public and private resources around objectives of economic growth and employment generation. It concludes by reflecting upon how the policy regimes in these two periods have led to the complex and evolving spatial patterns of race and residence in contemporary Birmingham.

Background

Birmingham is an archetypal industrial revolution city. Its industrial growth enabled it to overtake all but one of its British competitors in terms of its size and economic significance. There is a rich literature on this pattern of growth but one of the key questions about why Birmingham grew and how it became the centre for new industries and innovation relates directly to the reception of migration (Briggs, 1952; Cherry, 1994). Unlike larger, older, historical cities, it was not dominated by a conservative establishment intolerant of dissent and restrictive towards the activities of outsiders. Birmingham accepted and welcomed people with non-conformist religious backgrounds, including Quakers; some of the leading industrial families and economic pioneers in the city were migrants, seeking a more tolerant environment in which to develop their entrepreneurial gifts but also to be able to conduct their religious practices without the abuse and restrictions that they had experienced elsewhere. This is not to say that Birmingham was without its prejudices. For example, the Priestley riots were a notorious case of bigotry and attacks on the non-conforming scientific community. On balance, though, it is clear that Birmingham provided the environment in which enterprise and innovation flourished. Knowledge-based activities flourished with the direct application of scientific ideas to new manufacturing processes, and with an environment in which scientific and entrepreneurial skills were combined effectively.

The welfare state city

Birmingham by the 1950s had established itself as the UK's second city and its largest municipality. The municipal tradition was particularly strong. The city pioneered the growth of modern political parties and party politics at local government level. Alongside this was the development of municipal services such as gas, electricity, water and also a municipal bank. In the 1950s, Birmingham and the West Midlands region were experiencing a period of full employment and considerable prosperity. Manufacturing and the motor industry that was so strong in the region were booming. In relation to the rest of the UK the region only lagged behind the south-east in terms of prosperity. Indeed, this is indicated by the fact that regional policies of the time sought to steer industrial development away from the prosperous south-east and West Midlands towards the less prosperous regions of the north. Cities such as Glasgow, Newcastle, Liverpool and even Manchester were already some way down the route of economic decline as their basic industries and economic activities came under pressure from increased global competition.

Birmingham did not have as large a municipal housing sector as some of these other cities. Nevertheless, by the 1950s it had developed the pattern of tenure common in UK cities. The older, unplanned inner and central city neighbourhoods consisted principally of private (mainly rented) sector housing. These were often older, small terraced houses with enclaves of more expensive, more attractive housing within them. Some of these neighbourhoods were scheduled for slum clearance and became renewal areas and the locations for new council estates in the 1950s and 1960s – introducing a planned municipal element within these unplanned pre-1919 housing areas. Beyond the inner city there was a belt of inter-war suburbs which showed the evidence of planning, not just in the presence of large municipal housing estates alongside estates largely bought by home owners, but in terms of the separation of residential and industrial areas, the emphasis on well planned estates and neighbourhoods and the kinds of layouts which arose. Finally, beyond these areas was a new belt of council estates built in the reconstruction phase following World War II. These areas had higher concentrations of council housing than the inter-war suburbs because of the priority to meet housing needs in the early post-war period.

Thus by the mid-1950s there was a housing hierarchy in the city in which council housing was close to the top. Nearly all council housing was relatively modern and it consisted predominantly of well designed and well planned houses with gardens. The majority were three-bedroom houses which were much more spacious with larger rooms and more circulation space than the dominant property types within the private sector. Access to this housing was not determined by ability to pay but through a bureaucratic rationing process. The mass of the population lived in rented housing and council housing was generally considerably more modern and superior to the housing that was available in the privately rented sector.

All of this is the background to the study, *Race, Community and Conflict* (Rex and Moore, 1967), which identified the competition for housing not as a competition to obtain high status or positional goods, provided through the market or to occupy particular niches of market housing, but rather a competition to access the housing provided by the state. Rex and Moore referred to the notion of housing class to capture the nature of competition for housing.

The introduction of the concept of 'housing classes' by Rex and Moore (1967) marked a significant shift in the debate in housing. The dominant frameworks prior to Rex and Moore for the explanation of housing behaviour were ecological and economic or filtering approaches. The unsatisfactory nature of these approaches was particularly apparent where the development of not-for-profit and state housing through the twentieth century meant that there was high-quality housing which was accessible to people who did not have the wealth or income to gain access to such housing in the market. In these areas low social status was no longer associated with poor housing. The fundamental economic determinism of residential patterns was broken.

Rex and Moore acknowledged the influence of the ecological approach in talking about competitive processes in a zone of transition as recurrent urban phenomena. But they suggested that the competitive process was much more complex than assumed by the ecological school. In particular, and based on their empirical research in Sparkbrook in Birmingham, they emphasized the important role played by local authorities in the allocation of housing. To be allocated a dwelling in the public housing sector was seen as 'a considerable prize in a society where housing is a scarce resource. It can bring into being and help to sustain an especially privileged group' (Rex and Moore, 1967). Rex and Moore suggested that there was a class struggle over the use of houses that is the central process of the city as a social unit. Men in the same labour situation had differential degrees of access to housing and this determined the class conflicts of the city as distinct from the workplace. The fundamental reason was that state intervention had resulted in the provision of a high-quality council housing sector which provided desirable housing but where access to the housing was not determined by income or labour position.

As was argued at the time, the formulation of 'housing classes' appeared to assume a common value system and common status aspirations. For the competition to take the form that Rex and Moore (1967) suggested, implied that all households equally saw the benefits associated with council housing, there was a common view of the advantages associated with particular housing tenures and the particular neighbourhoods in which they were located. It was argued in this context that the delineation of classes arising from it was based on a static picture of a particular area. There was no consideration of the relevance of housing history and only guesses about future opportunities. The classification therefore rested on differences in the current situation (Haddon, 1970). It was partly because of the criticisms of the emphasis on current housing situation that Pahl (1975) highlighted the 'means of access' rather than current situation as central to competition within the city, but again Pahl's formulation tended to emphasize the situation of households with little or no choice. His reference was to households that 'must rent'.

What is remarkable about this debate is that it places state housing so near to the top of the hierarchy whilst significant parts of the private sector come very near to the bottom. The post-war welfare state had created a highly desirable commodity, which could not be accessed through ability to pay or price, but which because of its desirability had to be rationed. The demand for state housing was much greater than supply. The state actively pursued policies throughout to the supply of municipal housing including its activities to replace the slums with municipal dwellings. The management approach of the state reflected the perception that municipal housing was a privilege as well as responding to a need.

The practices that emerged included grading properties according to their desirability and applicants according to their level of need but also their perceived behaviour or housekeeping standards. The best housing would go to the best applicants who were in need, and the worst housing would go to households who were in the same levels of housing need but were not expected to be such good tenants or to look after their property in the same way. Because of the pressure of demand, local authorities took account of how long people had been waiting for a house and of the extent to which they had a local connection with the area. Residence qualifications were quite normal and people accumulated points (and so their bargaining power rose) if they had been on the waiting list for a longer period. The consequence of this was that the normal point of entry for new migrants to the city was the privately rented sector. They applied for council housing and waited for housing in that tenure. This had long been the experience of households moving to Birmingham and other cities from rural areas, or from Ireland or other parts of the UK.

The BME housing experience

It could be argued that the experience of the first wave of New Commonwealth immigrants coming to Britain in the 1950s was exactly the same as that of earlier waves of migrants. The booming economy and the high wage rates that could be generated in the automotive sector, for example, meant that Birmingham, in the same way as London, had shortages of people working in public transport and other sectors of the economy. New migrants from the British Commonwealth, initially the Caribbean, and subsequently the Asian subcontinent, were encouraged and found jobs in public services and elsewhere. It was not so easy to find housing.

There were reasons to suggest that the experiences of these groups were not just the same as earlier migrants. They were more likely to experience direct racial discrimination and there is evidence that this applied in the private rented sector as well as in public sector housing. Two other factors were particularly important. Firstly, the city was in the middle of a slum clearance programme, which meant that there was additional pressure on demand for council housing from people the council wanted to re-house. People in Urban Renewal Schemes tended to be put at the top of the waiting list and have overriding priority in order to progress clearance. Secondly, the city adopted racially discriminatory policies – the most notorious of which was the one in six rule designed to prevent concentrations of ethnic minority households. All of these aspects are fully discussed and described in Rex and Moore (1967) and in Henderson and Karn (1987).

The outcomes of the situation were clear. Minority ethnic groups found it very difficult to gain access to the high-quality municipal housing that was available. If they were eligible to go on the waiting list (which was only the case if they had established some degree of residence) they would be low on the list because they had not accumulated the waiting points that others had. Once they were on the list, and in a position to be offered properties, they would fall foul of other policy and practice elements. They would wait longer for an offer and they were more likely to be offered properties in the inner city in older and acquired properties rather than the newly built council housing.

Against this background, the 1962 Immigration Act was something of a watershed. The early migrants to the UK were almost exclusively economic migrants. They were young men who came to Britain in search of employment. The introduction of the 1962 Immigration Act, however, effectively limited immigration to 'key' workers but it carried with it the opportunity for the immediate relatives of migrants to accompany their 'menfolk'. The Act significantly changed the profile of migrants to Britain. Henceforth, their families often accompanied young males and this shifting profile of households implied a change in the form of accommodation they were seeking, away from shared accommodation for singles towards family accommodation. Originally they were looking for private rented accommodation but increasingly they began to generate their own solutions. Opportunities to buy older private sector property at low prices were considerable. Properties came on to the market with vacant possession and, although access to conventional mortgage loans was limited (both because of red lining of neighbourhoods and attitudes about lending to lower-income households), various mechanisms were effective in gaining access to owner-occupation. These included the provision of mortgages by the local authority, although other less conventional ways of borrowing were also in evidence (see for example Karn et al., 1985). All of this meant that by the time that households would have been in a position to gain access to the local authority sector, they were established in the owner-occupied sector and often their housing costs would be lower than if they moved to the rented sector.

The short-term strategies to cope with the lack of access to both municipal housing and suburban private sector housing merged into longer-term strategies which fixed the location of the BME groups in particular parts of the inner city. The outcome of welfare paternalism was a pattern of segregation and difference in housing market situation between working class white households and new working class immigrants.

The discussion in this section relates to developments that occurred within a local policy regime, described here as the welfare state city, that was most prevalent from the 1950s to the 1970s. By the late 1970s, the nature of housing policy intervention was changing. The era of slum clearance and new building for public sector renting had effectively come to an end. In Birmingham, as elsewhere, there was a major effort to retain and improve the older private housing stock. The City Council was particularly active, however, and over the next two decades there was a major programme of generously subsidized maintenance and repair of the inner-city private housing stock. The BME community, many of whom were by now homeowners, were major beneficiaries of this initiative. The 'urban renewal' programme as it was called in the city had an important influence in underpinning the concentration of minority communities in their original area of settlement. However, to understand the full complexity of contemporary spatial patterns of ethnicity and residence in Birmingham, it is necessary to consider how the effects of policies from this earlier era interact with the implications of a more recent and fundamental shift in the prevailing local policy regime. The characteristics of this new regime, and its impact on housing markets and policy in the city, are examined in the following section.

The post-welfare era – the competitive city

By the 1990s, the preoccupations of urban policy had shifted to economic restructuring and the competitiveness of cities. The economic changes of the 1970s and 1980s had undermined the economic base of Birmingham along with other industrial cities in Europe. Birmingham had experienced a considerable haemorrhaging of jobs and a dramatic deterioration in its economic position (Spencer et al., 1986). Further, city policy makers perceived a need to adapt to an emerging climate of international urban competition that presented Birmingham with significant challenges but also new opportunities in the context of an integrating European economic space. As a result, Birmingham emerged in the 1980s as an exemplar of the shift to urban entrepreneurialism, mobilizing public and private resources behind an ambitious programme of initiatives to strengthen its competitive position in this new context. In this respect, the turn in the city's governance reflected similar trends observed in many US cities in the 1970s and 1980s, where the pursuit of economic growth in perceived competition with other cities had been an established priority of public and private urban elites (Harvey, 1989). Indeed, many of the initiatives adopted in Birmingham were drawn directly from the precedents established by US cities such as Baltimore and Phoenix. The apparently successful waterfront and downtown transformation of the former city had a particularly strong influence on the thinking adopted in Birmingham and other British industrial cities in the 1980s.

Confronted by these new challenges, Birmingham policy makers responded with an ambitious programme to modernize and diversify the city economy, generate new employment, attract inward investment and enhance its external image through strategies to upgrade the city's physical fabric and strengthen its stock of economic assets. The spatial focus of this initiative was the city centre, which was seen to be under-developed (owing in large part to a poor-quality environment) but presented considerable opportunity for the generation of new economic functions for Birmingham. This response was spearheaded by extensive public investment in the late 1980s and early 1990s in a series of flagship projects, including the International Convention Centre, Symphony Hall, National Indoor Arena and Hyatt Hotel complex, alongside substantial environmental investments in the public realm. Subsequent high-profile mixed-use developments, such as Brindleyplace and the Mailbox, and later the Bull Ring retail complex (opened in 2003), were led by private capital but facilitated by a supportive spatial planning framework (Barber, 2001).

The prime purpose of these investments was to create the conditions for new economic activities, further private sector investment and the generation of new employment in growth sectors such as financial and professional services. However, an additional intended effect was to create opportunities for, and a market in, new city centre residential developments. The provision of this significant new type of housing was led by the private sector but the essential conditions were created by considerable initial public investments and the implementation of a supportive spatial policy framework. The imaginative use of canal-side locations to bring private residences back into the city centre, in what had previously been largely derelict and rundown former industrial areas, was also seen as important to the objective of attracting and retaining the highly skilled, and often young, individuals who are seen as essential to the expansion of the local economy.

In this context, then, the resources available to the city were no longer regarded as simply providing opportunities for redistribution or to meet the needs of households that were not able to access key services through the marketplace. Rather, the resources were being used to promote economic activity and to reshape the market. Unless market activity flourished and generated employment, the task of meeting social need would be overwhelming. Therefore, the appropriate approach was to put the resources of the city – financial and professional – into initiatives that would lead to a growth of employment. This did not mean that the city no longer had aspirations to meet housing needs, but they were not the primary ambitions. At the same time, especially in the housing sphere, other developments had changed the situation dramatically.

Transitions in the housing context

From the mid-1970s, especially after 1979, the amount of investment and new building of council housing had declined. The sale of council houses that had been pursued in Birmingham since the mid-1960s but was strengthened by legislation introduced in 1980 to provide tenants with a Right to Buy, had contributed to a change in the nature and reputation of the council housing sector. Much of the most attractive property had been 'creamed off' through sales to sitting tenants, and there was very little new building to maintain the competitive, attractive position relative to other tenures. Council housing began to be seen as less attractive and failures of management and repair services, together with the residualization of the tenure, contributed to a changed perception. It had ceased to be the aspirational tenure and was increasingly seen as the tenure of last resort.

In relation to the discussion earlier in this chapter, it would be fair to say that council housing had dropped down the hierarchy of housing choices and housing quality. Having undergone a major programme of refurbishment, some of the older, private sector housing was now seen as equally attractive as council housing, if not more so, and certainly had advantages in terms of asset values and wealth accumulation. It would be wrong to overstate this position. Some of the smaller back-of-pavement terraced housing in inner-city neighbourhoods lagged behind in the rate of appreciation of value, and in some neighbourhoods, the demand for these properties remains low. However, some parts of the market seen as second-best in the past, were now associated with considerable advantages.

At the same time, the second and third generations of BME households that had initially established themselves as owner-occupiers in these areas had moved on to more desirable neighbourhoods and expensive properties. This is partly about increased employability and educational qualifications leading to a different and improved economic situation for second and third generation households. But this is also associated with a maturation of the BME community or a cohort effect. One interpretation would be that some BME households were in a better position to take advantage of the operation of the market in the 1990s and early in the twenty-first century than the community left behind in the least desirable council estates. The housing booms of the late 1980s and after the mid-1990s did not affect all neighbourhoods or all parts of the private housing market to the same extent and there were parts of inner city Birmingham that did not share equally in the boom. However, those living in anything other than the bottom end of the owner-occupied sector almost certainly fared better than people in the municipal housing sector or in private rented accommodation.

The coping strategies of an earlier period had put some BME households in a position to take advantage of the way the market had developed and been promoted through public policy. Chapter 2 sets out the key elements of current residential patterns among BME groups, and highlights both concentration and dispersal effects.

While public policy in this period is preoccupied with other agendas, the effects of policy and market development have not consistently disadvantaged BME communities. As other welfare state objectives associated with redistribution and housing need have been subsumed within agendas about wealth creation and competitiveness, the results have not necessarily been disadvantageous. The reduction in the quality and privilege associated with welfare state housing provision has impacted most heavily on communities that have aspired to state housing or had previously benefited from it. Insofar as BME communities had not been made beneficiaries from this service, the reduction in its scale and quality had less impact on them. At the same time, some of the development of equal opportunities legislation had strengthened the position of BME groups. For some groups, access to council housing and housing association dwellings had increased, although the available evidence suggests the interaction between council housing allocation processes and practices, and the preferences and choices of BME communities, combine to mean that the council housing system continues to work less effectively for BME groups than for the white community. Black and minority ethnic groups are more likely to be in longer queues for types of housing and neighbourhoods where the demand is greater, and they are also less likely to obtain housing in their areas of choice. These factors are complicated by perceptions of safety and security, by the differential impact of the private sector housing renewal programme during the late 1970s and 1980s and the outcome in terms of the relative attractiveness of different neighbourhoods for ethnic groups.

Alongside these trends, the residential experience of BME groups in the contemporary city is affected by two significant dimensions of public policy relating to the growth of city living and the integration of asylum seekers and refugees. These are discussed in turn.

The growth of city living

The shift in Birmingham's urban policy priorities towards a competitiveness agenda, manifested in the extensive city centre investments of the 1980s and 1990s, has contributed to the emergence of a dynamic new housing market in the core of the city. Prior to the early 1990s, housing in the city centre was dominated by post-war municipal housing estates located mainly around the edge of the core commercial and retail area. A small number of housing association developments were completed in the 1980s and early 1990s. However there was no private market activity during this time, reflecting the wider economic and environmental decline affecting the city centre area in the post-war decades.

By the mid-1990s, the physical and economic changes in the city centre engendered by the public investments and subsequent private developments began to transform perceptions and confidence. The private market began tentatively with a handful of schemes near Brindleyplace in 1995–96, but their commercial success established viability of the market and generated significant developer interest. By the end of 2004, approximately 5500 new homes had been developed with a further 2650 under construction. This is not a large sum in the context of the city's overall housing stock, but the market is notable for the rapid rate of change that is occurring and its spatial concentration. In recent years, the relatively small city centre district has accounted for 35–40 percent of all new housing completions in Birmingham, and a greater proportion of all new private dwellings.

City council policy from the early 1990s was to encourage new residential development as part of the city centre ambitions, including the desire to attract and retain skilled individuals deemed to be essential for successful economic diversification. However, there were no specific policies to engender or shape this new housing supply beyond a more favourable land-use planning framework. The subsequent rise of city living has been driven by a combination of growth in the city centre economy and amenities, demographic and lifestyle shifts, and the 'push' factors of worsening transport and congestion problems.

Market characteristics

From 1998, there has been a sharp acceleration in the rate of new city centre housing development. This is increasingly dominated by private, mainly national, developers, with housing associations less able to participate due to rising land values that affect the financial viability of potential schemes. Developments are characterized by increasing scale and density, a dominance of apartments and a decrease in the average size of units.

This new city centre population is disproportionately comprised of young professional households without children attracted by proximity to city centre employment and amenities. The available research evidence suggests that the new housing is attracting people who may not have lived in the city otherwise. The market is therefore 'capturing' new residents for the city and is providing a desirable alternative to some of Birmingham's popular and high-priced suburbs. A significant proportion of residents use the city centre base as a second home. Research undertaken in early 2002 also suggests that BME households comprise around 15 percent of new residents – against a 29 percent share of Birmingham's total population – and the evidence suggests that these are more likely to be longer-term residents of the city centre rather than part of the more transient rental market (Blackaby et al., 2002).

As the market evolves, some critical issues are emerging. Most notable is the relatively high price of new and re-sale homes, well above the average for non-detached dwellings in Birmingham, and significantly beyond the purchasing power of most first-time buyers. This pattern reflects high land values and construction costs, the type of developer dominating the market and their continued success thus far in selling all of the new supply at these levels. This has generated concerns about new dimensions of polarization within the city centre along lines of economic class. The contrast of high-income households in new private homes alongside tenants in social housing is manifested in a 'middle market gap' (Barber and Blackaby, 2003). This is underpinned by a high rate of investor purchases (up to 50% of all new home sales) associated with a significant proportion of second homes and corporate lettings, and a related lack of community infrastructure.

Implications for spatial patterns of race and residence

The implications of the city living market for spatial patterns of housing and race in Birmingham are only beginning to emerge. It is a young, but rapidly maturing market that is highly distinctive in terms of its characteristics and current dynamism. But at this early stage it is possible to identify two dimensions of its impact.

The first concerns the nature of the new housing itself. The high value of the new private housing does not in itself exclude the city's BME households. Given the advances in terms of employment/income, education qualifications and accumulated household equity referred to earlier in the chapter, this housing may be more accessible to BME households than is often thought. The extent of these advances alongside the maturing of BME communities varies significantly between and within these communities. For example, indicators of labour market achievement and formal qualifications show that on average, Birmingham's Indian and Chinese communities fare better than the white community, while Caribbean, Pakistani and Bangladeshi communities trail in most respects (Birmingham Economic Information Centre, 2001). Issues of affordability will also be mediated by a host of other factors, including the degree to which these groups are represented in city centre employment, cultural values and perceptions of the city centre generally, and even the channels and means through which most of this distinctive type of housing is marketed. A further, related factor is the predominance of apartments in this new development and the relatively small size of the units being constructed – there are very few new homes with more than two bedrooms. In one respect, this new supply is ill suited to the demands of large households that comprise a significant segment of housing demand from some BME communities. At the same time, however, the BME population in Birmingham is disproportionately young (under 25), and this factor, combined with the growing affluence and role in professional occupations, suggests that city centre housing geared to small households at the beginning of their housing cycle could be appropriate to large elements of the city's BME communities.

The second dimension of impact relates to processes of change in adjacent districts of the city. As the market expands and matures, it is generating increasing pressure on adjacent residential areas. Thus far, city centre housing development does not represent a case of gentrification, at least in the classic sense of the term, and the displacement effects have concerned industrial and commercial uses. But the city living momentum is beginning to enter contemporary 'zones of transition', some of which are characterized by large municipal housing estates with significant BME populations. One such area, Attwood Green (previously known as the Lea Bank municipal estate), is already experiencing a restructuring and diversification of tenure that introduces substantial private housing alongside refurbished and new-build social housing. This process, in which the city council is a lead partner alongside a registered social landlord and a private developer, may well be replicated in other areas such as Ladywood and Highgate.

Elsewhere, the spatial expansion of city living could have implications for the role of areas just beyond the city centre that are characterized by older, mixed stock including high levels of private rented homes as well as significant tranches of declining industry. These areas are beginning to attract considerable private developer interest and early plans envisage a wider mix of new housing types, including 3–4 bedroom townhouses. The ultimate implications of this process remain to be seen, but it may involve displacement of some existing BME populations on the city centre fringes. It is just as likely that it could generate significant new alternatives for BME households that are located in adjacent districts and may wish to move into high-quality new housing relatively close to their existing place of residence, rather than having to shift further away from the centre as has been the case in recent years.

The final question concerns city council policy towards this process and its potential implications. In the early years of city living, policy makers were content to let market forces operate without constraint, as the priority was to encourage developer involvement and to facilitate a rebalancing of tenure in the city centre away from the dominance of social housing. But as the market evolves and critical issues emerge, there is greater recognition of the rationale for more pro-active engagement to guide and shape future change. The motivation in this respect concerns not only a desire to broaden access but also to ensure the long-term sustainability and growth of what is in many ways a positive process for the city.

The main intervention available is the city-wide affordable housing policy which aims for all major developments to include 35 percent of units for a mix of low-cost sale, shared ownership or social rent. But in practice, the norm is to achieve 10–15 percent of low-cost market or shared ownership homes owing to the economic viability of schemes due to land costs and site conditions. Given the limitations of this policy, there remains a need for more effective means to diversify the supply of housing by type, size and price (including the approach to the other municipal estates in or adjacent to the city centre). The intention is that this would broaden the potential city centre population, although recent policy debate concerns provision for households on more moderate incomes, families, key workers, young graduates and home-workers – issues of ethnicity have not been explicit in the discourse. There are doubts that market forces alone will generate this diversification in the medium term, and policy makers are therefore exploring possibilities for innovative forms of intervention and partnerships with private interests.

To conclude, it appears that a simplistic impression – that the narrow range and character of new city centre housing thus far necessarily excludes BME households and contributes to new dimensions of ethnic segregation in the city's housing markets – does not necessarily hold up. The survey evidence thus far suggests that BME households form a significant, though not proportional, share of the new city centre population. Future patterns will be shaped by the complexities of this market (especially the role of investors/rental properties), the nature and impact of new policy interventions and trajectories of economic change among BME communities (including the growing incidence of dual heritage households in the most recent census). This market is, and is likely to continue to be, led by private development, albeit supported and mediated by policy interventions. For the time being, policy relating to widening city centre access involves attempts to harness the momentum of market forces and the private actors that drive them for some wider social benefits. It is clear that much more understanding is needed, but it may be the case that this significant dimension of recent housing market change combines with the maturation of BME communities to create unexpected new patterns of residential growth in the heart of Birmingham.

Refugees and asylum seekers

References to BME households relate to established households whose families initially moved to Birmingham over 30 years ago. We are talking about subsequent generations and about later stages in household and family housing histories. To this extent, we are not strictly comparing how the housing system in the city functions for BME communities in different phases. A more comparable insight with the experience of new arrivals in the UK in the 1950s is the experience of refugees and asylum seekers newly arriving in Birmingham in the twenty-first century.

During the 1990s, together with other western European countries, the UK has experienced a rapid growth in refugees and asylum seekers. In order to deal with this growing 'problem', successive governments have introduced a series of legislative measures aimed at improving efficiency in identifying 'genuine' cases (as opposed to those seeking access as 'economic migrants'); clarifying their rights in respect of access to employment, housing and social welfare benefits; and their rights of citizenship and nationality. One of the main aims of government policy has been to 'disperse' applicants to the major conurbations and other provincial cities in order to ease the problem of concentration in London and the south-east and a National Asylum Seekers Support Service (NASS) was established in 2000 to manage this dispersal programme.

Both the NASS scheme and the informal arrangements established before it have relied heavily on private landlords to provide accommodation for asylum seekers and refugees. The circumstances in Birmingham have been no exception. Letting arrangements under the NASS scheme have been relatively generous compared with the standard system of housing allowances available to those on Housing Benefit. The scheme has also operated differently in that it has required landlords to provide accommodation to a high standard of internal repair in return for leasing arrangements of up to 5 years. This has proved very attractive to local landlords and has also ensured an enormous injection of investment into some of the older, inner-city areas of pre-1919 privately rented stock and into properties that, prior to this, were in poor condition and occasionally void. Indeed, compared with other rental opportunities, the returns have been so lucrative that there has been a veritable scramble to purchase similar properties becoming available for sale on the open market in order to rent to asylum seekers. In some inner-city areas, landlords formerly letting to those on Housing Benefit have also moved away from the sector to cater for asylum seekers.

There are no authoritative statistics accounting for the numbers of asylum seekers and refugees that have come to the city in the last 5 years or so, but the most recent estimates suggest that by October 2003, the figure was about 8250 (Phillimore et al., 2004). It will have expanded sharply since, owing to an amnesty introduced by the government at that time for those having been here for over 3 years. The main concentrations of asylum seekers are in areas of pre-1919, terraced accommodation in the inner city and especially in three wards in the north-western quadrant of the city. These are all areas that have traditionally provided low-cost rental accommodation to successive waves of migrants to the city. But the area to the north-west is a highly multicultural area in which indigenous white residents were already in a minority. From a recent survey undertaken by Phillimore et al. (2004), the main countries of origin of asylum seekers and refugees in the city are Iraq (27 percent), Iran (10 percent), Somalia (8 percent), Afghanistan (7 percent), Albania (5 percent), Kosovo (5 percent), Congo (4 percent), the former Czechoslovakia (3 percent) and Zimbabwe (3 percent).

The outcome of this new wave of immigration to the city is both to consolidate the process of residential segregation which already exists between the areas of indigenous white settlement and those of ethnic minorities, but also to enhance the complexity of a layering system within areas of existing ethnic concentration. In the latter areas, as we have seen, second and third generation migrants tend to be more mobile. To improve their residential circumstances they move either to the periphery of their existing residential communities or seek new locations further afield but invariably still within the conurbation. The newer migrants tend to take their place, not only in the poorest and oldest residential stock, but also often in accommodation now rented to them by a previous generation of migrants. The residential areas discussed thus far in the chapter, and their spatial relationship, are illustrated in Figure 5.1.

Conclusion

This chapter has sought to explain how two distinct local policy regimes in Birmingham's recent history have shaped the spatial configuration of ethnicity and residence in the contemporary city. The main characteristics of these regimes, their implications for housing policy and provision, and the associated impacts for BME communities are summarized in Table 5.1. A core theme linking these two regimes was the restructuring and erosion of the welfare state since the 1960s. This contrasts a 'golden age' of welfare, in which the role of the state was more interventionist, more generous and more re-distributive, with a more liberal welfare state which is less generous, less redistributive and contributed to social exclusion. Some of the critiques of the golden age argue that the post-war welfare state was not so generous and redistributive to key groups including women and minority households (e.g. Ginsburg, 1992). The foundations of the welfare state had not been sensitive to new challenges that emerged after the founding principles were put in place. Housing is a good example of this. Some accounts identify housing as the most redistributive element of the post-war welfare state (Le Grand, 1982) but others identify the

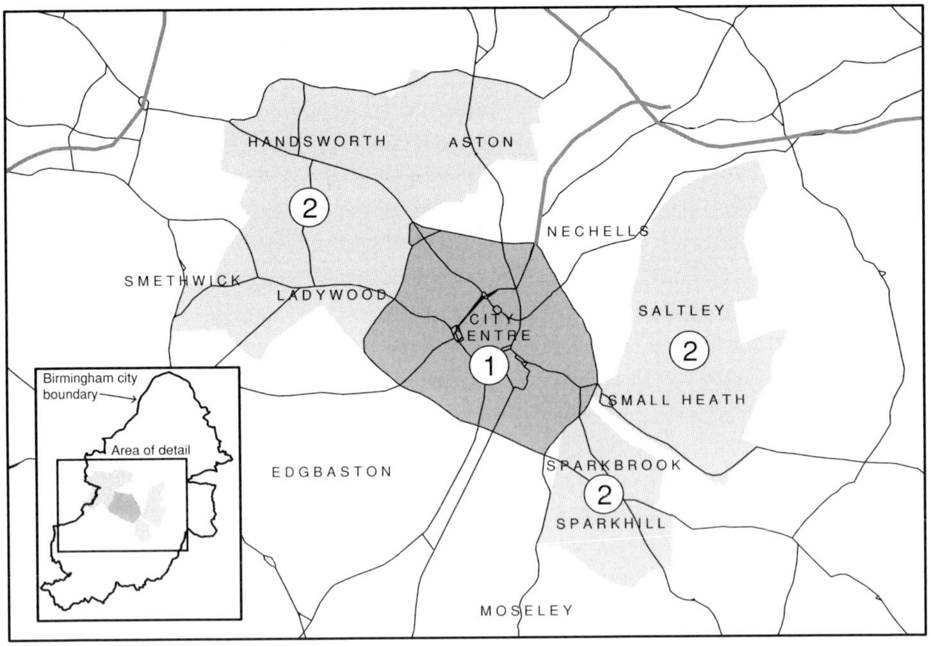

Figure 5.1 Main residential areas relating to the spatial patterns of BME households in Birmingham.
Area 1: City Centre. Contains several post-war council housing estates with large BME populations. These estates are beginning to undergo substantial restructuring including tenure diversification. From the late 1990s, this area has seen a rapid growth in new private market housing with a significant proportion of BME households.
Area 2: Inner City Districts. These consist largely of pre-1919 terrace dwellings that housed large numbers of post-war migrants to Birmingham, first in private rented accommodation and, later, in owner-occupation supported by Urban Renewal programmes. They retain important BME communities today. From the 1990s, some of the districts, particularly Ladywood, Aston and Handsworth, have become home to large numbers of asylum seekers and refugees occupying rented accommodation in houses owned by BME owners who have since moved further afield within the city.

paternalistic and exclusionary assumptions underlying council housing and relating to household type, housing need and merit. More fundamentally, the practices that operated in the allocation of council housing tended to involve stereotypes of households. Black and minority groups fell foul of these stereotypes and prejudices.

One perspective on the re-structuring of the British welfare state is that the generation and the communities that benefited most from it were active in withdrawing the welfare state from the cohort who would next benefit. In this perspective, BME communities which were not major beneficiaries of council housing provision in the period before the 1980s would, if there had been continuity of policy, have been more likely to

Table 5.1 Housing in Birmingham in the welfare state and post welfare state eras.

Feature	Welfare state city era	Competitive city era
Dominant mode of city governance	Urban managerialism Meet social needs of local population Focus on provision of public services and amenities including housing Development of economic assets and employment generation left to the market	Urban entrepreneurialism Improve city competitiveness in international context Direct and indirect public development of economic assets and infrastructure Cooperation with private interests through formal and informal mechanisms
Dominant housing policy discourse	Housing need Families with children Decentralization Slum clearance, redevelopment and new peripheral development Enforcement and bureaucratic management	Urban renaissance Attract and retain young, skilled individuals City centre focus High density, brownfield development Choice and consumer focus Homelessness
Role of the local state	Direct provider of council housing Rationing and determining access to housing Management of slum districts and council estates	Enabling and facilitating roles Provider of housing of last resort Management of residual mixed tenure neighbourhoods
Key relationships influencing policy	Central government financial and subsidy regimes affecting housing	Central government regeneration initiatives and planning policy Private developers and owners Voluntary organisations
Housing provision	Direct provision by the state through council housing in overspill and new housing areas and to replace slums Market provision to meet other needs	Public policy objectives met principally through market provision with attempts to steer the market and secure planning gain Housing associations to meet the new housing needs market misses Stock transfer
Significance of council housing	High quality, high demand Expanding aspirational sector	Mixed quality, mixed demand Declining low aspiration, transitional sector
Position of BME communities	Segregated and disadvantaged Excluded from state housing estates and upper end of market	Segregated but represented in most parts of the city Differentiation between BME groups related to ethnicity and family length of residence – parts of BME community and new migrants concentrated in least desirable housing Significant presence in new city centre markets linked with increasing affluence and employment in professional sectors
Position of new migrants	Benign neglect and exclusion Exclusion from council housing through allocation policy and practice Access to private renting and low price home ownership through the market	Policies for refugees and asylum seekers Managed schemes to access parts of private renting Limited access to home ownership, access for some to least attractive council housing and private renting

benefit in the subsequent period. But there was not continuity of policy. The opportunities declined in number and quality and the 'drawbridge' was pulled up before BME communities were able to cross. However, this perspective underestimates the extent to which the changing impact of housing policy to assist low-income homeowners and strategies adopted by BME communities, had actually put these groups in a better position in the housing market than some of those who had benefited from state housing. Exclusion from public sector housing led people to adopt strategies, which in the long run and with a subsequent focus of policy on home ownership, have proved advantageous, and have contributed to increased opportunity and wealth accumulation.

Legislation on homelessness and the particular policies adopted towards refugees and asylum seekers provide more protection for BME households settling in Birmingham for the first time today than in the earlier phase of policy. The shift in emphasis towards a post welfare state agenda produces a more complex picture than one that is simply about 'pulling up the drawbridge'. It is much more difficult to talk about such wide categories of the population than in the past. The rules and regulations affecting access to housing in the 1950s and 1960s were much more likely to operate as a blanket exclusion of BME households from council housing and presented common difficulties in accessing mortgage loans and house purchase. In the post welfare state era, the pattern is more complicated. Access to council housing is more open, although it is access to poor-quality council housing that is most easily negotiated. There is still difficulty in gaining access to quality council housing. At the same time, households in different ethnic groups with different community and family linkages and with different levels of resources of different kinds are able to negotiate different routes through the housing system. Contemporary accounts of the operation of the housing market in the city of Birmingham would acknowledge the outcomes are much more complex than in the past.

There are a number of layered influences on this. The welfare state and the structure of the city have changed, but so has the composition and nature of the BME community. Households are in different stages in their housing careers and are in different generations. Many households are in a stage of secondary, rather than primary, settlement. Public policies have changed and the bargaining power of different minority ethnic groups has changed. Both because of this, and because of their different economic and demographic characteristics, they are more able to take advantage of different opportunities in the city. The fast-growing city centre housing market and its imminent expansion into fringe districts in particular offer new alternatives that are highly accessible for many of the second and third generation BME households, particularly those comprising young, skilled individuals. At the same time, however, newcomers to the city are just as restricted in their points of entry and their opportunities. They are concentrated in particular locations and parts of the housing market. Unequal bargaining power is now apparent within the BME community rather than low bargaining power simply being a distinguishing feature of the whole.

In conclusion, can we say something about whether the different nature of the welfare state and the organization of housing provision in Birmingham as an example of a British city accounts for differences in the nature and extent of segregation? British cities are less segregated than American cities, although the analyses carried out suggest that they are generally more segregated than other European cities (Musterd et al., 1998). There is a tendency to argue that this latter effect is because the British welfare state is weaker and more comparable to the American welfare state than is true in other European cities. Ironically, however, this may not be the most important explanation. Instead it may be more appropriate to focus upon the significance of council housing at a key stage in the establishment of the BME community in the UK. In general, the high quality and desirability of council housing has been argued to have moderated the importance of income and social class on residence in British cities. Access to council housing was not determined purely by income and class position and the desirability for council housing meant that people did not move on as their circumstances changed or improved in every case. The decommodification of council housing meant that there is a greater degree of social mix in good-quality council estates or neighbourhoods, such as Bourneville, than would have occurred in a purely market-based system.

However, this positive effect of municipal housing on socio-spatial patterns within the city does not stand up so well to scrutiny when it comes to BME communities. For the reasons outlined earlier in this chapter, BME groups were largely excluded from council estates, especially in the early phase of residence. As a result they were concentrated in affordable and accessible parts of the private sector. The strength of municipal housing, rather than minimizing segregation, has contributed to it. The reason why British cities are more segregated than their European counterparts is then more to do with the strength of council housing and the welfare state than the weakness of the welfare state. The shift in public policy away from state housing provision towards a more market-based approach has further consolidated the tendency towards segregation.

The housing market in the UK has generally seen a sustained high level of real house price increases, and households in the owner-occupied sector have benefited compared with others. However, not all parts of the owner-occupied sector have performed in the same way. Within Birmingham it is clear that BME communities have been more likely to be located in some of the areas where the performance has fallen behind. Even in this perspective, the position of BME communities in the hierarchy of opportunity represented by the housing market has shifted compared with the 1950s and 1960s.

The argument might be advanced that this understates the position of those within BME communities who, by accident or design, did move to parts of the market that subsequently have performed more strongly, or, because of achievements in education or employment, have moved to stronger economic situations. These households have then moved within the housing market, reflecting the opportunities that their housing or their economic position has given them. Situations that were associated with lack of choice in the past, have contributed to increased capacity to exercise choice at a later stage. As we have seen, this ability to exercise choice includes the possibility to access dominant new forms of housing development associated with the wider shift in role of local state to one of city competitiveness. In this context, access to new markets in the city centre is determined more by considerations of economic class than ethnicity. As municipal estates have become mixed-tenure estates, following privatization, opportunities to move into council estates (as purchasers on the open market) have increased for BME households and, in some areas, it is clear that BME households have moved into the larger properties in former council housing estates. While this, as well as the ability to move out of areas of original settlement, contributes to some change in the location of BME communities within Birmingham, it does not alter the continuing tendency for many BME groups to reject council housing because of tenure, neighbourhood and especially safety reasons.

At the same time, there are signs of complex new spatial patterns of residence emerging in the heart of the city. The character of the emerging city living market and its potential to catalyse large-scale regeneration in adjacent inner-city districts suggests that there may be a significant BME presence in emerging communities in the core areas of Birmingham in the years ahead, confounding some fears that this high-profile development may generate new dimensions of exclusion and segregation by ethnicity in the central area. And yet while this dynamic private market process evolves, there remain strong forces for concentration of BME groups, including the residential experiences of new migrants to the city. These issues are examined in a separate chapter of this book but the overall pattern indicates that, although the dominant policy approach has moved more towards an American model in recent years, the evolving spatial patterns of residence and race continue to be path dependent. They reflect specific historical factors and in particular the legacies of the specific role of the local state in relation to housing provision and planning.

References

Akhtar, J. (ed.) (1996) *Pakistanis in Britain: In the 1990s and Beyond*. Pakistan Forum, Birmingham.

Anwar, M. (1996) *British Pakistanis: Demographic, Social and Economic Position*. Pakistan Forum, Birmingham.

Barber, A. (2001) *The ICC, Birmingham: A Catalyst for Urban Renaissance?* CURS, The University of Birmingham, Birmingham.

Barber, A. (2003) City living and new urban policy challenges: the implications of city centre housing development in England's regional cities. In: Kang, M. (ed.) *Urban Squatter Policies (III): The Cases of Korea and United Kingdom*. Korean Research Institute for Human Settlements.

Barber, A. and Blackaby, B. (2003) *Filling the Gap: The Need for Middle Market Housing in Birmingham City Centre*. Centre for Urban and Regional Studies, Birmingham.

Birmingham Economic Information Centre (2001) *Social Exclusion and Employment*. BEIC, Birmingham.

Blackaby, B., Murie, A. and Barber, A. (2002) *City Living in Birmingham: An Independent Review*. Birmingham City Pride, Birmingham.

Briggs, A. (1952) *History of Birmingham (Vol. 2)*. Oxford University Press, Oxford.

Cherry, G. (1994) *Birmingham*. Wiley, Chichester.

Ginsburg, N. (1992) *Divisions of Welfare*. Sage, London.

Groves, R. and Niner, P. (1998) *A Good Investment? The Impact of Urban Renewal on the Inner City Housing Market*. Policy Press, Bristol.

Haddon, R. (1970) A minority in a welfare state society: the location of West Indians in London. *New Atlantis* **2** (1): 80–133.

Harvey, D. (1989) From managerialism to entrepreneurialism: the transformation of urban governance in late capitalism. *Geografiska Annaler* **79**(B) (1): 3–17.

Henderson, J. and Karn, V. (1987) *Race, Class and State Housing*. Gower, Aldershot.

Karn, V., Kemeny, J. and Williams, P. (1985) *Home Ownership in the Inner City – Salvation or Despair?* Croom Helm, London.

Le Grand, J. (1982) *The Strategy of Equality*. George Allen and Unwin, London.

Musterd, S., Ostendorf, W. and Breebaart, M. (1998) *Multi-Ethnic Metropolis: Patterns and Policies*. Kluwer Academic Publishers, Dordrecht.

Pahl, R.E. (1975) *Whose City?* Penguin, Harmondsworth.

Phillimore, J., Ferrari, E. and Fathi, J. (2004) *The Housing Needs and Aspirations of Asylum Seekers and Refugees Living in the Birmingham Sandwell Pathfinder Area*. Birmingham Sandwell HMRA, CURS, The University of Birmingham.

Rex, J. and Moore, R. (1967) *Race Community and Conflict: A Study of Sparkbrook*. Oxford University Press, Oxford.

Spencer, K., Taylor, A., Smith, B., Mawson, J., Flynn, N. and Batley, R. (1986) *Crisis in the Industrial Heartland*. Clarendon Press, Oxford.

6

Race and Public Housing Transformation in the United States

Susan J. Popkin

Introduction

A massive transformation of public housing is under way in the US, one that is dramatically reshaping inner-city neighborhoods. Driven by the $5 billion federal HOPE VI program[1], US public housing authorities are demolishing their worst public housing properties and replacing them with new, mixed-income developments. Since 1992, 166 housing authorities have received HOPE VI grants and more than 60,000 units of distressed public housing have been demolished. In place of traditional public housing, federal policy now emphasizes replacing distressed housing with mixed-income developments and dispersing very low-income tenants throughout metropolitan areas using Section 8 Housing Choice Vouchers, which allow recipients to rent in the private market[2]. And, in a departure

---

[1] HOPE VI stands for 'Housing Opportunities for People Everywhere'. Originally called the Urban Revitalization Demonstration Program, HOPE VI was intended to comprehensively address the physical and social problems of severely distressed public housing communities. Congress first provided funding for the program through the Departments of Veterans Affairs and Housing and Urban Development and Independent Agencies Appropriations Act of 1993. However, the program did not receive congressional authorization until the passage of the Quality Housing and Work Responsibility Act of 1998 (Holin et al., 2003).

[2] Under the federal Section 8 Housing Choice Voucher program, recipients pay 30 percent of their income for rent and the housing authority pays the remainder. HUD sets what it considers Fair Market Rents for each jurisdiction. Recipients must apply for assistance and must meet income eligibility criteria and a criminal background check. In most US cities, there are long waiting lists for assistance; families relocated under the HOPE VI program receive special vouchers that have been allocated for relocation. Once the participants receive their vouchers, they must find a private market landlord willing

from traditional housing policy, the policies and programs that underlie the transformation of public housing explicitly seek to use federal assistance to promote economic self-sufficiency for very low-income families.

After a decade, this effort has achieved substantial successes; it has demolished some of the most distressed and destructive housing environments and replaced them with higher-quality housing, often in mixed-income communities. Many residents who relocated with vouchers are living in higher-quality housing in safer neighborhoods. However, it is also clear that the transformation effort has not yet achieved its full potential to improve the lives of poor, minority families. There is evidence that original residents have suffered as a result of inadequate relocation planning, have ended up in other troubled public housing developments or been 'lost' during the relocation process. Others are struggling to maintain stable housing in the private market. And most are still living in communities that are very poor and racially segregated.

Notably missing from the policy dialogue about how to transform federal housing assistance in the US has been the question of race – how segregation contributed to the social distress in public housing and how racial discrimination might affect outcomes for these vulnerable families, many of whom would now have to negotiate the private rental market. But the reality is that race must play a very central role in this debate. Racial segregation is one of the major causes of the problems of crime and concentrated poverty in the US that public housing transformation was intended to address. Because of decades of housing authorities' discriminatory policies and practices, nearly all the residents of distressed public housing are African-American or Hispanic. Under HOPE VI, a substantial number of these households are being relocated to make room for higher-income, sometimes white households. And, even in cities without tight rental markets, residents' choices for replacement housing are constrained by discrimination and segregation in the private housing market.

In this chapter, I consider the implications of the policy decision to avoid making remedying racial segregation an explicit goal of US public housing transformation. I argue that this choice has had serious consequences, not only for the way that the policy debates over the impact of public housing transformation are framed, but for the families who endured the deplorable conditions in distressed public housing, and are now the ones most affected by revitalization efforts. Finally, I raise the question of whether, without addressing directly racial prejudice, discrimination and segregation, public

---

to lease the unit for what the housing authority determines is an acceptable rent. Under current regulations, participants can continue to receive assistance as long as they are in compliance with program rules and their household incomes do not exceed eligibility guidelines.

housing transformation can fully realize its potential to improve the lives of poor, minority families.

In the first section, I review the research and policy debates that led to the current transformation of public housing and the role of racial segregation in creating the problems in distressed public housing. Then I provide an overview of the policies that reshaped US housing policy during the 1990s. I review the research evidence to date on outcomes for original residents of developments targeted under HOPE VI and how the failure explicitly to address racial discrimination may have played a role in bringing about the mixed results for these families. I conclude with a call for a bold policy approach that will directly address the problems of segregation and discrimination in public and assisted housing and in the broader housing market.

The rationale for transforming public housing

Public housing transformation was a response to policy debates in the US during the 1980s over the purported failure of federal social welfare programs to improve the lives of low-income families. Researchers and advocates, particularly those on the right, argued that these programs discouraged work and reinforced 'deviant' behaviors such as single motherhood and teen pregnancy (Murray, 1984). Public housing was regarded as one of the biggest and most visible failures of social welfare policy; the forbidding high-rises of Chicago came to symbolize all that had gone wrong with this federal effort (National Commission, 1992).

In the housing arena, the debate was further complicated by the accusation that the US Department of Housing and Urban Development (HUD) was so corrupt and badly managed that the agency was incapable of carrying out its basic mission. HUD earned this reputation in large part due to the financial scandals that plagued the agency during the Reagan administration, but also because of the visible failures of distressed public housing. By the early 1990s, Congress was actively considering eliminating HUD altogether (Popkin et al., 2004b).

At the same time that policy makers were debating the role of the federal government in social welfare, there was increasing concern about the negative consequences of the increasing concentration of extremely poor households in inner-city communities (Massey and Denton, 1993; Goetz, 2003). Wilson (1987) argued in the late 1980s that the concentration of poverty was responsible for a range of social ills in inner-city communities, including high-school drop-out, teen parenthood, delinquency, drug and alcohol use, and weak labor force attachment. Wilson attributed the rise of what he called the 'underclass culture' to the increasing economic isolation of the poor in inner-city communities, and the loss of middle- and working-class role models. Many other scholars (Briggs, 1997; Ellen and Turner, 1997; Leventhal and Brooks-Gunn, 2000) argued that neighborhoods have profound effects on residents' life chances, although the mechanisms that bring about these effects were poorly understood.

Nowhere were the negative consequences of concentrated poverty more evident than in distressed public housing, epitomized by Chicago's high-rise developments like Cabrini-Green and Robert Taylor Homes. By the end of the 1980s, living conditions in the nation's most dilapidated public housing were intolerable. A complex layering of problems had left these developments mired in the most destructive kind of poverty (Blank, 1997). The situation was so bad that Congress created a National Commission on Severely Distressed Public Housing to develop a plan for addressing the problems (National Commission, 1992). The level of racial and economic segregation in these developments was extreme, and public services, particularly police, schools and sanitation, were inadequate. The neighborhoods that surrounded these developments generally had few services or stores and even fewer employment opportunities. Most residents were unemployed, depending on public assistance or the underground economy (Popkin et al., 2000b). Ineffective housing authority management and inadequate federal funding had left these developments with huge backlogs of repairs, creating hazardous conditions that placed residents at risk of injury or disease (Rosenstreich et al., 1997). Exacerbating the problems, many of these developments were dominated by violent criminals and drug dealers, and residents lived in constant fear.

The role of racial segregation and discrimination

Public and assisted housing represent only a small proportion of the overall rental market in most US cities[3] and patterns of racial segregation in public housing reflect the overall trends in the broader housing market. Although there have been some improvements in the past three decades, the level of racial segregation in the overall housing market – particularly black–white segregation – remains extremely high (Lewis Mumford Center, 2001). Likewise, while discrimination against African-Americans in the rental housing market has declined slightly since 1989, the evidence from the most recent national discrimination study indicates that it still continues at a high rate, as does discrimination against Hispanics (Turner et al., 2002).

[3] Even in Chicago, where the housing authority now has more than 34,000 vouchers and 25,000 public housing units, assisted housing represents just 10 percent of the total rental market.

Moreover, racial segregation and discrimination persists across all income levels in the US. As whites and Asians earn more, they tend to move to neighborhoods with higher levels of public services. But, because of racial discrimination, higher income African-Americans are not able to move to neighborhoods commensurate with their economic status (Massey and Denton, 1993; Lewis Mumford Center, 2002). Middle- and upper-class African-American neighborhoods are generally less class segregated than white neighborhoods, and, as a result, more vulnerable to crime and other social problems (Patillo-McCoy, 1999; Cashin, 2004). Because of racial segregation, these neighborhoods have lower property values, weaker schools, higher levels of crime and lower public services than comparable white communities (Massey and Denton, 1993). Racial segregation is closely linked with economic segregation. The majority of high-poverty neighborhoods in urban areas – census tracts where more than 40 percent of the residents are poor – are predominantly African-American and, in contrast, the majority of very low-poverty neighborhoods are predominantly white (Jargowsky, 1996). Recent research shows that the persistence of these problems has given rise to potent stereotypes: Americans of all races tend implicitly to associate predominantly African-American neighborhoods with high rates of social disorder regardless of whether these neighborhoods are objectively more dangerous than other communities (Sampson and Raudenbush, 2004).

Segregation and public housing

There is no question that racial segregation and discrimination played a major role in creating the terrible conditions that existed in public housing in America's inner cities (Bickford and Massey, 1991; Massey and Kanaiapuni, 1993). Race and subsidized housing had been explicitly intertwined since the inception of New Deal housing initiatives (Olion-Chandler, 1992). For example, before World War II, 236 of the 261 developments subsidized by the United States Housing Authority and 43 of the 49 supported by the Public Works Administration (or 90 percent of the total) were completely segregated racially, usually by restricting admissions to developments (Coulibaly et al., 1998). Discriminatory public housing practices continued during and after the war (Rossi and Dentler, 1961; Hirsch, 1983; Goldstein and Yancey, 1986). In the three decades following, approximately 700,000 units of public housing were constructed, generating a potent legacy of segregation (Goering and Kamely, 1997; Goetz, 2003; Popkin et al., 2003).

In addition to discrimination in admission practices, in many US cities, public housing was deliberately located in poor, historically minority neighborhoods. The most notorious case was Chicago, where Mayor

Richard J. Daley and the Chicago City Council overrode the objections of housing authority administrators and razed hundreds of slums in black communities, replacing them with enormous high-rise developments (Bowly, 1978; Cohen and Taylor, 2000; Popkin et al., 2000a). As in many cities, having constructed these massive developments to house poor black families, state and local officials used federal highway funds to isolate them further by constructing expressways that cut through the communities (Fosburg et al., 1996).

Title VI of the US Civil Rights Act of 1964 officially prohibited discrimination in the administration of federally assisted housing. However, in practice, the Act had little impact on racial segregation in public housing, primarily because housing authority practices changed little (Goering, 1986). But another contributing cause was that white families increasingly chose not to remain in or apply for public housing family developments. As a result, by 1977, over half of the family developments were predominantly occupied by African-Americans; in contrast, nearly half the developments for seniors were predominantly occupied by whites (Coulibaly et al., 1998; Popkin et al., 2003).

The net effect of years of discriminatory policies in public housing was 'hypersegregated' developments in deteriorating neighborhoods that were almost equally racially segregated (Bickford and Massey, 1991; Massey and Denton, 1993). By the 1990s, the average African-American public housing resident lived in a development that was 85 percent black, while white residents lived in developments that were predominantly white (Goering and Kamely, 1997). Further, two-thirds of public housing developments in the US had 80 percent or more tenants in one racial-ethnic group, and 15 percent had all tenants in one group (Coulibaly et al., 1998)[4]. The 'hypersegregation' in these developments exacerbated the racial segregation and decline in the surrounding neighborhoods (Massey and Denton, 1993).

The economic segregation in distressed public housing was also staggering. By the 1990s, more than 80 percent of public housing residents lived below the poverty threshold, and most of the households earned less than 20 percent of what non-public housing residents in the same communities earned[5]. More alarming still, there was an eight-fold increase in the share of the very poorest families living in public housing between 1981 and 1991 (National Commission, 1992)[6].

---

[4] Of these homogeneous developments, 57 percent are occupied by African-Americans, 43 percent by whites.

[5] This figure excludes elderly households.

[6] In 1981, 2.5 percent of families in public housing earned 10 percent less than the area median income; by 1991, the proportion earning less than the median income had increased to 20 percent.

Poverty 'deconcentration' and self-sufficiency

The best policy response to the problems of distressed public housing, many scholars and advocates argued, was to enact bold new strategies that would emphasize 'poverty deconcentration'. As noted above, by the 1990s, virtually all distressed public housing in the US was racially and economically segregated; there was little hope for being able to desegregate these communities in their current form. These new initiatives would both revitalize these distressed communities in hopes of attracting higher-income residents and offer current residents vouchers that they could use to move to better, less segregated neighborhoods. Politically, proponents hoped that poverty-based strategies would prove easier to sell to policy makers who had soured on affirmative action and other race-based strategies.

Beyond the political advantages, proponents hoped that poverty deconcentration would have substantial benefits for very low-income families and would help to undo the negative effects of living in concentrated poverty. The theory was that mixing residents of different income levels – through place-based strategies like mixed-income and mobility strategies such as vouchers – would both expose very low-income public housing residents to working- and middle-class role models and provide them access to neighborhoods that offered greater opportunity. Based on surprisingly little hard evidence, proponents claimed that a range of benefits could be gained from aggressively pursuing these strategies, including: improved job and educational opportunities for low-income families; positive role models; more stable communities; better public services; better management; and investments in the larger neighborhood (Schwartz and Tajbakhsh, 1997; Popkin et al., 2000b; Goetz, 2003).

The influence of Gautreaux

The high expectations for the potential self-sufficiency effects of 'deconcentration and dispersal' approaches were based in large part on largely positive research findings on Chicago's Gautreaux Housing Desegregation Program. Initiated in 1976, the Gautreaux program was created as a result of the settlement of the seminal *Gautreaux* case, a major public housing desegregation case initiated in the late 1960s[7]. In *Gautreaux*, the courts found that the Chicago Housing Authority and HUD had discriminated against African-American tenants, concentrating them in large-scale developments located in poor, African-American neighborhoods. The

[7] *Gautreaux* v. *Chicago Housing Authority*, 304 F. Supp, 736 (N.D. Ill 1969) *enforcing* 296 F. Supp. 907 (N.D. Ill 1969) and *Gautreaux* v. *Landrieu*, 523 F. Supp. 665, 674 (N.D. Ill 1981).

decision against the housing authority in 1969 called for the creation of new public housing at 'scattered sites' in predominantly white communities. The case against HUD eventually moved to the US Supreme Court and was settled in 1976. The court ordered relief in the form of 7100 Section 8 certificates – subsidies that were to be provided to current and former Chicago Housing Authority residents to use in neighborhoods that were less than 30 percent African-American. Practically, this requirement meant that many Gautreaux participants had to move to suburban areas. The agreement was later amended to permit some households to move to 'revitalizing' African-American communities, which meant that about a third of the participants ended up moving within the city of Chicago (Rubinowitz and Rosenbaum, 2000).

Although the Gautreaux program was a desegregation remedy, not a social welfare program, research on outcomes for families participating in the mobility program seemed unusually encouraging. Studies suggested that adults who had moved to white suburban communities were more likely to be employed than those who remained in the city and that their children were doing better in school (Kaufman and Rosenbaum, 1992; Popkin et al., 1993; Rubinowitz and Rosenbaum, 2000). The fact that these studies had a number of significant limitations, particularly small sample sizes and retrospective design, was often overlooked.

In the policy debate over how to help very poor families, the Gautreaux findings were often cited as definitive rather than suggestive of the benefits of deconcentration and mixed-income strategies (Popkin et al., 2000a). Policy makers looking for models of successful social welfare efforts seized on the results, hoping that they offered an effective way to dramatically improve the life circumstances of poor minority families. As a result, the Clinton administration and the then HUD Secretary, Henry Cisneros, made addressing the problem of concentrated poverty a major focus of the effort to 'reinvent' HUD, promoting innovative strategies such as mobility counseling and mixed-income housing (Popkin et al., 2004b).

However, these new policies differed from the Gautreaux model in a significant way: Gautreaux was explicitly designed to be a remedy for the *racially discriminatory* practices of the Chicago Housing Authority. In contrast, many of the new housing policies focused solely on '*deconcentrating poverty*'. This emphasis was a political decision, intended to enhance the prospects for a new initiative that sought to funnel massive amounts of federal funds into a program widely regarded as a failure (Goering and Feins, 2003). Policy makers and researchers acknowledged the central role that racial segregation had played in creating the conditions in distressed public housing in many cities. However, proponents of the new federal initiatives feared that drawing attention to the problem of racial segregation would generate considerable opposition both from those who opposed desegregation initiatives and affirmative action as unfair and harmful to whites and those who viewed mobility efforts as 'social engineering'[8]. Because of the clear relationship between racial and economic segregation, proponents hoped that poverty deconcentration would also accomplish the goal of desegregation and would help former residents move to communities which would offer economic opportunity.

Pursuing public housing desegregation

At the same time that policy makers pursued poverty deconcentration through new federal initiatives, there was a renewed effort to address aggressively the problem of racial discrimination in public and assisted housing through litigation. During the 1990s, HUD departed from its previous strategy of contesting public housing lawsuits such as *Gautreaux*. Instead, the agency acknowledged the historical role of the government in neglecting housing authority segregation and discrimination, and developed a strategy of settling the cases by supporting solutions to ensure 'fair housing' for the plaintiffs. Minority households were to be given expanded opportunities to live in higher-quality, desegregated residential environments by changing the character of existing public housing developments and surrounding communities, building new developments in scattered-site settings, and expanding the use of vouchers in areas with low percentages of minority households. Other than the specific emphasis on race, many of these remedies were similar to the larger federal programs that were simultaneously reshaping public housing.

These efforts to bring about public housing desegregation through litigation were only modestly successful. An early assessment of these initiatives (Popkin et al., 2003) showed that housing authorities faced significant obstacles in implementing these consent decrees, including the fact that housing authority resident populations and waiting lists were almost exclusively minority; there were inadequate supplies of affordable rental housing; and there was a lack of public transportation to low-poverty, white neighborhoods. Two of the most difficult obstacles were minority public housing residents' understandable reluctance to attempt to move to unfamiliar areas that they perceived as potentially hostile to African-Americans and Hispanics and organized resistance to integration efforts by white residents in communities targeted for scattered-site

---

[8] Indeed, one of the major objections to the Gautreaux program was that the mobility program constituted social engineering. See Rubinowitz and Rosenbaum (2000) for a full discussion of the debate over mobility as a strategy for remedying discrimination against public housing residents.

housing[9]. In the case with the most far-reaching implications, a homeowners' association in Dallas successfully challenged the *Walker* decree[10]. This precedent-setting suit alleged that the construction of public housing in a predominantly white community denied the residents 'equal protection' under the Fourteenth Amendment. Essentially, the plaintiffs argued that only the 'least intrusive' race-based remedy should be used to correct racial discrimination, and that, since rental assistance was less intrusive than building public housing, funds for scattered-site public housing construction should be used instead to fund vouchers. The Fifth Circuit Court of Appeals ruled in favor of the homeowners in 1999; the Supreme Court declined to hear the case, thereby upholding this ruling.

Even with these obstacles, some of these efforts ultimately succeeded in achieving at least some of their objectives. Dallas, where the white community put up such effective resistance to scattered-site housing, has actually become one of the best examples of what can be accomplished through litigation. While the scattered-site housing program failed, the housing authority was able successfully to implement a mobility program using Section 8 vouchers and replace its deteriorated, segregated developments with mixed-income housing. After 19 years of litigation, the *Walker* case was declared settled in late 2004 and the housing authority was released from its obligations[11]. Likewise, in Minneapolis, a regional effort succeeded in convincing a number of suburban communities to accept several hundred units of affordable scattered-site units as part of the settlement of the *Hollman* case (Goetz, 2003). Although the Bush administration is no longer actively pursuing settlements in these cases, litigation continues in a few cities. In February 2005, a federal judge in Baltimore found HUD liable for the segregation of Section 8 tenants and ordered a new mobility program (Rusk, 2005).

Public housing transformation – federal initiatives

At the federal level, the effort to reduce the concentrations of poverty in public and assisted housing took two forms. The first approach was

---

[9] This problem affects many public housing residents searching for housing with vouchers. Several studies have found that residents often voice fears about being forced to move to the suburbs and of limiting their housing searches due to the fear of encountering racial discrimination (Popkin and Cunningham, 2000; Turner et al., 2000; Smith, 2002).

[10] *Walker* vs. *The US Department of Housing and Urban Development, Highlands of McKamey VI, and Community Association, Preston Highlands Homeowners' Association, et al.*

[11] *Walker* vs. *The US Department of Housing and Urban Development, Dallas Housing Authority, et al., Agreed Final Judgement*, US District Court, December 21, 2004.

treating the public housing developments themselves – demolishing distressed properties and attempting to create new, healthier communities on the same site. The $6 billion HOPE VI program was the centerpiece of this effort. Since 1992, HUD has awarded 446 HOPE VI grants in 166 cities. To date, 63,100 severely distressed units have been demolished and another 20,300 units are earmarked for redevelopment (US Department of Housing and Urban Development, 2003). As of the end of 2002, 15 of 165 funded HOPE VI programs were fully complete (US General Accounting Office, 2003). The billions of federal dollars allocated for HOPE VI have leveraged billions of dollars more in other public, private and philanthropic investment.

Enacted by Congress in 1992, the HOPE VI program had extremely ambitious goals. The program explicitly intended to move beyond 'bricks and mortar' to address the social and economic needs of residents and the health of the surrounding neighborhood. Officially, the program's major objectives were:

- To improve the living environment for residents of severely distressed public housing by demolishing, rehabilitating, reconfiguring or replacing obsolete projects in part or whole.
- To revitalize the sites of public housing projects and help improve the surrounding neighborhood.
- To provide housing in ways that avoid or decrease the concentration of very low-income families.
- To build sustainable communities.

In addition to these explicit goals, the creators of the HOPE VI program believed that these revitalization efforts might have implications for the surrounding neighborhoods, spurring broader reinvestment and development activities.

To support this effort, the government made changes in a number of federal laws to help reduce the concentration of extremely poor households in all, not just distressed, public housing. These changes included the repeal of the one-for-one replacement rule[12], and an emphasis on promoting self-sufficiency and employment for public housing residents. Congress repealed an array of federal admission rules that had required local housing agencies to give preference to very poor families (and even homeless families) in resident selection. Housing authorities were given greater flexibility in setting resident selection preferences, based on local housing needs

[12] The one-for-one replacement rule prohibited housing authorities from demolishing any units unless they also constructed new replacement units. See Popkin et al. (2000a) for a full discussion of the regulatory changes during the 1990s that affected public and assisted housing.

and priorities – thus making possible a more diverse mix of incomes among public housing residents. The Quality Housing and Work Responsibility Act of 1998 allowed housing authorities to take a number of measures to attract higher-income residents, particularly reinstituting 'ceiling rents,' which keep rents at a set level even as tenants' income increases[13].

The focus on reducing poverty concentrations and promoting self-sufficiency rather than explicitly addressing racial segregation is evident in all of these efforts. The report of the National Commission on Severely Distressed Public Housing (1992), which laid the groundwork for the HOPE VI program, describes the magnitude of problems in distressed public housing and provides a fairly comprehensive list of the factors that contributed to the crisis situation in many developments. However, there is no mention of racial segregation; indeed, there is no discussion of the racial composition of the population of distressed public housing anywhere in the report, nor in the final authorizing legislation.

Promoting choice and mobility

The other major approach to transforming public housing was to shift HUD's emphasis from project-based assistance to greatly expanded use of tenant-based assistance through the Section 8 program. Section 8 participants receive certificates or vouchers that allow them to search for housing in the private rental market[14]. During the 1990s, HUD instituted a number of reforms to expand the use of vouchers and make the program more acceptable to the private market[15].

The objective of these changes was to use Section 8 to further the goal of deconcentration, using strategies to shift the poorest assisted households from high-poverty public housing to the private market. As part of this effort, the Quality Housing and Work Responsibility Act of 1998 set aside 75 percent of all new and turnover Section 8 vouchers for participants whose incomes were less than 30 percent of the area median. New Section

---

[13] The Quality Housing and Work Responsibility Act of 1998 (Section 513.d.2) requires that at least 40 percent of a housing authority's units made available in a year must be occupied by families with incomes at or below 30 percent of the area median income. If more than 75 percent of the new or turnover Section 8 vouchers are used by families with incomes below 30 percent of the area median income, then this 40 percent requirement can be reduced to as low as a 30 percent share.

[14] All units must meet HUD's housing quality standards and fall within what HUD has determined to be fair market rents; under recent regulations, participants must also pay a full security deposit for their unit.

[15] Provisions requiring landlords to give tenants additional notices were repealed. Also repealed were provisions that (a) required landlords accepting one voucher family to accept all qualified voucher families (the 'take one, take all' rule) and (b) prohibited term leases.

8 program rules included a performance measure on how well housing authorities were doing on dispersing their Section 8 populations throughout 'low-poverty' areas of their jurisdictions[16]. And, in 1998, reflecting the new emphasis on mobility and choice, the Section 8 program was renamed the 'Housing Choice Voucher program'. By the end of the decade, the voucher program had surpassed the public housing program to become the largest housing assistance program in the United States.

Several new initiatives during the 1990s involved providing mobility counseling to voucher holders to encourage families to move to low-poverty areas that policy makers hoped would offer greater economic opportunities. Again, although these programs were modeled on Gautreaux, they emphasized poverty deconcentration rather than desegregation. In particular, HUD's major mobility initiative – the Moving to Opportunity (MTO) Demonstration – used a poverty-based rather than a race-based strategy. MTO was designed to test whether the positive effects reported in the Gautreaux research would hold up in a rigorous experiment (for details see Goering and Feins, 2003; Orr et al., 2003). Instead of using a racially-based criterion as in Gautreaux, the program provided participants with vouchers they could only use to move to areas where less than 10 percent of the households *were below the poverty line.*

According to Goering and Feins (2003), the designers of MTO hoped that by making the inherently political decision to sell the program as focused on economic opportunity rather than desegregation, the program would be less controversial. Further, they believed that using the poverty criterion would likely result in most participants moving to predominantly white suburban areas comparable to those in Gautreaux, and thus would offer the same types of benefits for low-income families. However, this approach was only partially successful as a political strategy; shortly after the program was announced, local politicians in the Baltimore suburbs (one of the five MTO sites) protested the program as a threat to the stability of their communities (Galster et al., 2003). The Baltimore controversy resulted in

[16] Under the 10 September 1998 proposed rules (24 CFR Part 185) for the Section 8 Management Assessment Program (SEMAP), housing authorities get up to five bonus points if they meet one of the following criteria for Section 8 families with children: 50 percent live in low-poverty census tracts; the percentage of movers who move to low-poverty census tracts in the current year is at least two percentage points higher than the percentage in the previous year; or the percentage of movers who move to low-poverty census tracts in the last 2 years is at least two percentage points higher than the percentage who moved to low-poverty census tracts in the next to last year. A low-poverty census tract is defined as: a poverty rate at or below 10 percent or a poverty rate below the average in the housing authority's operating area. SEMAP also awards points to housing authorities that have policies that encourage participation of landlords with units located outside areas of poverty or minority concentration (Pistilli, 1998).

Congress killing the second year funding for the MTO Demonstration (Goering and Feins, 2003).

Other HUD efforts during the 1990s to provide mobility counseling were lower profile and less controversial. The Regional Opportunity Counseling Initiative provided funds for a small number of housing authorities to experiment with mobility counseling programs to expand choice for Section 8 holders. HUD tried to encourage mobility counseling for public housing relocatees under its Vacancy Consolidation Program and, to some extent, through HOPE VI (Turner and Williams, 1998). The Welfare to Work Voucher program also encouraged housing authorities to consider mobility counseling as part of their efforts (Smith and Johnson, 2001). Finally, HUD authorized a few housing authorities to experiment with mobility approaches to meet the goal of deconcentration. For example, Chicago's Housing Opportunities Program (HOP) offered mobility counseling and assistance in locating housing in low-poverty neighborhoods to current Section 8 participants (Cunningham and Popkin, 2002)[17].

The impact of public housing transformation

Now, more than a decade after the beginning of the effort to transform public housing, enough evidence is accumulating to allow an initial assessment of the effects of public housing transformation. A review of the research to date suggests that these initiatives have had some real benefits for public housing developments and neighborhoods. Most notably, revitalizing the distressed public housing site has often led to substantial decreases in neighborhood poverty and crime.

In practice, the goal of providing housing in 'ways that would avoid or decrease the concentration of very low-income households' has meant both replacing public housing with mixed-income developments and relying heavily on 'soft units' or vouchers for replacement housing. Although many sites are still under construction, the mixed-income developments completed to date have varied considerably, from sites that are a mix of low and very low income to some with a mix of market rate, affordable and public housing units. Generally, however, the new developments are considerably lower density than what they replaced, and have a relatively small number of public housing units on site (Holin et al., 2003). Some of these designs have been truly innovative and have dramatically transformed their communities, replacing high-density, high-rise and barracks-style housing with lower-density townhomes and low-rise dwellings

[17] The HOP program was created by CHAC, the private organization that manages Chicago's Section 8.

(Popkin et al., 2004b). In many sites, housing authorities have succeeded in attracting market-rate tenants to the new development; in some cases, these higher-income tenants are white, meaning there has been at least some minimal racial integration at these sites (Holin et al., 2003).

There is only limited evidence available on how public housing transformation has affected the neighborhoods around formerly distressed public housing developments. However, the studies that have been done to date suggest generally positive effects (Holin et al., 2003; Zeilenbach, 2003; Turbov and Piper, 2004). In particular, there is no doubt that demolishing the high-density distressed public housing developments has reduced poverty and crime in these communities, and may have stimulated additional development. However, there is little evidence of impact on racial segregation at any site (Holin et al., 2003), and virtually no evidence of a reduction in racial segregation in the surrounding communities.

Further, we know little about how public housing transformation has affected 'receiving communities'. Receiving communities are neighborhoods where either scattered-site housing has been constructed or voucher holders have moved, and, while the issue of potential impact is highly politically charged, there is little hard evidence. In the only systematic study to date, Galster et al. (2003) argue that in small numbers, Section 8 has no negative impact, but that large concentrations of Section 8 households in a single neighborhood may destabilize moderate-income communities and trigger declines in property values.

Outcomes for residents

While the findings on neighborhood impact are generally positive, the evidence to date on how residents have fared is much more mixed. While the mixed-income strategy has made for more livable – and, in a few cases, more racially integrated – communities, there is an obvious downside to not requiring developers to replace all the public housing units that are demolished. HOPE VI has led to a real loss of hard units for very low-income families. As of mid-2003, 115,000 units were earmarked for demolition but only 66,000 new housing units were proposed, with only a portion of these units being set aside for public housing residents[18]. Other, less well documented HUD initiatives such as the 'demo-dispo' rule that requires housing authorities to demolish units where the cost of rehabilitation exceeds the cost of demolition and the conversion of thousands of former Section 8 properties to market rate housing, account for the loss of an even larger number of units (Popkin et al., 2000b).

---

[18] So far, 49,828 units have actually been demolished and 21,000 have been created.

One of the most controversial issues facing the HOPE VI program is the small number of original residents who have returned to revitalized HOPE VI sites. Critics have pointed to the small numbers of returning tenants – as low as 11 percent nationwide in 2002 – as evidence that public housing transformation is another form of 'urban renewal' which is displacing poor, minority households from gentrifying neighborhoods to make room for higher-income whites (Keating, 2000; Wexler, 2001; National Housing Law Project, 2002). Defenders of the program counter that many HOPE VI sites are only partially reconstructed and even fewer are reoccupied, so the number of original residents who will ultimately return to the revitalized sites is unknown[19].

However, simply looking at the number of returning tenants provides little information about how these former residents are faring overall. Many may have gladly taken the opportunity to receive a voucher and be settled in a new unit in a better neighborhood. Others may be living in different public housing developments, while still others may have left assisted housing altogether. The HOPE VI Tracking Study (Buron et al., 2002), a retrospective survey of former residents from eight early HOPE VI sites, found that 19 percent of the households surveyed were living in a revitalized HOPE VI development, 29 percent were living in other public housing properties, 33 percent were renting units using housing vouchers and 18 percent had left assisted housing altogether. Similarly, an analysis of HUD administrative data for 73 HOPE VI sites found that few of the original residents had returned to a revitalized development; about one third of former residents had received vouchers, half had relocated to other public housing developments and the remainder had left subsidized housing (Kingsley et al., 2003)[20].

The HOPE VI Panel Study, which is tracking residents from five HOPE VI sites where redevelopment activities began in 2001, found that the majority of residents who had been relocated by the follow-up in 2003 had received vouchers. None of the sites had constructed replacement housing, so returning to the new development was not yet an option. However, the majority of residents, both prior to relocation and at the 2003 follow-up, expressed interest in returning to the site when it was complete (Popkin et al., 2002; Cunningham, 2004). The contrast between this finding and the evidence about actual numbers of returning tenants to date suggests that

---

[19] One study of HOPE VI (Holin et al., 2003) found wide variation in the numbers of returning tenants, depending on the characteristics of new development. Developments that remained 100 percent public housing, not surprisingly, had large numbers of returning tenants, while the numbers living in developments that had become mixed-income were substantially smaller.

[20] This analysis of HUD administrative data included former residents at 73 HOPE VI sites in 48 cities.

some residents who would like to return are not being allowed to, either due to the reduction in public housing units or not meeting the screening criteria for new mixed-income developments. However, it is also possible that once residents have moved, they may decide they prefer living in the private market with a voucher or simply that they are happy with their new community and do not want to move again. The long delays between relocation and reoccupancy at many sites contribute to the likelihood that former residents may choose to stay where they are rather than risk moving again.

Impact on racial segregation

Former residents have ended up in at least somewhat lower-poverty neighborhoods – in the HOPE VI Panel Study the average census tract poverty rate for those who had been relocated dropped from 40 percent at baseline to 28 percent at follow-up (Buron, 2004). In addition, 30 percent of movers now live in low-poverty neighborhoods, where less than 20 percent of the population lives in poverty. Findings from the national analysis of HUD administrative data were similar, showing residents still in assisted housing living in neighborhoods with considerably lower poverty rates than their original HOPE VI developments. While these findings are encouraging, they are tempered by the fact that residents of HOPE VI developments were living in extremely high-poverty neighborhoods, therefore, most moves would bring improvements in poverty rates. Further, in sites with tight rental markets or where demolition outpaced the production of new units, many former residents ended up in other distressed communities (Buron et al., 2002; Popkin et al., 2003).

Public housing transformation has had much less impact on racial segregation. While former residents are living in neighborhoods that are substantially less poor, the majority of former residents are still living in census tracts that are predominantly minority (Buron et al., 2002; Kingsley et al., 2003; Buron, 2004). The HOPE VI Panel Study found that after relocation, 76 percent of relocatees still lived in neighborhoods that were at least 80 percent minority and just 24 percent had relocated to communities that could be characterized as racially diverse. Likewise, MTO participants, those who received special assistance and those who did not, are generally living in lower-poverty areas than the distressed public housing developments they came from[21]. However, relatively few MTO participants, even among the experimental group, initially moved to white or racially integrated areas; by 2002, nearly 90 percent of all participants were living in areas of high minority concentration (80 percent

[21] Because of HOPE VI, even many of those in the control group (who did not receive a voucher through MTO) are also living in lower-poverty communities.

or more) (Orr et al., 2003). The experience of HOPE VI and MTO demonstrates the limitations of relying on poverty rather than race as a criterion for 'deconcentration' and desegregation. The reality is that African-American and Hispanic public housing residents who receive vouchers face the same discriminatory patterns as other non-subsidized tenants. And their situation is compounded by discrimination against Section 8 – often a proxy for racial discrimination (Popkin and Cunningham, 2000; Turner et al., 2000; Galster et al., 2003).

Impact on well-being

Even while public housing transformation has not resulted in many former residents moving to the types of opportunity-rich communities that the planners envisioned, all of the research to date suggests that former public housing residents perceive substantial improvements in their quality of life. Most notably, original residents report much less crime in their communities than those still living in public housing developments awaiting revitalization. For example, the percentage of HOPE VI Panel Study respondents reporting 'big problems' with shootings and violence in their communities dropped from 67 percent prior to relocation in 2001 to just 20 percent at the follow-up in 2003. Nearly three quarters of respondents reported that they felt safer in their current neighborhood than in their original public housing development (Buron, 2004).

Former residents also report better housing conditions (Buron et al., 2002; Popkin and Cunningham, 2002; Orr et al., 2003). Nearly two thirds of HOPE VI Panel Study relocatees described their current housing as being in good or excellent condition, with voucher holders reporting significantly better housing than those who moved to other public housing developments (Comey, 2004). Movers also reported substantially fewer problems with their housing, such as mold, broken heating and peeling paint or plaster[22].

Perhaps the most striking finding has been the impact of relocation on former residents' physical and mental health. Residents of distressed public housing suffer alarmingly high rates of health problems like obesity, hypertension, arthritis, diabetes and depression (Harris and Kaye, 2004). The Chicago study documented almost immediate improvements in respondents' mental health, likely a result of living in a safer neighborhood (Popkin and Cunningham, 2002). The MTO Interim Assessment found major improvements in adult mental health overall, as well as improvements in mental health for adolescent girls. In addition, MTO documented a significant reduction in adult obesity, which researchers attribute to the

[22] However, even with this substantial improvement, voucher holders reported more housing problems than other poor renters nationally (Comey, 2004).

healthier environment (Orr et al., 2003). Preliminary, single-site studies on MTO also documented dramatic reductions in asthma (Goering and Feins, 2003).

Impact on self-sufficiency

While public housing transformation has had a generally positive impact on former residents' quality of life, it appears to have had none of the hoped-for impacts on self-sufficiency. As noted earlier, policy makers and researchers had high hopes for the potential benefits of moving residents to neighborhoods that offered greater opportunity, helping to overcome spatial mismatch; providing residents with positive role-models; and getting children into better, more effective schools. However, at least so far, these hopes for public housing transformation have not been realized. No study of resident outcomes has found any indication that moving to lower-poverty communities has had any significant impact on household income, adult employment or welfare use. Employment increased overall for very low-income households because of welfare reform and the economic boom of the late 1990s, but there is no evidence of any independent benefits from moving to a lower-poverty community. Further, there have been no sustained effects on educational outcomes for youth. Both the MTO and HOPE VI studies find that children are attending schools that are less poor and parents perceive as safer, but there have been no sustained impacts on achievement (Orr et al., 2003; Popkin et al., 2004b).

On the other hand, there is no doubt that even those residents who have benefited have paid a price for public housing transformation. They have experienced the stress of involuntary relocation, and the resulting disruption of their social networks and their children's schooling. Most worrying, there is some evidence that at least some residents are worse off as a result of public housing transformation: a substantial proportion of former residents appear to be experiencing hardship, living in neighborhoods that are still very poor, extremely segregated, and suffer very high crime. This problem may intensify as the pace of relocation increases and a substantial proportion of the remaining residents of distressed developments – those with the most complex personal problems – may have great difficulty making successful transitions to either the private market or revitalized developments, and even more trouble becoming self-sufficient.

Public housing transformation: the next stage

The record for public housing transformation is mixed and the ultimate impact for communities and former residents is still unclear. Removing distressed public housing developments that were blighting neighborhoods has had some important benefits for surrounding communities and produced new and innovative mixed-income housing. It is less clear how these changes have affected the receiving communities and it is possible some moderate-poverty neighborhoods have been 'tipped' into high poverty as a result of poor relocation planning. However, it is abundantly clear that there has been little to no improvement in the racial segregation of either public housing communities or surrounding neighborhoods.

Former residents seem to have benefited in some important ways, but there is, thus far, no evidence of larger impacts on their economic well-being, the types of effects policy makers hoped would come of these deconcentration and dispersal efforts. The reasons why public housing transformation has had so little impact on residents' socio-economic status are not clear. It is possible that not enough time has elapsed for these kinds of changes to occur and we may see more positive effects as families adjust to new neighborhoods and schools. It is also possible that other factors, such as welfare reform or the economic boom of the late 1990s, have masked any effects. It may also be that these expectations were simply unrealistic, given the many health barriers and other challenges residents face (Levy and Kaye, 2004).

However, we must also consider whether the failure to address racial segregation directly has limited the effect of these ambitious reform efforts. One piece of evidence from MTO is particularly suggestive: unlike the Gautreaux experience, children in the MTO Demonstration generally did not change school districts as a result of relocation (Orr et al., 2003). In other words, they did not end up in the white, suburban, resource-rich environments that policy makers envisioned. Because of the pernicious effects of racial segregation, even more affluent African-American neighborhoods lack the types of public services and resources found in comparable white communities (Patillo-McKoy, 1999; Cashin, 2004). Further, because black neighborhoods are less segregated by class than white communities, they are more vulnerable to crime and other social problems. Perhaps simply moving to a safer, slightly better school in a moderate-income, minority urban neighborhood was simply not enough to bring about even the modestly encouraging effects seen in the earlier research on Gautreaux participants.

The reality is that, whether or not it is acknowledged explicitly, race overlies virtually every aspect of the effort to transform America's public housing and the failure to address racial segregation directly may have undermined even the most ambitious efforts to improve the lives and well-being of public housing residents. Racial segregation was a major factor in creating the humanitarian disaster in distressed public housing that led to the call for a bold new approach to federal housing assistance. But the architects of public housing transformation felt strongly that focusing on deconcentrating poverty was a more politically viable approach and that it would lead to reductions in racial segregation and improved socio-economic outcomes for poor, minority families.

Policy makers must grapple with the implications of the fact that the residents who are now experiencing both the modest benefits and the myriad hardships created by the transformation of public housing are almost all very low-income women of color – African-American and Hispanic single parents – and their children. The residents who are benefiting from the new, mixed-income developments are generally new higher-income and, in some cases, white – not the people who endured the terrible conditions in the distressed housing these communities replaced. Thus, the reality is that question of race must overlie any discussion of public housing transformation's ultimate impact on residents and communities.

In order for public housing transformation to realize its potential to improve the lives of poor, minority families, policy makers should now address some very tough questions about the impact of race and class and the implications of the political decision to focus federal housing initiatives during the 1990s on deconcentrating poverty. It is clear that addressing poverty alone has not been sufficient to substantially improve the lives of many very low-income, minority public housing residents. Fearing discrimination, former public housing residents may not even try to move to more affluent or less segregated communities (Popkin and Cunningham, 2000). And housing authorities attempting to replace their developments with scattered site units must face the likely prospect of organized resistance from community residents. Policies that focus on poverty deconcentration alone will not address these fundamental problems.

Future efforts *must* explicitly address racial segregation. Policy approaches that could potentially bring about better outcomes for public housing residents and promote desegregation include:

- Hands-on help with housing search in low-poverty, predominantly white communities and long-term follow-up support.
- Requiring housing authorities to recruit landlords in low-poverty, predominantly white areas for their voucher programs.
- Requiring housing authorities to market all new mixed-income developments to whites as well as minorities.
- Providing financial incentives to landlords in low-poverty, predominantly white communities to accept voucher holders and/or incentives for voucher holders to make mobility moves to these areas.

In order truly to improve outcomes for public and assisted housing residents, the federal government must address racial discrimination and segregation in the broader housing market; aggressive fair housing enforcement is essential. As part of this effort, the federal government should explicitly pursue discrimination against Section 8 holders and assess the extent to which it is being used as a proxy for racial discrimination. To address opposition to scattered-site and assisted housing, policy makers must develop approaches to build support and readiness in white communities. The experience of the successful effort to convince white, suburban communities near Minneapolis to construct scattered-site units as part of the settlement of the *Hollman* decree provides an excellent model (Goetz, 2003). Finally, the federal government must pursue policies to increase the production and preservation of affordable housing stock, especially in predominantly white communities.

It is crucial that policy makers, and Americans in general, begin a new dialogue about race and poverty. Until our society, and public policy, comes to terms with the long-term and wide-ranging effects of segregation, discrimination and prejudice, even our most ambitious social welfare efforts are likely to have only modest effects on the lives of poor, minority families.

References

Bickford, A. and Massey, D. (1991) Segregation in the second ghetto: racial and ethnic segregation in American public housing, 1977. *Social Forces* **69** (June): 1011–36.

Blank, R.M. (1997) *It Takes a Nation: A New Agenda for Fighting Poverty*. Princeton University Press, Princeton.

Bowly, D. (1978) *The Poorhouse: Subsidized Housing in Chicago 1895–1976*. Southern Illinois University Press, Carbondale.

Briggs, X. (1997) Moving up versus moving out: neighborhood effects in housing mobility programs. *Housing Policy Debate* **8**: 195–234.

Buron, L. (2004) *An Improved Living Environment? Neighborhood Outcomes for HOPE VI Relocatees. A Roof Over Their Heads Policy Brief No. 3.* The Urban Institute, Washington, DC.

Buron, L., Popkin, S., Levy, D., Harris, L. and Khadurri, J. (2002) *The HOPE VI Resident Tracking Study*. The Urban Institute, Washington, DC.

Cashin, S.D. (2004) *The Failures of Integration: How Race and Class are Undermining the American Dream*. Public Affairs, New York.

Cohen, A. and Taylor, E. (2000) *American Pharaoh*. Little, Brown, and Company, Boston.

Comey, J. (2004) *An Improved Living Environment? Housing Quality Outcomes for HOPE VI Relocatees. A Roof Over Their Heads Policy Brief No. 2.* The Urban Institute, Washington, DC.

Coulibaly, M., Green, R.L. and James, D.M. (1998) *Segregation in Federally Subsidized Low-Income Housing in the United States*. Praeger, Westport.

Cunningham, M.K. (2004) *An Improved Living Environment? Relocation Outcomes for HOPE VI Relocatees. A Roof Over Their Heads Policy Brief No. 1.* The Urban Institute, Washington, DC.

Cunningham, M.K. and Popkin, S.J. (2002) *CHAC Mobility Counseling Assessment: Final Report*. The Urban Institute, Washington, DC.
Ellen, I.G. and Turner, M.A. (1997) Does neighborhood matter? Assessing recent evidence. *Housing Policy Debate* **8** (4): 833–66.
Fosburg, L., Popkin, S. and Locke, G. (1996) *An Historical and Baseline Assessment of HOPE VI, Volume 1*. US Department of Housing and Urban Development, Washington, DC.
Galster, G.C., Tatian, P.A., Santiago, A.M., Pettit, K.L.S. and Smith, R.E. (2003) *Why NOT in My Back Yard? The Neighborhood Impacts of Assisted Housing*. Center for Urban Policy Research, Rutgers University, New Brunswick.
Goering, J. (ed.) (1986) *Housing Desegregation and Federal Policy*. University of North Carolina Press, Chapel Hill.
Goering, J. and Feins, J. (2003) *Choosing a Better Life? Evaluating the Moving to Opportunity Social Experiment*. The Urban Institute, Washington, DC.
Goering, J. and Kamely, A. (1997) Recent research on racial segregation and poverty concentration in public housing in the United States. *Urban Affairs Review* **32** (5, May): 723–45.
Goetz, E.G. (2003) *Clearing the Way: Deconcentrating the Poor in Urban America*. The Urban Institute, Washington, DC.
Goldstein, I. and Yancy, W. (1986) Public housing projects, blacks, and public policy: the historical ecology of public housing in Philadelphia. In: Goering, J.M. (ed.) *Housing Desegregation and Federal Policy*. University of North Carolina Press, Chapel Hill.
Harris, L. and Kaye, D. (2004) *How Are HOPE VI Families Faring: Health. A Roof Over Their Heads Policy Brief No. 5*. The Urban Institute, Washington, DC.
Hirsch, A. (1983) *Making the Second Ghetto: Race and Housing in Chicago, 1940–1960*. Cambridge University Press, New York.
Holin, M.J., Buron, L., Locke, G. and Cortes, A. (2003) *Interim Assessment of the HOPE VI Program Cross-Site Report*. Prepared by ABT Associates for the US Department of Housing and Urban Development, Washington, DC.
Jargowsky, P.A. (1996) *Poverty and Place: Ghettos, Barrios, and the American City*. Russell Sage Foundation, New York.
Kaufman, J.E. and Rosenbaum, J.E. (1992) The education and employment of low-income black youth in white suburbs. *Educational Evaluation and Policy Analysis* **14** (3): 229–40.
Keating, L. (2000) Redeveloping public housing; relearning urban renewal's immutable lessons. *Journal of the American Planning Association* **66** (4, Autumn): 384–96.
Kingsley, G.T., Johnson, J. and Pettit, K. (2003) Patterns of Section 8 Relocation in the HOPE VI Program. *Journal of Urban Affairs* **25** (4): 427–47.
Leventhal, T. and Brooks-Gunn, J. (2000) The neighborhoods they live in: the effects of neighborhood residence on child and adolescent outcomes. *Psychological Bulletin* **126** (2): 309–37.
Levy, D.K. and Kaye, D. (2004) *How Are HOPE VI Families Faring: Employment. A Roof Over Their Heads Policy Brief No. 4*. The Urban Institute, Washington, DC.
Lewis Mumford Center (2001) *Ethnic Diversity Grows, Neighborhood Integration Lags Behind*. http://mumford1.dyndns.org/cen2000/WholePop/WPreport/page1.html
Lewis Mumford Center (2002) *Separate and Unequal: The Neighborhood Gap for Blacks and Hispanics in Metropolitan America*. http://mumford1.dyndns.org/cen2000/SepUneq/SUReport/SURepPage1.htm

Massey, D.S. and Denton, N.A. (1993) *American Apartheid: Segregation and the Making of the Underclass.* Harvard University Press, Cambridge.

Massey, D.S. and Kanaiapuni, S. (1993) Public housing and the concentration of poverty. *Social Science Quarterly* **74**: 109–122.

Murray, C. (1984) *Losing Ground*. Basic Books, New York.

National Commission on Severely Distressed Public Housing (1992) *The Final Report of the National Commission on Severely Distressed Public Housing*. US Government Printing Office, Washington, DC.

National Housing Law Project (2002) False HOPE: A Critical Assessment of the HOPE VI Public Housing Redevelopment Program. http://www.nhlp.org.

Olion-Chandler, M. (1992) 'Public housing desegregation: what are the options? *Housing Policy Debate* **3** (2): 509–534.

Orr, L., Feins, J.D., Jacob, R., Beecroft, E.S., Katz, L., Lawrence, F., Liebman, J.B. and Kling, J.R. (2003) *Moving to Opportunity Interim Impacts Evaluation.* US Department of Housing and Urban Development, Washington, DC.

Patillo-McCoy, M. (1999) *Black Picket Fences: Privilege and Peril Among the Black Middle Class*. University of Chicago Press, Chicago.

Pistilli, L. (1998) Successful Section 8 Programs: How I Learned to Stop Worrying and Love SEMAP. *Housing Strategies*, **V1** (4): 1–3.

Popkin, S. and Cunningham, M. (2000) *Searching for Rental Housing With Section 8 in Chicago.* The Urban Institute, Washington, DC.

Popkin, S.J. and Cunningham, M.K. (2002) *CHA Relocation Counseling Assessment Final Report*. Report prepared for the John D. and Catherine T. MacArthur Foundation. The Urban Institute, Washington, DC.

Popkin, S.J., Buron, L., Levy, D.K. and Cunningham, M.K. (2000a) The Gautreaux: what might mixed-income and dispersal strategies mean for the poorest public housing tenants. *Housing Policy Debate* **11** (4): 911–42.

Popkin, S.J., Cove, E. and Eiseman, M. (2004a) *How Are HOPE VI Families Faring: Children. A Roof Over Their Heads Policy Brief No. 6.* The Urban Institute, Washington, DC.

Popkin, S.J., Cunningham, M.K. and Burt, M. (2005) Public Housing Transformation and the Hard to House. *Housing Policy Debate* **16** (1): 1–24.

Popkin, S.J., Galster, G.C., Temkin, K., Herbig, C., Levy, D.K. and Richer, E. (2003) Obstacles to desegregating public housing: lessons learned from implementing eight consent decrees. *Journal of Policy Analysis and Management* **22** (2): 179–99.

Popkin, S.J., Gwiasda, V.E., Olson, L.M., Rosenbaum, D.P. and Buron, L. (2000b) *The Hidden War: Crime and the Tragedy of Public Housing in Chicago*. Rutgers University Press, New Brunswick.

Popkin, S.J., Katz, B., Cunningham, M.K., Brown, K.D., Gustafson, J. and Turner, M.A. (2004b) *A Decade of HOPE VI: Research Findings and Policy Challenges.* The Urban Institute, Washington, DC.

Popkin, S.J., Levy, D., Harris, L.E., Comey, J., Cunningham, M.K. and Buron, L. (2002) *HOPE VI Panel Study: Baseline Report*. The Urban Institute, Washington, DC.

Popkin, S.J., Rosenbaum, J.E. and Meaden, P.M. (1993) Labor market experiences of low-income black women in middle-class suburbs: evidence from a survey of Gautreaux Program participants. *Journal of Policy Analysis and Management* **12** (3): 556–73.

Rosenstreich, D.L., Eggleston, P., Kattan, M. et al. (1997) The role of cockroach allergy and exposure to cockroach allergen in causing morbidity among inner-city children with asthma. *New England Journal of Medicine* **336**: 1356–63.

Rossi, P. and Dentler, R. (1961) *The Politics of Urban Renewal: The Chicago Findings*. Free Press, New York.

Rubinowitz, L.S. and Rosenbaum, J.E. (2000) *Crossing the Class and Color Lines: From Public Housing to White Suburbia*. University of Chicago Press, Chicago.

Rusk, D. (2005) The regional solution. *The Baltimore Sun*, 18 January.

Sampson, R.J. and Raudenbush, S.W. (2004) Seeing disorder: neighborhood stigma and the social construction of 'broken windows'. *Social Psychology Quarterly* **67** (4): 319–42.

Schwartz, A. and Tajbakhsh, K. (1997) Mixed-income housing: unanswered questions. *Cityscape: A Journal of Policy Development and Research* **3** (2): 71–92.

Smith, R.E. (2002) *Housing Choice for HOPE VI Relocatees*. The Urban Institute, Washington, DC.

Smith, R. and Johnson, J. (2001) *Welfare to Work Housing Voucher Program: Early Implementation Assessment*. Department of Housing and Urban Development, Washington, DC.

Turbov, M. and Piper, V. (2004) *HOPE VI as a Catalyst for Neighborhood Change*. The Brookings Institution, Washington, DC.

Turner, M. and Williams, K. (1998) *Housing Mobility: Realizing the Promise*. Report from the Second National Conference on Assisted Housing Mobility. The Urban Institute, Washington, DC.

Turner, M.A., Popkin, S.J. and Cunningham, M.K. (2000) *Section 8 Mobility and Neighborhood Health: Emerging Issues and Policy Challenges*. The Urban Institute, Washington, DC.

Turner, M.A., Ross, S.L., Galster, G.C. and Yinger, J. (2002) *Discrimination in Metropolitan Housing Markets: Phase 1*. US Department of Housing and Urban Development, Washington, DC.

US Department of Housing and Urban Development (2003) *HOPE VI Quarterly Management Report for September 30, 2003*. US Department of Housing and Urban Development, Washington, DC.

US General Accounting Office (2003) *Public Housing – HUD's Oversight of HOPE VI Sites Needs to be More Consistent*. US General Accounting Office, Washington, DC.

Wexler, H.J. (2001) HOPE VI: market means/public ends – the goals, strategies and midterm lessons of HUD's Urban Revitalization Demonstration Program. *Journal of Affordable Housing* **10** (3): 195–233.

Wilson, W.J. (1987) *The Truly Disadvantaged: the Inner City, the Underclass, and Public Policy*. University of Chicago Press, Chicago.

Zeilenbach, S. (2003) Assessing Economic Change in HOPE VI Neighborhoods. *Housing Policy Debate* **14** (4): 621–55.

7

Living Parallel Lives? Housing, Residential Segregation and Community Cohesion in England

David Robinson

Introduction

Until recently little attention had been paid to the issue of ethnic residential segregation in England. In the US racial residential segregation has long been recognized as a constituent factor in racial inequality and regarded as a problem to be solved. Minority ethnic population clusters in English towns and cities, in contrast, have been considered an inevitable consequence of the tendency for new migrants to gravitate towards population clusters of people from similar backgrounds, for reasons of mutual support and security in the face of hostility from the majority ethnic population, as well as the availability of cheap and accessible accommodation (Johnston et al., 2002). Through time, it was assumed, these population clusters break down and new populations become residentially assimilated. This assimilation scenario is supported by evidence revealing that the longer a population group has been resident in the UK the greater the degree to which it is residentially assimilated with other population groups. Set against this backdrop, the recent shift in public policy, from a position of relative indifference and towards the problematizing of minority ethnic settlement patterns in English towns and cities, appears a little bemusing. Policy makers, however, have become increasingly convinced that the residential segregation of certain minority ethnic groups is proving a persistent aspect of the urban landscape. These patterns of settlement have been problematized for allowing identities, values and principles perceived as being at odds with the dominant moral order to threaten the supposed neutrality of the public cultural domain (Robinson, 2005). This story line was evident in the policy response to both the violent disturbances that rocked northern England during 2001 and the London bombings of 2005 and has been most clearly articulated in the community cohesion agenda.

During the summer of 2001 violence erupted on the streets of a number of towns and cities in northern England, most notably Bradford, Burnley and Oldham. Commonly portrayed as 'race riots' and characterized as pitching young south Asian (Bangladeshi, Pakistani and Indian) men against the agents of 'mainstream' society (principally the police), the disturbances sparked an intense period of contemplation and reflection regarding issues of race, ethnicity, identity, segregation and social harmony in English urban society. The government sanctioned various local and national reports in a bid to explain the disturbances and to devise a response (Home Office, 2001; Independent Review Team, 2001; Oldham Independent Review Team, 2001; Burnley Task Force, 2002). Most influential was the report of the Community Cohesion Review Team, which was commissioned by the Home Office and directed to capture the views and opinions of local residents and different community interests in the affected towns, as well as other parts of England. A much quoted section of the Independent Review Team report captures the essence of the collective conclusion arrived at by these various reports:

> 'Whilst the physical segregation of housing estates and inner city areas came as no surprise, the team was particularly struck by the depth of polarisation of our towns and cities. The extent to which these physical divisions were compounded by so many other aspects of our daily lives, was very evident. Separate educational arrangements, community and voluntary bodies, employment, places of worship, language, social and cultural networks, means that many communities operate on the basis of a series of *parallel lives*. These lives often do not seem to touch at any point, let alone overlap and promote any meaningful interchanges.' (Independent Review Team, 2001, p. 9, emphasis added)

The physical separation borne out of residential segregation was identified as the principal cause of tension and unrest. Physical isolation was concluded as having led to isolation in service use and social life and as thereby having fostered ignorance, suspicion and intolerance between different communities. These divisions were further exacerbated by problems of deprivation and disadvantage in isolated communities and perceived injustices in the distribution of area-based resources (such as regeneration funding). Add to this the inflammatory activities of far-right extremist groups and the insensitive, inaccurate and provocative reporting of local news media and the result was social disharmony and unrest.

Implicit in this analysis was the perceived self-segregation of the south Asian population in many towns and cities. Housing policy and provision was also singled out as a major determinant of the shape of communities and criticized for contributing to high levels of segregation. The Independent Review Team (2001), for example, suggested that some minority ethnic groups are choosing to live within their own communities, although some residential choices are not freely made and reflect the outcome of housing policy and provision. Difficulties accessing better housing in more sought-after areas, either because of poverty or racial harassment, were identified as fostering frustration and resentment, as well as serving physically to isolate communities. The challenge for policy was therefore identified as being to counter the increasing residential segregation of different groups, which are 'retreating into "comfort zones" made up of people like themselves' (Ouseley, 2001, p. 3), and to forge unity through a common sense of place and shared sense of belonging (Home Office, 2001).

Policy therefore emerged from the period of contemplation, reflection and analysis following the disturbances in 2001 convinced that ethnic enclaves, promoted by the self-segregating tendencies of certain minority ethnic groups and reinforced by the vagaries of the housing system, were a persistent feature in many towns and cities. The result was the undercutting of social cohesion and promotion of social disharmony and unrest. In response, government invoked the need to foster community cohesion; to promote shared identities, values and principles through the social interaction borne out of residential integration. The logic of this position was subsequently reinforced for politicians and policy-makers by the London bombings of July 2005. As Trevor Phillips, Chair of the Commission for Racial Equality, commented: 'the aftermath of 7/7 forces us to assess where we are. And here is where I think we are: we are sleepwalking our way to segregation. We are becoming strangers to each other, and we are leaving communities to be marooned outside the mainstream' (Phillips, 2005).

This chapter takes these fundamental assumptions of the contemporary policy agenda as the starting point for an exploration of ethnic residential settlement patterns in England. First, the fundamental assumption of the community cohesion agenda that certain minority ethnic groups are increasingly segregating into ethnic enclaves within England's towns and cities is critically examined. Revealing that segregation is often a product of a gap between housing aspirations and outcomes, rather than an active choice, discussion then moves on to consider the factors bounding the housing choices and shaping the residential settlement patterns of minority ethnic groups. In doing so, both the benefits and challenges often associated with living in segregated neighbourhoods are considered. Finally, a response to segregation is proposed that focuses on challenging the injustices restricting the residential choices of many minority ethnic households, rather than setting up segregated communities as a problem to be solved.

Residential segregation: separating myth from reality

Public policy now assumes that ethnic residential segregation is a persistent presence in many towns and cities. These settlement patterns are understood as the direct consequence of the self-segregating tendencies of certain minority ethnic groups, compounded by the institutional procedures and operation of the housing system. Segregation is perceived to be limiting social interaction, undercutting the effective promotion of shared values, principles and norms of behaviour and fostering disharmony and unrest. Yet, reviewing demographic statistics of race and ethnicity it is difficult to uncover any substantial evidence to back up this discourse. In fact, analysis of the 1991 Census of Population has led geographers to conclude that there are no ghettos in England; no polarized enclaves where a large proportion of an ethnic group's population reside (Champion, 1996; Peach, 1996; 1998; Phillips, 1998; Johnston et al., 2002). Analysis also suggests that the assimilation model best summarizes the process of ethnic segregation in English cities, with the clustering of certain groups reflecting the early stages of concentration followed by dispersal. Johnston et al. (2002), for example, map ethnic segregation in 18 large English urban areas. A correlation is revealed between different migration streams into the UK and patterns of residential settlement, with cultural, religious and linguistic particularities being recognized as important contingent factors. The Irish population, for example, one of the oldest migration streams into the UK and involving people whose culture is described as similar to that of the white British majority, is revealed to be the least concentrated minority ethnic population. The South Asian population, in contrast, a more recent migration stream involving people who are more culturally and linguistically distinct, is revealed as being more residentially segregated. The African-Caribbean population, meanwhile, is described as occupying an 'intermediate position' (p. 606).

Of course, it would be wrong to assume that just because the Irish population has become spatially assimilated through time that other ethnic groups with their own distinct cultures, arriving into the UK at different points in time and encountering very different social, economic and political circumstances, will inevitably follow in their footsteps. Analysis of the 2001 Census appears to support this scenario, however, although residential assimilation has taken different forms and proceeded at different rates for different groups.

Dorling and Thomas (2004) analyse the 2001 Census with a view to deriving an atlas of social divisions expressed through changing patterns of settlement. Reflecting on the changing geographies of ethnicity at the district level, they conclude that 'while many aspects of society in the UK became more polarised geographically over the course of the 1990s, this was generally not the case for people's ethnicity'. In most cases, they suggest, 'minority ethnic groups became slightly less geographically concentrated over time' (p. 36). More specifically, they reveal that minority ethnic groups, including the Indian, Chinese and black Caribbean populations, are moving out of cities to more distant suburbs and small towns. The active dispersal of these particular groups is consistent with changing settlement patterns hinted at by analysis of the 1991 Census (Peach, 1996, 1998; Rees and Phillips, 1996). More noteworthy is evidence of the active dispersal beyond traditional population centres of the Pakistani and Bangladeshi populations. These are two groups predominant within the minority ethnic populations of the towns and cities of northern England where disturbances broke out in 2001. The 1991 Census had suggested that Pakistani and Bangladeshi households were rarely present among the 'spatial pioneers' moving out of inner-city areas and beyond traditional minority ethnic population clusters (Phillips, 1998). By 2001 public policy was assuming that these groups were actively retreating into neighbourhoods made up of people like themselves. The analysis of the 2001 Census presented by Dorling and Thomas, however, suggests that both groups, although still concentrated in areas of initial settlement, 'became more dispersed out of these areas during the 1990s' (p. 36).

Local demographic studies support the impression of gradual dispersal of minority ethnic populations out from inner-city population clusters. Of particular interest are findings to emerge from Bradford, one of the locations where street disturbances erupted in the summer of 2001 and a location commonly held up as the archetypal polarizing city. The Bradford Metropolitan District comprises the city of Bradford, a number of smaller satellite towns and a rural hinterland that stretches north to the borders of the Yorkshire Dales National Park. In 2001 the district had a population of approximately 500,000. One in five people were recorded as belonging to a minority ethnic group, the largest single minority ethnic group being the Pakistani population, which accounted for 15 percent of district population in 2001.

In the immediate aftermath of the disturbances in 2001, a report examining 'community fragmentation' in Bradford was published (Ouseley, 2001). Although commissioned before the disturbances, the findings of the report became the focus of the official search for an explanation and response to the violence. The review team's terms of reference focused on 'the worrying drift towards self-segregation [and] the necessity of arresting and reversing this process' (Ouseley, 2001; foreword). Very much in tune with the various local and national reports sanctioned by the Home Office in the aftermath of the disturbances, the report took as a given that relations between different cultural communities were deteriorating, attitudes were hardening and intolerance of difference was growing. This situation was concluded as being fuelled by 'white people's resentment about a perceived dominant presence of visible minorities with strong religious affiliation' resulting in the Muslim (Pakistani) community drawing on 'the comfort and security derived from staying together, retaining its strong culture, religious affiliation and identity, to live in self-contained communities and maintain strong links with Pakistan' (Ouseley, 2001, p. 17).

No demographic evidence was presented to support these claims. The report spoke of dominant notions regarding the self-segregation and cultural isolationism of the South Asian population and was largely accepted without question by local and national politicians, policy makers and the media. Bradford City Council, however, has been compiling demographic statistics with a racial dimension since the 1970s, allowing the validity of this story line to be tested. What emerges is a very different and far more complex picture than that presented in the Ouseley Report or portrayed in local and national media reports following the 2001 disturbances and the London bombings of July 2005.

Drawing on detailed time-series population statistics with a racial dimension to analyse segregation within Bradford, Simpson (2004) reveals that overall segregation between South Asian and 'other' populations actually remained unchanged in the city during the 1990s. This apparent stability, however, masks significant dynamism. First, the city did experience an increase in the number of majority South Asian areas. Rather than driven by the movement of South Asian households into traditional population clusters, this growth was due to the concentration of immigration and natural population growth in existing areas of settlement. Between 1981 and 2001 the South Asian population of the district grew from 48,800 to 94,250, compared to a reduction in the white population from 404,500 in 1981 to 374,100 in 2001. As the population grew, so new households formed and moved into housing that was in some cases vacated by white households. The net result was an increase in the number of majority South Asian areas in the inner city. The net migrationary trend for *both* white and South Asian households alike, however, was out of the city. The South Asian population actively dispersed from traditional inner core areas to the outer edges of the city and beyond. Rural-town wards located beyond the city fringe tripled their South Asian population during the 1990s. There were therefore much fewer mono-racial areas in the city at the beginning of the twenty-first century than 10 years earlier.

Simpson's (2004) demographic findings corroborate evidence to emerge from qualitative studies of minority ethnic housing preferences. The aspiration and intention of, especially younger, people from different minority ethnic groups to move beyond established areas of settlement and into ethnically mixed neighbourhoods has been repeatedly revealed (Birmingham City Council, 1998; Hawtin et al., 1999; Cole and Robinson, 2003; Robinson et al., 2005). One such study in Bradford involved interviews with 182 South Asian households (Ratcliffe et al., 2001). One third (62) of these households indicated that, at the time of the survey, they were looking to move house. When asked where they would want to move, only 14 households specified their current neighbourhood. In total, three out of five South Asian households expressed an interest in moving to a different area, most wanting to move to specific neighbourhoods located adjacent to established areas of South Asian settlement. Asked about their preference for these areas, respondents referred to their intrinsic qualities (clean, quiet areas, away from 'inner city problems' providing larger houses with gardens), rather than their proximity to existing areas of South Asian settlement. Questioned further about the outward movement of south Asian households from traditional areas of settlement, respondents identified younger people as the catalyst for this change and suggested that the process would gather momentum as children got older and married. Rather than self-segregating into ethnic enclaves, Bradford's South Asian communities are presented as having acknowledged and accepted that these changes are under way and will continue over the coming years (Ratcliffe et al., 2001).

Minority ethnic households commonly aspire to move, and often are relocating, beyond established areas of settlement. By no means a new trend, this process appears to have gathered pace during the 1990s. This is not to say that strong continuities do not remain in the geography of minority ethnic residential settlement in England. Indeed, England's minority ethnic populations remain concentrated in locations originally settled during the large-scale migration of the post-war period. Almost half of all minority ethnic people live in Greater London (47.6 percent), one eighth (13.6 percent) live in the West Midlands, while smaller concentrations are found in Yorkshire (7.6 percent) and the north-west (6.8 percent) (Strategy Unit, 2003). According to Dorling and Thomas (2004), in 2001 49.0 percent of the Pakistani population lived in just 13 of the 459 local authority districts in Britain, over half a million (47.5 percent) Indian people were concentrated in just 13 districts, a total of 53.4 percent of African-Caribbean people lived in 13 districts and 58.0 percent of the Bangladeshi population were clustered in just 10 districts. Even at the local level, where spatial changes in settlement patterns have been most pronounced, the well established pattern of minority ethnic settlement that is characterized by residential concentration, segregation and deprivation remains a persistent feature of the residential landscape. Set against the apparent loosening of ties to particular traditional locations and aspirations toward suburbanization, these continuities in minority ethnic settlement patterns beg an obvious question, to which we now turn; what factors are constraining the residential mobility of minority ethnic households?

Housing, neighbourhoods and residential mobility

At the micro-level, residential mobility is commonly understood as a manifestation of individual or household choice, what Mandic (2001, p. 53) refers to as 'an adaptive mechanism by which a household can adjust its current housing consumption to a preferred one'. The expressed desire of many minority ethnic households to venture beyond traditional residential locations might therefore be understood as an attempt to secure an improved residential environment and access accommodation better suited to their particular needs, given household size, structure and life-cycle position. The living conditions endured by many minority ethnic households certainly appear to give good reason for people wanting to improve their accommodation situation. A relatively large proportion of virtually all minority ethnic groups in England live in deprived housing situations[1], overcrowding being a particularly acute problem, as revealed in Table 7.1. The deprived industrial areas where many minority ethnic populations are concentrated have also been revealed to provide some of the most unpopular residential areas in the country (Burrows and Rhodes, 1998). But if these residential situations explain the expressed desire and intention of minority ethnic households to move beyond areas of traditional settlement, what factors explain the apparent failure of many households to fulfil these aspirations? To answer this question we need to consider the distinct package of opportunities, compromises and constraints that inform the housing outcomes of different minority ethnic groups in different locations and examine the importance of place as a determinant of residential preference and settlement patterns.

Housing and residential mobility

Housing and, in particular, desirable housing is a scarce resource and, as Rex and Moore (1967) observed, different groups are differentially placed to access accommodation. Financial resources (capital, size and security of income, access to loan finance) are particularly important determinants of the strength of a household in the housing market. The key to financial

[1] Deprived housing is defined as accommodation that is either overcrowded, or is in a shared dwelling, or does not have sole use of bath/shower and toilet, or has no central heating.

Table 7.1 People in overcrowded accommodation in England, by ethnicity (2001).

Ethnic group		Percentage of ethnic group	Number of people
White	British	5.6	2,335,050
	Irish	9.7	73,021
	Other white	16.7	220,392
Mixed	White and black Caribbean	17.5	18,618
	White and black African	28.3	14,367
	White and Asian	16.9	17,082
	Other mixed	21.1	19,297
Asian or Asian British	Indian	21.2	218,370
	Pakistani	26.4	184,893
	Bangladeshi	47.9	129,961
	Other Asian	32.8	83,340
Black or black British	Black Caribbean	20.5	127,246
	Black African	44.7	214,624
	Other Black	27.7	20,592
Chinese and other ethnic group	Chinese	23.5	46,908
	Other ethnic group	29.7	200,397

Note: Overcrowding, as defined by the Census of Population, relates to the actual number of rooms compared to the number of rooms 'required' by the members of the household (based on the relationships between household members and their ages).

power and the opportunity to choose is, in large part, linked to labour-market position. Put simply, people in a relatively good labour-market position are in an advantageous position in the housing market (van Kempen and Ozuekren, 1998). Lower-income households might therefore be expected to live in worse housing conditions and be more restricted in their spatial choices than higher-income households. The continued concentration of minority ethnic groups in traditional locations can therefore be understood as, in part, a reflection of the income and labour-market position of minority ethnic populations, who are more likely than the majority ethnic, white British population to live in low-income households. For example, 68 percent of the one million Pakistani and Bangladeshi people in Britain in 2001 were living in low-income households, according to the 2001 Census, compared to 21 percent of the white population. Low incomes reflect low employment rates; the working age employment rate for non-white ethnic groups in England in 2002/03 was 57.4 percent, compared to 76.3 percent for white people, a situation consistent across all regions and local areas of the UK (Brook, 2004).

These experiences are indicative of the disadvantages that all minority ethnic populations in England encounter within the labour market. Demand-side factors include low levels of business activity in areas where minority ethnic people live, employment practices that fail to engage with, recruit and maintain minority ethnic staff, and deprivation and poor levels of public infrastructure. Supply-side factors include the relatively low skills base within certain minority ethnic populations and difficulties accessing services that help people secure and sustain a place in the labour market, including childcare, transportation and training. The result, as the Cabinet Office (2003) has recognized, is that 'after taking into account the key variables that can be quantified, all ethnic minorities remain disadvantaged in terms of employment and occupational attainment' (p. 61). It is important to note, however, that disadvantage within the labour market is not the only constraint limiting the power of minority ethnic households to effect a change in their residential situation, a point illustrated by the fact that many economically successful minority ethnic households have remained in areas of community settlement (Harrison and Phillips, 2003). So what other factors constrain the housing choices of minority ethnic households in England?

Minority ethnic households are persistently disadvantaged by the actions of key individuals and institutions within the housing system, including landlords, letting agents, estate agents and mortgage finance companies. Market segmentation by private sector institutions, including racial steering by estate agents, inequalities in access to housing finance and private landlord discrimination have all been recognized as determinants of minority ethnic segregation (Harrison and Phillips, 2003). For example, the discriminatory practices of lender and financial services and the reluctance of building societies to fund mortgages for Pakistani families have been revealed (Bowes et al., 1998). Estate agents have also been found to be operating discriminatory practices, including the direction of minority ethnic buyers away from white areas and toward areas of minority ethnic settlement, in some cases in response to instructions from vendors not to sell properties to anyone who is black (CRE, 1990). Changing market conditions do appear to be forcing lending institutions and estate agents to be less discriminatory and more responsive to minority ethnic buyers. Evidence of the discriminatory practice of some estate agents continues to accumulate, however (Ahmed et al., 2001; Harrison and Phillips, 2003).

Labour market disadvantage and practices of private sector institutions have played a major role in restricting the mobility and reproducing the concentration and segregation of minority ethnic groups. But England is a country with a long history of state intervention in housing supply. The state-sponsored house building programme might be long past its post-war peak, but the social rented sector still accounts for one fifth of all housing in England. With allocations based on the concept of need, rather than ability to pay, the sector represents a key mechanism through which households in a relatively weak housing market position can effect a change in their residential situation. So why did the South Asian households surveyed in Bradford by Ratcliffe et al. (2001) suggest that the only people likely to be moving out of the inner city were 'middle class professionals', 'business people' and 'those who can afford it'? Why does greater locational flexibility remain, for so many minority ethnic households, closely associated with their ability to buy into the opportunities provided by the owner-occupied sector in locations beyond traditional areas of settlement? Why is the social rented sector so limited in its ability to provide significant opportunities for residential mobility among minority ethnic populations? The answer lies in the exclusion of minority ethnic groups from the opportunities offered by the social rented sector. Three broad explanations can be drawn from the extensive evidence base that has accumulated over the last 40 years to explain this situation.

First, the supply, allocation and management of social rented housing has largely failed to recognize that the situations and experiences of minority ethnic households differ in degree and kind from the white British population. An obvious example is the mismatch between the particular accommodation wants and needs of different ethnic groups and the location, design and size of housing available in the social rented sector. Even schemes developed with the express intent of meeting minority ethnic housing needs have been found to provide inappropriately or insensitively located, designed and sized housing (Robinson et al., 2002).

Policy and provision have also disadvantaged minority ethnic households in more subtle ways. Minority ethnic people in England are more likely than the rest of the population to live in deprived neighbourhoods, have low incomes, be unemployed, live in inadequate or unsuitable accommodation, experience poor health and be the victims of crime (Social Exclusion Unit, 2000). Rather than recognizing and responding to the impact of these inequalities on minority ethnic households across the whole range of supply, management, allocation and service delivery activities within the social rented sector, policy has assumed that racialized inequalities can be tackled through the eradication of racial discrimination. Policies that appear non-discriminatory, however, can serve to exacerbate racialized inequalities. A prime example is the Right-to-Buy programme, through which council houses have been sold to sitting tenants. With lower incomes and limited access to the suburban council estates most likely to provide a source of capital accumulation following purchase, minority ethnic households have been less able to take advantage of Right-to-Buy policies. Through the sale of council housing, meanwhile, combined with limited spending on new build and rehabilitation and maintenance of the remaining stock, households unable to buy have been presented with housing opportunities declining in size and quality

(Robinson, 2002). In particular, many larger properties were lost to the sector through Right-to-Buy, while housing associations, the providers of new social housing for rent, have struggled to build the larger dwellings required by some minority ethnic populations. The social rented sector has therefore been increasingly unable to respond to the problem of overcrowding within, for example, the Pakistani and Bangladeshi population.

The second broad explanation for the restricted opportunities for residential mobility provided by the social rented sector is that, even though race relations legislation has outlawed more overt forms of discrimination, the policies and practice of social landlords (local authorities and housing associations) have continued to have discriminatory consequences. The impact of the actions or inaction of key individuals or gatekeepers within the social rented sector (subjective racism) were recognized as far back as the 1960s by Rex and Moore (1967). Stereotyping, fed by ignorance or prejudice, continues to inform the actions of housing officers and shape the opportunities provided by the social rented sector. One example is the efforts of housing officers to make 'sensitive' lettings by matching minority ethnic applicants to areas of traditional settlement (Robinson et al., 2002). Perhaps driven by good intentions – limiting the likelihood of harassment and maximizing the possibility of long-term housing outcomes – the consequence is to restrict choice and limit the opportunities available to minority ethnic applicants.

The discriminatory consequences of the policies and administrative processes of social landlords (institutional racism[2]) have also been revealed as central to explaining why the proportion of certain minority ethnic populations living in social rented housing is relatively small and why those minority ethnic households that do live in the sector are more likely to reside in less desirable properties in less popular neighbourhoods (Robinson, 2002). Numerous studies stretching back over the last 30 years have documented how institutionally racist practices are created and maintained. Many of these policies have been exposed, reformed or abandoned, following formal investigations that revealed race to be a key factor

---

[2] The concept of 'institutional racism' became the subject of heated debate during the inquiry into the murder of Stephen Lawrence. The Commissioner of the Metropolitan Police Service appealed against the conclusion that the Service was institutionally racist. In response the report of the inquiry took care to distinguish between individual racism and institutional racism, and in doing so provided a useful definition of institutional racism as 'the collective failure of an organisation to provide an appropriate and professional service to people because of their colour, culture or ethnic origin. It can be seen or detected in processes, attitudes and behaviour which amount to discrimination through unwitting prejudice, ignorance, thoughtlessness and racist stereotyping which disadvantage minority ethnic people' (Macpherson, 1999, para. 6.34).

in shaping differential housing outcomes[3]. Their effects are still being felt, however, by households condemned to live in less attractive residential settings or excluded from the sector altogether.

Research also continues to highlight common practices that might appear fair and impartial but serve to disguise the needs of minority ethnic households and inform the geography of housing opportunities and potential for residential mobility provided by the social rented sector. Minority ethnic applicants, for example, have been found to be disadvantaged in the allocation process because they are over-represented in the homeless stream, where choices have traditionally been more limited (Jeffers and Hoggett, 1995). The corrosive influence of the fear or experience of racial harassment is also a major influence on the residential opportunities of minority ethnic households. The tendency for some allocation officers to steer minority ethnic applicants away from traditional white areas, for fear of the likelihood of racial harassment, has already been noted. This policy only appears rational because of the ongoing failure to adequately support minority ethnic households moving beyond traditional areas of settlement and effectively to tackle harassment as and when it occurs. Similarly, the reluctance of housing associations to develop new schemes targeted at meeting the needs of minority ethnic households beyond traditional areas of settlement reveals much about the failure of landlords to tackle problems with racist abuse, harassment and violence (Robinson et al., 2002).

The third broad explanation for the limited opportunities for residential mobility provided by the social rented sector resonates across all sectors of the housing system (social and private) and relates to the various intangible resources that households call on when seeking to move house. Awareness and knowledge of the local housing market, points of access and bureaucratic procedures and administrative processes of key agents (cognitive resources), for example, can prove critical in negotiating access to a desired housing situation. So too can direct and indirect access to key individuals or gatekeepers within the housing system and membership of networks of influence and power (social resources). These resources, in turn, are reflective of the formal and informal rights secured locally and nationally by different groups (political resources).

Recent research examining the housing situations and experiences of Somali households in England (Cole and Robinson, 2003) found access to these resources to be an important influence on the residential settlement patterns of Somali households across the five case study districts. Somali

---

[3] See, for example, the Commission for Racial Equality's report on the London Borough of Hackney (CRE, 1984).

people, including recent arrivals to the UK as well as second and third generation community members, were typically living in social rented accommodation clustered in inner-city neighbourhoods characterized by poor housing conditions, deprivation, high levels of unemployment, crime problems, poor-quality services and limited local amenities. Although respondents often talked at length about the benefits of living in areas of established Somali settlement, the majority also expressed a desire to move in an attempt to secure an improved residential situation. Many Somali people, in particular respondents who had settled in the UK in recent years, reported limited knowledge and awareness of available provision and access routes into different sectors of the housing market. The absence of Somali people in either front-line or senior positions within key institutions of the housing market (social landlords, private landlords and letting agencies, estate agents, mortgage finance companies), meanwhile, meant that social networks within the Somali population rarely included persons or extended into groups able to advise households about housing opportunities or to influence policy and provision to ensure greater sensitivity to the requirements and aspirations of Somali households. These factors were reported to be compounded by the limited political power and influence that Somali communities wield locally and nationally and the difficulties encountered focusing attention on the population's requirements and needs. The limits of Somali power and influence were also apparent in the politics of resource allocation and agenda setting within minority ethnic networks, with Somali community group members commenting that other minority ethnic groups were often better organized and more able to articulate their needs, influence policy and secure access to key resources.

The opportunities for residential mobility are clearly bounded by the particular resources that people have at their disposal. Evidence suggests that minority ethnic people typically have a more limited package of resources available to them. The operation of institutions within the housing system and the actions of individuals within these institutions can also impact on housing opportunities and spatial outcomes for different groups. It would be wrong, however, merely to regard minority ethnic households as passive recipients of constrained consumption opportunities. Far from being unable to resist, shape and control the forces shaping their lives, minority ethnic people are active agents capable of devising strategies involving elements of avoidance, accommodation and resistance, even within constrained circumstances (Law, 1996; Harrison and Phillips, 2003). Place – the location in which people live out their everyday life – is integral to the development and mobilization of these strategies and it is to the importance of place as a determinant of residential settlement patterns of minority ethnic households that we now turn.

Place and residential mobility

Locality or place provides the context within which people act and a focus for social organization. It represents the environment created through people's individual and collective actions. As such, place represents a repository of difference, with the nature of places reflecting cultural variations. Place also provides a territorial focus for the politics of identity and political action. Residential mobility involves leaving one place, with all its attendant supports, opportunities, potentials and problems, and settling in another, with its own distinctive milieu. Any attempt to understand the continuing concentration of minority ethnic groups in certain neighbourhoods in particular towns and cities must consider the specifics of localities that people would be leaving if they were to move and the potential risks and benefits associated with the places to which they might move.

People are often reluctant to leave an area with which they have developed an affinity and grown accustomed. Familiarity with the local environment, facilities and amenities, the existence of established social networks and nearness to friends and relatives can all serve to bind people to existing areas of settlement. There is no doubt that the segregated ethnic communities that contemporary public policy seeks to problematize have their difficulties, often associated with poverty and social exclusion. There is also little doubt that they have their benefits. They can provide a territorial focus for a sense of identity and belonging, fostering ontological security in the face of persecution and oppression. Minority ethnic people have also talked about the defence against harassment and violence and the greater sense of comfort and security that can stem from living in an area where they do not readily stand out for not being white (Robinson et al., 2005).

These locales can also serve as a repository for key social, economic and political structures and opportunities. Social networks can represent an important source of advice and assistance when engaging with service providers, as well as providing a substitute service for people unable to access formal provision, for example, with social care needs. Community-led businesses can provide job opportunities for local people. The clustering of minority ethnic populations can also provide a critical mass of demand required to warrant the development of key facilities, such as community-led services, religious amenities and shopping opportunities, as well as the visible presence required to prompt statutory service providers to recognize and sensitize service provision to the needs of group members in the area (Robinson et al., 2005). Population clustering can also allow a group to exert greater power through the structures of representative and participative governance.

If residential mobility is perceived to involve leaving behind this arena of predictable encounters (Kearns and Parkinson, 2001), where people are secure in the knowledge that the resources they require to negotiate the discrimination, disadvantage and exclusion they experience from wider society are readily available, then a certain degree of reticence about moving out is only to be expected. If, in addition, risks are perceived to be associated with living in alternative locations, then the idea of moving beyond existing areas of settlement can prove threatening.

Knowledge of the opportunities available elsewhere is acquired from a variety of local sources, including family and friends, local media, estate agents and housing officers. Based on lived experience, received wisdom and rumour, perceptions can have a strong influence on the residential strategies negotiated by households. In particular, research has revealed how existing patterns of segregation are actively sustained by the strategies of avoidance adopted by minority ethnic households in response to heavily racialized notions of space. Nearly all of the South Asian respondents interviewed during a study in Bradford, for example, identified areas of the city that were perceived as 'white' or 'Asian' and out of bounds to one group or another (Phillips et al., 2002). Similarly, respondents to a study in the rural district of north Lincolnshire referred to certain neighbourhoods in the town of Scunthorpe as being 'hostile', 'white' or 'racist' and considered these areas to be off their 'residential radar'. In both of these studies, these perceptions were most often associated with social housing estates, thereby further limiting the opportunities available to minority ethnic households through this sector.

Benefits are often associated with living in traditional areas of settlement and problems can be encountered moving beyond these locations into new areas of settlement. However, it is important that life in traditional areas of settlement is not portrayed as problem-free or that aspirations and housing outcomes are not presented as fixed and never changing. The aspirations, preferences and actions of minority ethnic groups are shifting in response to changing needs, requirements, aspirations and opportunities. This general point is illustrated by the north Lincolnshire study. Interviews with local people revealed a shared aspiration across all age groups and different minority ethnic communities (Pakistani, Bangladeshi and Somali) to move beyond the Crosby neighbourhood of Scunthorpe, where the local minority ethnic population has traditionally been clustered (Robinson et al., 2005). Three factors emerged as driving shifting aspirations and the changing profile of local demand. First, although respondents often talked at length about the benefits of living in Crosby, many reported dissatisfaction with the quality of key services and public goods and a perceived decline in the quality of life in the neighbourhood. In particular, respondents talked about a series of issues encountered on an increasingly frequent basis, including drug-related activities, prostitution, crime and anti-social behaviour, racial harassment and poor environmental conditions, that were impacting on quality of life and well-being. Second, the population was growing rapidly, largely as a consequence of the relatively young age profile. Overcrowding was a major problem, with demand for family housing outstripping supply and concerns were voiced about the quality of housing in the neighbourhood, the majority of which was Victorian terraced accommodation.

A third factor identified as driving a change in residential settlement patterns in north Lincolnshire was a generational shift in aspirations and intentions. Older people often expressed concerns about the quality of life in the neighbourhood and observed that the area was 'on the slide', but rarely indicated any intention to move elsewhere. Younger people, in contrast were often both concerned about the quality of life in the area and intent on moving out. To understand this shift it is important to realize that issues other than ethnicity were informing the aspirations of these young people. Younger respondents talked explicitly about their multiple identities, pointing out that as well as being Pakistani or Somali, they had been born, raised and educated within the UK. Whether university graduates pursuing professional careers or currently unemployed and seeking work, they conveyed a personal identity that reflected not only the cultural particulars of their parents and their ethnic community, but also school friends, work and business colleagues and wider society. These respondents also referred to the greater portfolio of resources (financial, cognitive and social) at their disposal that placed them in a stronger position to act on their aspirations to move out of the neighbourhood:

> 'It's like, you've got two different perspectives as well, you've got to look at the actual, the older generation and people that have come into the country now, but also the third and fourth generations we've got, they're adapting more, they know about services, they've got their own transport, they're able to access everything and they're just living day to day lives.' Omar Ali, 22-year-old Somali man (quoted in Robinson et al., 2005, p. 28).

Even so, many of the younger people, like Omar Ali, still expressed a degree of reluctance about moving beyond the traditional area of settlement, often rationalizing their situation through reference to some kind of cost–benefit analysis:

> 'I'm quite a family person, do you know what I mean, I like to be close to my Mum and things like that so obviously, you know, she's getting older and things like that so obviously I want to be there for her, but I don't think long term I could stay in the Crosby area, I do know that. I'm a professional myself so I'm looking to get out. But then again, also, you've

> got the community ties here as well; it's a real struggle you see. Whether to go outside and people thinking 'ah he thinks he's moving out of the way' or stay in the close knit community but then again the standard of housing in this area, its quite, its not really that good. There's certain areas, certain few houses that you know, quite a good standard of building and that's like that but also the area, for the money you're paying for them you could have get a slightly similar sort of house but in a different sort of area, less crime, less things like that . . .' Omar Ali, 22-year-old Somali man (quoted in Robinson et al., 2005, p. 28).

Place, as a realm of associational activity, is clearly a key determinant of the residential preferences of many minority ethnic households. Equating commitment to existing areas of settlement with voluntary self-segregation fails to appreciate the extent to which locational preferences are bounded up in the strategies that people devise for managing the exclusion from wider society that they both perceive and encounter. Segregated communities can serve as an important resource and refuge from exclusion. Alternative locations, meanwhile, are often perceived and experienced as threatening places to live.

One way forward: a role for social housing?

The assumption that concentrations of non-white households represent a challenge to be measured and treated as if a pathological problem is increasingly encroaching into mainstream thinking in England (Harrison et al., 2005). Certain minority ethnic groups are being blamed for self-segregating into ethnic enclaves of people from similar cultural and religious backgrounds. This physical separation is assumed to lead to people living 'parallel lives', that rarely overlap or interconnect, allowing disharmony to flourish. The response is to promote social cohesion through the interaction borne out of residential integration. Clustering, however, is not necessarily problematic. Segregated neighbourhoods can prove a vital resource for households faced with the disadvantage, discrimination and exclusion visited upon minority ethnic populations by wider society. The problematizing of ethnic clusters also serves to deny the multiple identities of minority ethnic people and the diversity that exists within such communities. There is also evidence to suggest any attempt to socially engineer ethnically mixed neighbourhoods is likely to meet with problems, difficulties and failure. As Amin (2002) observes, many mixed estates are riddled with racism, interethnic tension and cultural isolation, while many areas dominated by a single ethnic group are relatively trouble free, a point well made by the chief executive of a minority ethnic-led housing association quoted in

Robinson et al. (2004), who observed that 'the majority of (our) tenants are happily segregated and my most sustainable schemes are mono-cultural' (p. 15).

Contemporary policy is right to be concerned about the segregation of minority ethnic communities, but rather than promoting fallacious arguments about the ills of segregation and the rich rewards that will inevitably flow from residential integration, the priority is to tackle the injustices that bound the residential choices of minority ethnic households. So what might be done to 'break down the barriers'?

The social rented sector is often portrayed as a tired brand; a relic from a bygone era of state welfarism; an anachronism in an age of personalized, private provision. Yet it represents one of the few points of direct leverage that policy now possesses in a housing system characterized by increasing institutional complexity and diversity (Robinson, 2003). It might only account for one-fifth of the housing stock in England, but it remains a key instrument for achieving social objectives through housing policy. But, are social landlords committed to the objective of widening choice for minority ethnic households? The sector certainly has much to do before it can claim to provide a culturally competent service across its full range of activities. Survey evidence also suggests that opening up minority ethnic choices is not on the agenda of most local authorities and housing associations (Phillips and Unsworth, 2002). Recent research, however, has uncovered commitment among senior social housing managers to the principle of extending choice for minority ethnic households, respondents pointing out that not doing so would represent a failure to deliver on their general duty to promote equality of opportunity and to avoid discrimination (Robinson et al., 2004). The housing managers interviewed also suggested that integration could well flow from extending the historically restricted choices of certain minority ethnic groups, given latent demand and the broadening aspirations of younger minority ethnic people.

The social rented sector certainly has the potential to serve as a catalyst for change, providing a bridgehead into new areas of settlement through the supported relocation of minority ethnic tenants. Social housing can still provide housing of a high standard and better quality than many minority ethnic people experience in the private sector, particularly given the large sums of money being ploughed into stock renewal in recent years in pursuit of the government's Decent Homes target[4]. The allocation of housing on the principle of need serves to counter the disadvantages in the labour market that restrict the power of minority ethnic households to exercise choice in the private sector. Tenancy support activities designed

---

[4] In 2000 the government set a 10-year target of ensuring that all social housing meets set standards of decency by 2010.

to help vulnerable tenants sustain their tenancy can help support tenants new to the area, by countering the fear of harassment and attack, while community development work can introduce and foster interaction between new and long-standing residents.

Social landlords are also on the frontline of the government's assault on anti-social behaviour and neighbourhood nuisance and are well armed to tackle head on the hostilities and harassment that minority ethnic people can encounter when venturing into new areas of settlement. Under the Anti-Social Behaviour Act (2003), for example, housing associations and local authorities have the power to seek an injunction against persons engaged in behaviour that is capable of causing nuisance or annoyance to any person with a right to reside in or occupy housing in the neighbourhood. Local participatory structures, meanwhile, such as tenant bodies or resident partnerships engaged in master planning exercises, can prove a powerful vehicle for drawing people together into the kind of 'micro-publics' in which the 'prosaic negotiations' can occur that allow people to come to terms with ethnic difference (Amin, 2002).

The social rented sector also benefits from the contribution of a strong minority-ethnic led housing movement that has increasingly worked in partnership with white-led agencies to improve cultural competence across the whole sector, as well as meeting the needs of minority ethnic households through direct provision. Indeed, the involvement of minority ethnic-led housing associations was identified as crucial to the success of local (marketing, lettings, tenancy support, tenancy management and community involvement) initiatives that were reported to have succeeded in extending the locational choices of minority ethnic households (Robinson et al., 2004). Finally, all public bodies, including social landlords, are duty bound, under the Race Relations Amendment Act (2000), to promote equality of opportunity and good relations between different groups and to eliminate direct *and* indirect discrimination in all aspects of their activities.

Realizing this potential represents a major challenge for a sector still grappling to address the racialized inequalities borne out of shortcomings in institutional policy and practice. Bringing about a change in residential settlement patterns via the social rented sector will also be a slow and resource-hungry process. A recent Chartered Institute of Housing study provides working examples of the practical ways in which social landlords can help to widen locational choices, resulting in the movement of minority ethnic households out from traditional areas of settlement (Robinson et al., 2004). It is questionable, however, whether many landlords are likely to be willing to commit the resources and staff time found necessary to secure even modest gains, particularly if under no explicit obligation to so. It is therefore vital that national policy commits to the principle of widening choice for minority ethnic households. Policy makers should, in theory, have no problem signing up to this objective, given the recent proselytizing of government ministers about the importance of promoting choice within public service provision. At this point, however, we are confronted with an irreconcilable contradiction in contemporary policy.

On the one hand, government is promoting the virtues of introducing greater choice into the rented sector. 'It's about choice' as the Minister of Housing recently asserted, the aim being to secure 'a shift away from a system which is based entirely around meeting what administrators perceive as the needs of people, to one that is much more geared to meeting the applicant's own aspirations' (Hill, 2005). On the other hand, government is increasingly convinced of the need to engineer residential integration for the purposes of controlling minority ethnic communities and fostering social change. Implicit within this agenda is the denial of choice; in particular, the denial of the right of minority ethnic (but not white British) households to choose to live in segregated communities. Just how government will square this policy circle is currently unclear. The evidence presented in this chapter is unequivocal in the picture it presents; there is little or no evidence to support the thesis that minority ethnic groups are actively self-segregating into areas dominated by people from similar ethnic and cultural backgrounds, while there is a wealth of evidence regarding the barriers restricting the locational choices and residential opportunities of minority ethnic households. Whether policy acknowledges and responds to these insights in this supposed era of evidence-based, 'what-works' policy making will be interesting to observe, but judging by the pervading rhetoric in the aftermath of the 2001 disturbances and the London bombings, the omens do not look good.

References

Ahmed, N.M., Bodi, F., Kazim, R. and Shadjareh, M. (2001) *The Oldham Riots Discrimination, Deprivation and Communal Tension in the United Kingdom.* http://www.mediamonitors.net/mosaddeq6.html.

Amin, A. (2002) Ethnicity and the multicultural city: living with diversity. *Environment and Planning A* **34**: 959–80.

Birmingham City Council (1998) *Black and Minority Ethnic Communities' Access to Outer City Housing.* Birmingham City Council.

Bowes, A., Dar, N. and Sim, D. (1998) *Too White, Too Rough, and Too Many Problems: A Study of Pakistani Housing in Britain.* Research Report No. 3. Department of Applied Social Science, University of Stirling.

Brook, K. (2004) *Labour Market Data for Local Areas by Ethnicity.* Office for National Statistics, Labour Market Trends, October 2004, 405–416.

Burnley Task Force (2002) *Report of the Burnley Task Force.* Chaired by Lord Clarke.

Burrows, R. and Rhodes, D. (1998) *Unpopular Places? Area Disadvantage and the Geography of Misery in England.* The Policy Press, Bristol.

Cabinet Office (2003) *Ethnic Minorities and the Labour Market: Interim Analytical Report.* Cabinet Office, London.

Champion, T. (1996) Internal migration and ethnicity in Britain. In: Ratcliffe, P. (ed.) *Social Geography and Ethnicity in Britain: Geographical Spread, Spatial Concentration and Internal Migration (Ethnicity in the 1991 Census, Vol. 3)*, pp. 135–73. HMSO, London.

Cole, I. and Robinson, D. (2003) *Somali Housing Experiences in England.* Centre for Regional Economic and Social Research, Sheffield Hallam University, Housing Corporation, Sheffield.

Commission for Racial Equality (CRE) (1990) *Racial Discrimination in an Oldham Estate Agency. Report of Formal Investigation into Normal Lester and Co.* CRE, London.

Commission for Racial Equality (2004) *Race and Council Housing in Hackney: A General Investigation.* CRE, London.

Dorling, D. and Thomas, B. (2004) *People and Places. A 2001 Census Atlas of the UK.* The Policy Press, Bristol.

Harrison, M. and Phillips, D. (2003) *Housing and Black and Minority Ethnic Communities: Review of the Evidence Base.* Office of the Deputy Prime Minister, London.

Harrison, M., Phillips, D., Chahal, K., Hunt, L. and Perry, J. (2005) *Housing, Race and Community Cohesion.* CIH, Coventry.

Hawtin, M., Kettle, J. and Moran, C. (1999) *Housing Integration and Resident Participation: Evaluation of a Project to Help Integrate Black and Minority Ethnic Tenants.* Housing Corporation, London.

Hill, K. (2005) *Speech by the Minister of Housing to the ODPM/CIH Conference on Choice Based Lettings Schemes.* 26 January 2005.

Home Office (2001) *Building Cohesive Communities: A Report of the Ministerial Group on Public Order and Community Cohesion.* Home Office, London.

Independent Review Team (2001) *Community Cohesion. A Report of the Independent Review Team.* Chaired by Ted Cantle. Home Office, London.

Jeffers, S. and Hoggett, P. (1995) Like counting deckchairs on the Titanic: A study of institutional racism and housing allocations in Haringey and Lambeth. *Housing Studies* **10** (3): 325–44.

Johnston, R., Forrest, J. and Poulsen, M. (2002) Are there ethnic enclaves/ghettoes in English cities? *Urban Studies* **39** (4): 591–618.

Kearns, A. and Parkinson, M. (2001) The significance of neighbourhood. *Urban Studies* **38** (12): 2103–110.

Law, I. (1996) *Racism, Ethnicity and Social Policy.* Harvester Wheatsheaf, Hemel Hempstead.

Macpherson, W. (1999) *The Stephen Lawrence Inquiry.* Command Paper 4262-I. The Stationery Office, London.

Mandic, S. (2001) Residential mobility versus in-place adjustments in Slovenia: Views from a society in transition. *Housing Studies* **16** (1): 53–73.

Oldham Independent Review (2001) *One Oldham, One Future.* Panel Report, chaired by David Ritchie. Oldham Metropolitan Borough Council, Oldham.

Ouseley, H. (2001) *Community Pride not Prejudice. Making Diversity Work in Bradford.* Bradford Vision, Bradford.

Peach, C. (1996) Does Britain have ghettos? *Transaction of the Institute of British Geographers* **21** (1): 216–35.
Peach, C. (1998) South Asian and Caribbean ethnic minority housing choice in Britain. *Urban Studies* **35** (10): 1657–80.
Phillips, D. (1998) Black minority ethnic concentration, segregation and dispersal in Britain. *Urban Studies* **35** (10): 1681–702.
Phillips, D. and Unsworth, R. (2002) Widening locational choices for minority ethnic groups in the social rented sector. In: Somerville, P. and Steele, A. (eds) *Race, Housing and Social Exclusion*, pp. 77–93. Jessica Kingsley, London.
Phillips, D., Butt, F. and David, C. (2002) The racialisation of space in Bradford. *The Regional Review* July: 9–10.
Phillips, T. (2005) *After 7/7: Sleepwalking to Segregation. Transcript of speech given on 22 September 2005 to the Manchester Council for Community Relations* (http://www.cre.gov.uk).
Ratcliffe, P., with Harrison, M., Hogg, R., Line, B., Phillips, D. and Tomlins, R., and with Action Plan by Power, A. (2001) *Breaking Down the Barriers: Improving Asian Access to Social Rented Housing*. Chartered Institute of Housing, on behalf of Bradford MDC, Bradford Housing Forum, The Housing Corporation and FBHO, Coventry.
Rees, P. and Phillips, D. (1996) Geographical spread: the national picture. In: P. Ratcliffe (ed.) *Ethnicity in the 1991 Census: Volume 3. Social Geography and Ethnicity in Britain*, pp. 23–109. HMSO, London.
Rex, J. and Moore, R. (1967) *Race, Community and Conflict*. Oxford: Oxford University Press.
Robinson, D. (2002) Missing the target? Discrimination and exclusion in the allocation of social housing. In: P. Summerville and A. Steele (eds) *Race, Housing and Social Exclusion*. Jessica Kingsley, London.
Robinson, D. (2003) Housing governance in the English regions: emerging structures, limits and potentials. *Housing Studies* **18** (2): 249–68.
Robinson, D. (2005) The search for community cohesion: key themes and dominant concepts of the public policy agenda. *Urban Studies* **42** (8): 1411–28.
Robinson, D., Coward, S., Fordham, T., Green, S. and Reeve, K. (2004) *How Housing Management Can Contribute to Community Cohesion*. Chartered Institute of Housing, Coventry.
Robinson, D., Iqbal, B. and Harrison, M. (2002) *A Question of Investment: From Funding Bids to BME Housing Opportunities*. The Housing Corporation, London.
Robinson, D., Reeve, K., Bennington, J. and Coward, S. (2005) *Minority Ethnic Housing Experiences in North Lincolnshire*. Centre for Regional Economic and Social Research, Sheffield Hallam University, North Lincolnshire Council, Sheffield.
Simpson, L. (2004) Statistics of racial segregation: measures, evidence and policy. *Urban Studies* **41** (3): 661–81.
Social Exclusion Unit (2000) *Minority Ethnic Issues in Social Exclusion in Neighbourhood Renewal*. HMSO, London.
Strategy Unit (2003) *Ethnic Minorities and the Labour Market*. Cabinet Office, London.
van Kempen, R. and Ozuekren, A.S. (1998) Ethnic segregation in cities: new forms and explanations in a dynamic world. *Urban Studies* **35** (10): 1631–56.

Institutions and Change Agents

8

Resistance and Change: Political and Community Mobilization in Harlem

Lucille McEwen

Introduction

Community-based organizations play a crucial role in black communities in the US. Blacks had fought for and won basic civil rights after World War II, but any steady progress stopped abruptly with the death of Martin Luther King, Jr in 1968. Urban centers lost significant ground as civil unrest followed his death and inner cities' fragile economies were destroyed through riots and waves of housing abandonment. When President Johnson declared the War on Poverty, this precursor to the growth of community development organizations – the Model Cities Program – organized local offices that with federal funding began to rebuild blighted inner cities through urban renewal. This new production of housing, while removing physical blight, often destroyed the basic support systems among residents that had existed amid the blight. As a result community activists began to shift from calls for black political power via increased numbers of African-American elected officials to local control of schools. Economic issues that were the result of racial discrimination and unequal treatment emerged and low-income residents, in partnership with community-based organizations, began to work towards improving their communities.

An overview will be provided of a community-based organization – the Harlem Congregations for Community Improvement (HCCI) – its history, services and current priorities. In the course of that overview we will focus on some of the limitations of a not-for-profit operation. The focus will then turn to wealth-building strategies that are being undertaken as a means to reduce the impact of gentrification and facilitate more equitable development of the Harlem community.

Harlem: a symbol for African-Americans

Harlem, the historic African-American community, is known throughout the world. Dutch settlers established Nieuw Haarlem in 1658, as a farm community; Morris Jumel Mansion, which was George Washington's headquarters during the revolutionary war, remains standing from that period. It has been home to many new immigrants. The area was overbuilt in the early 1900s and that paved the way for Jewish and Italian immigrants to populate Harlem. St. Phillip's Episcopal Church purchased the apartment house complex between Lenox and Seventh Avenues (later renamed Malcolm X and Adam Clayton Powell Jr, Boulevards) and the first African-American enclave was established in Harlem in 1910 (Hickman, 2001, p. 44).

Subsequent waves of African-American migration firmly established Harlem as an African-American community. It gained prominence in the 1920s during the period of the Harlem Renaissance when arts and culture thrived and a cultural surge was led by prominent writers, such as Zora Neal Hurston and Langston Hughes, who took up residence in Harlem. However, this area's cultural and economic vitality began to diminish with the onset of laws prohibiting housing discrimination and by the late 1970s many remaining Harlem residents were living in substandard housing and often lacked basic services like heat and hot water.

As the situation worsened, the community looked to the Church for solutions. With the prophetic leadership of Martin Luther King, Jr, the leaders of HCCI had been leaders in the civil rights struggle and the anti-apartheid movement. They knew that they could not 'wait for an end of racist domination to create conditions under which we [could] live lives of sustained well being' (Hooks, 1996, p. 254). They now wanted to address the redevelopment of the community in a way that would 'share the liberatory power of self-sufficiency' (Hooks, 1996, p. 254), by allowing residents and local institutions to take the leadership role. Harlem had been romanticized during the 1920s and 1930s; it was subsequently demonized as the community deteriorated and opportunities for the black middle class to move to the suburbs increased. Today, Harlem is hot and the real estate market's escalating prices reflect new interest of the middle class across all races!

Harlem Congregations for Community Improvement overview

Founded in 1986, Harlem Congregations for Community Improvement, Inc. (HCCI) is a coalition of more than 90 inter-faith congregations dedicated to the physical, economic and cultural revitalization of the Harlem community, with a focus on the 40-block area called Bradhurst[1]. The origins of HCCI were fraught with struggle as a group of ministers was inspired to say they were not going to let the community slip any further away. This group included Rev. Dr Wyatt Tee Walker who served in the civil rights struggle with Dr Martin Luther King, Jr and other prominent ministers of the Harlem community. This group recognized that power concedes nothing without a struggle, and they were ready for the struggle that would ensue, as they demanded a say in the plans to redevelop Harlem. Self-determination, a core principle that permeates HCCI's programs and service, drove their effort.

Black self-determination was critical for the ministers of HCCI. In reaction to the civil rights struggle against institutional racism, the ministers wanted to ensure that redevelopment would create economic opportunity for Harlem residents, and that the residents would be participants in the process and would not be treated as victims being taken care of in a patriarchal system that often equates human worth with material privilege. As Bell Hooks notes, 'at the heart of black self-determination is the political awareness that we must assume responsibility for constructively transforming our lives' (Hooks, 1996, p. 254). While economic self-sufficiency was an important goal, the most important priority was to bring hope to the hopeless, to demonstrate that it is 'possible for black people to create meaningful lives irrespective of their material conditions' (Hooks, 1996, p. 261). We wanted a self-determination process that would 'first engage the hearts and minds of black folks' (Hooks, 1996, p. 261) and encourage coalition building across religions. Numerous federal programs had been directed to 'remove blight' from areas of significant distress. These urban renewal programs often resulted in the destruction of communities of color, and were commonly referred to by the residents as 'Negro Removal' programs. The community did not trust government to include the community in any significant way. They expected that government would contract with outsiders to come in and prescribe what they thought best for the community, without any significant community input. In the past this had resulted in programs that allowed contractors to profit significantly on construction of housing that resulted in high concentrations of poverty with an excessive number of special needs populations and municipal services that no other community would accept.

Ironically, 1986 represented the first opportunity fully to celebrate the Martin Luther King official holiday, and the main theme of this King

[1] The group includes African Methodist Episcopal (AME), African Methodist Episcopal Zion (AMEZ), Baptist, Church of God in Christ (COGIC), Christian Methodist Episcopal, Episcopal (CME), Interdenominational, Islamic, Pentecostal, Presbyterian, Roman Catholic and United Methodist (HCCI corporate office's records).

celebration called upon all Americans to 'rededicate themselves to work toward the kind of society that King envisioned' (Cruse, 1987). Rev. Dr Preston Washington, the President of HCCI, emphasized the importance of self-determination when he noted that it was important for Harlem-based groups 'to signal to the rest of the city that we have the genius and the expertise to develop Harlem ourselves' (Washington, 1989).

For more than 16 years, HCCI has worked to build affordable housing and provide a comprehensive array of services to address the needs of low-income residents of central Harlem. Programs have proven their ability to create affordable housing and safe streets; offer opportunities for individuals and groups to become and remain economically independent; increase understanding of, and access to, health care; and provide substantive education programs for adults and young people. Through alliances with other community organizations, elected officials and local residents, HCCI has also helped reduce crime in the community, increase public sanitation, and preserve and transform open space.

As not simply a faith-based institution, but as an interfaith-based institution, HCCI does not promote any particular faith. Instead, values that cut across all religions, such as love thy neighbor as thyself, are emphasized and integrity, justice, equity and self-sufficiency are promoted. The organization seeks to reinforce core values and support for residents as they act on core values to build community. This community is an organic system, and so the work of HCCI is not limited to physical redevelopment, but includes action to empower young people and adults, with the tools to be successfully self-reliant. Encouragement is given to help them to look beyond the confines of their apartment to improving their building, their block and their outdoor environment.

The black church has always been independent of the majority owned and operated power institutions in the US. It is the place where each of us, of whatever station in life, could come and gain the respect of our friends and neighbors. For example, Mr Jones, who throughout the week may be washing dishes, can put on his Sunday best and serve the church as an usher or deacon. It has also been the place where Mr Jones or Sister Johnson can come to get advice and assistance with any issue. Because of its unique position, the black church has always dealt with the disparities in our communities. This includes housing, education, health care, technology and access to economic capital. As an organization, HCCI deals with these disparities in our community through its programs and community-development strategies.

As Dr Wyatt Tee Walker says 'the black church's mission has been to provide hope in a hopeless situation' (Walker, 1994, p. 3). He acknowledges that the African-American church has been 'critical in the process of creating wholeness in black life' (p. 11) and further notes that the first banks and insurance companies were started in churches and with the onset of Jim Crow laws, the black churches and their congregations were the primary support of black entrepreneurship. The church also provided elementary education and supported the secondary schools and colleges (Walker, 1994, pp. 19–20). Since World War II, fueled by the growth of earning power of black congregations, and the opportunities created after the civil rights era, the black church has grown exponentially. As the church has developed it has continued to address a comprehensive array of needs of its people, which includes its membership and the residents of the surrounding, often poor and disenfranchised communities.

Planning for neighborhood redevelopment

The initial slogan of HCCI '*Empowering the Church and Community to Make Housing A Right for All*' was descriptive of the mission that created the Bradhurst Redevelopment Plan (Scotland, 1999). The Redevelopment Plan was developed in concert with Columbia's Urban Technical Assistance Project, HUDC, and the CCNY's School of Architecture. The Redevelopment Plan called for mixed-income housing with rental and ownership opportunities throughout a 40-square block area of blight from 140th Street to 155th Street, A.C. Powell, Jr Boulevard to Edgecombe and Bradhurst Avenues. HCCI envisioned an opportunity to create a mixed-income community, while the city's initial plan proposed to rehabilitate 1200 units as all low-income rental housing with maximum incomes of approximately $25,000 (Scotland, 1999). Most of the recommended sites in the Redevelopment Plan proposed by HCCI more than a decade ago were eventually authorized for redevelopment in the Bradhurst Urban Renewal Area Plan of the city's housing development agency (NYC Department of Housing Preservation and Development).

There was significant political turmoil as HCCI joined with Consortium for Central Harlem Development (CCHD), a group of other business associations and professionals, and lobbied the city to adopt the Bradhurst Plan. But as Lionel McIntyre, Head of Columbia's Urban Technical Assistance Unit and primary author of the Bradhurst Redevelopment Plan, noted, he had 'yet to see a project without political turmoil' (Scotland, 1999). Initially, HCCI's focus was to advocate for the housing and the quality of life of the community, with CCHD providing the technical assistance needed successfully to rehabilitate the physical conditions in the community.

The city was reluctant to provide such large development responsibilities to a group of churches (Moore, 1988, 1989), but eventually consented providing that a consortium, which included the New York Urban Coalition, would have responsibility for development and that the properties would be turned over to HCCI only upon completion and issuance of occupancy approvals from the NYC Department of Buildings. Time proved that HCCI in fact was the group with staying power. By 1997, the New York Urban Coalition had folded, and CCHD had only two employees. By 2002, the CCHD Board of Directors voted to close down the corporation. In contrast, HCCI remains a growing corporation with 1400 housing units under ownership and management and another 500 units in various stages of development[2].

From its inception, CCHD was doomed to fail because of its faulty infrastructure. The organization was supposed to rely on the expertise of the organizations represented on its board[3]. While each of these organizations wanted to develop Harlem for Harlemites, the focus was inevitably on running their non-profit organization and not on stepping up to provide the necessary management support for the understaffed CCHD. HCCI, while headed by a visionary minister, who was also senior pastor of Memorial Baptist Church, had the benefit of full-time staff to manage its daily operation.

Nearly all of the remaining undeveloped sites in the Bradhurst Urban Renewal Plan are due for construction over the next 2 years. Now that the Bradhurst Plan has been almost fully implemented to address the city-owned properties, HCCI recognizes a need for another round of planning to focus on the blighted privately owned vacant lots and properties where residents are living in substandard conditions.

HCCI Services

HCCI's Department of Real Estate Development and Property Management has developed more than 1400 units of low- and moderate-income rental housing, and 30 commercial spaces; 400 rental units, 300 moderate, middle-income and limited equity co-op units and a 45,000 sq ft PathMark supermarket are in development[4]. Since 1986, HCCI has leveraged more than $200 million in support from federal, state and city agencies. In addition, HCCI provides scattered-site housing for 60 families and individuals living with HIV/AIDS.

---

[2] HCCI corporate records.

[3] Black Agency Executives, Greater Harlem Real Estate Board, Greater Harlem Chamber of Commerce, Harlem Artists and Scholars Interest Consortium, Harlem Athletic Association, Harlem Business Alliance, Harlem Churches for Community Improvement, Harlem Urban Development Corporation, New York Coalition of Black Architects (NYCOBA), and New York Urban Coalition, HCCI corporate records.

[4] See Appendix A, HCCI Property Listing.

The slogan changed from 'Empowering the Church and Community to Make Housing A Right for All' to 'Empowered Congregations Rebuilding Harlem' as it evolved into a comprehensive community development organization that addresses both the physical redevelopment of the Harlem community, and the development of its residents to be prepared to take greater advantage of economic opportunities. With the development of the first properties, an 85/85 site[5], immediately followed by a series of openings of four projects totalling 301 units that were designated Bradhurst Phase I, HCCI recognized the need for human services for these tenants. The Bradhurst Social Service Delivery Plan was created, primarily funded with tax credit income generated through the federal Low Income Housing Tax Credit Program[6]. Today, HCCI participates in a citywide Welfare to Work training and placement program (formerly Charitable Choice)[7]. Other adult education programs include soft skills, literacy training (in conjunction with Literacy Partners), and computer software training – all linked with job placement. To date we have placed more than 750 residents in employment opportunities.

HCCI also runs two after-school programs. The organization improves scholastic achievement of elementary students through a focus on maths, reading tutorials and homework assistance at the Bradhurst Academy of Excellence. In addition, HCCI runs a Computer Clubhouse that provides young people with access to cutting-edge graphic, computer editing and robotics programs. The Clubhouse includes a one day per week program that focuses on introducing girls to enhanced maths and science training.

HIV/AIDS is a serious problem in Harlem. To tackle this, HCCI's Health and Wellness Strategies (HWS) include operation of 60 units of scattered-site AIDS housing, as well as HIV/AIDS outreach, education and prevention services. Recognizing the importance of replicating best practices in service delivery, in 2002, HCCI secured funding to train five congregations to provide faith-based HIV/AIDS education and outreach services. To avoid repetitive services and prevent undue competition, HCCI was one of the founding members of the Harlem Directors Group that coordinates services among several HIV/AIDS providers serving Upper Manhattan. Other

[5] The 85/85 Program was initiated by the City of New York and the State of New York, each contributing $85 million towards the development of housing that was 51 percent homeless and 49 percent very low income.

[6] See Section 42 of the Internal Revenue Code.

[7] From fall 2001 to spring 2003, HCCI was the lead agency coordinating a citywide pilot program called Charitable Choice. Although the program was successful in bringing welfare recipients that had been sanctioned into compliance for reinstatement and training for employment, the program was discontinued as the current administration set other priorities for the Human Resources Administration. In 2003 HCCI joined another Citywide Welfare to Work employment effort as a subcontractor to Seedco, a national community development intermediary.

health-related services of HWS include a community-based medical library and a Start Right 2010 Coalition with other Upper Manhattan organizations, to ensure that babies get all of their immunization shots by 3 years of age. The organization also provides enrolment in WIC (a supplemental food program for pregnant women, infants and children) and is facilitating enrolment for food cards, a federal food subsidy program.

Community engagement

HCCI hosts a monthly meeting known as the Community Improvement Group with city agencies, community residents and other institutions to discuss community quality-of-life issues in Bradhurst and throughout Central Harlem. In addition, HCCI has initiated and nurtured resident associations for each housing development and initiated the Jackie Robinson Park Conservancy, which is now an independent non-profit organization that raises funds and supports programming for Jackie Robinson Park. The park in its blighted condition was an emblem of the poor quality of life that existed in the area. This ten-block long park should have been the primary recreational resource for the community, but it had become a battleground for local gangs and was unused for its intended purpose. It created a barrier that seemed to wall in the poorer families on the east side of the park and physically separated them from the more affluent who lived along Edgecombe Avenue in the area known as Sugar Hill.

In recent years, HCCI has recognized the critical importance of building individual and community wealth to enable low-income families to move from poverty to prosperity. Working closely with national intermediaries – PolicyLink and the Conservation Law Foundation – we embarked on a comprehensive planning process that culminated in the vision of the Harlem Equitable Development Initiative (the 'Harlem EDI').

Limitations of a non-profit

> 'Let us throw off everything that hinders . . . and let us run with the perseverance the race marked out for us.' (Hebrews 12:1)

Non-profits are often seen as weak and all too often capabilities are not judged by competency of service delivery, but in the ability to raise funds. Stephanie M.M. Smith notes, 'Some people believe that faith is a mask that covers up a fundamental weakness' (Smith, 2004, p. 166). She states that the non-profit church group is thought to have little to no bargaining position in the real estate deal. Outside of faithfully tithing with their chosen religious institution blacks have not acquired the wealth for organizational-sustaining philanthropy. Non-profits must secure their foundation and their internal capacity. That foundation must be carefully laid, and regularly reinforced. Reinforcement is achieved through ongoing monitoring and rewarding and strengthening best practices by infusing the institution with new ideas by employing qualified experts to fulfil long-term needs, and contracting consultants for temporary or specialized project needs. Being mission driven does not mean that the organization has the expertise to create successful outcomes. And while many black-run non-profits have the technical skills, they have not made the necessary connections to create an organizational-sustaining donor base.

Generally, non-profits struggle with program and organizational management because there is a lack of infrastructure or resources common to larger organizations. Despite this, community organizations are consistently 'in the fray', and punching above their weight. Creating the right performance measures, identifying the necessary resources and partners, as well as creating and vigilantly monitoring a realistic budget require a focus on purpose and accountability, without blame. Simply providing a service is not enough. A large portion of the work that non-profits need to finance is not appealing to funders. Most notably the areas of marketing, development and communications are almost universally under-funded. As a result HCCI is always fundraising. Enough cannot be said about how the right relationships are essential to success. For a faith-based institution there is an even greater limitation on the sources of money raised. For example, HCCI has not accepted money from tobacco and alcohol companies. As mergers continue, many conglomerates that are potential funders may be eliminated because the parent company owns tobacco interests.

HCCI projects

Brownfield redevelopment

In the early spring 2002, with support from PolicyLink and CLF Ventures, HCCI began to identify appropriate non-residential properties for development. Two promising prospects were the former Film Storage Warehouse at 203 West 146th Street and PS 90, an abandoned public school property on a block that is currently being revitalized by HCCI. Both buildings are city-owned and have been abandoned for decades.

In early 2002, HCCI focused on obtaining control for the sites, gathering additional assessment information that HCCI would need to undertake site remediation and redevelopment. HCCI was able to garner support on the Board of Education for a sale/lease-back arrangement for PS 90 to renovate the site for a 600-student high school[8]. Changes within the Board of Education in late spring 2002 postponed further development of the project. Then in September 2003, the Board of Education advised that the project had been removed from the Five Year Capital Plan; it would not be reconsidered prior to 2009[9]. HCCI has obtained a new site control authorization letter to permit the property's redevelopment as market-rate housing. The plan is to develop the school site that is on West 148th Street, where, in partnership with for-profit developers, HCCI has created home ownership and rental opportunities affordable to low- and moderate-income families through the preservation of approximately 300 units[10].

Rehabilitation of the film storage warehouse

Development plans often take many years to be realized. With limited resources, a non-profit must rely on funding cycles and coordinate them as effectively as possible to facilitate each incremental step of the development process. HCCI has also focused on the development of the film storage warehouse as the priority site for the relocation of its offices. The organization is currently pursuing the redevelopment of 203–209 West 146th Street into a consolidated office space for programs and staff, and additional rental space to help cover debt service. The film storage warehouse is a 23,600 square foot, two-story brick building. It was constructed in the early 1900s. During its operation, the film storage warehouse served as a storage facility for cinematographic purposes. It was abandoned in the 1970s and has remained unoccupied ever since.

After presentations by HCCI staff in June 2002, Community Board 10 (the community planning board for the Bradhurst area) recommended that the city give site control to HCCI so that it could pursue clean-up/development strategies. In July, the New York City Department of Housing Preservation and Development (HPD) granted HCCI site control.

In late July, HCCI submitted a letter and supporting documentation to the New York State Department of Environmental Conservation (DEC) requesting resources to conduct environmental studies and perform an assessment of the site. HCCI received verbal approval for a $125,000 grant

[8] The proposed sale/lease-back arrangement was based on a model under way between the Board of Education and the Abyssinian Development Corporation.

[9] Based on preliminary findings, it appears the building does not have sufficient environmental contamination to meet the federal Environmental Protection Agency (EPA) funding criteria. HCCI, therefore, will continue to pursue this development opportunity and seek alternative funding for site redevelopment.

[10] See Appendix A, HCCI Properties Listing.

from the DEC in early November[11]. The assessment was completed in August 2004.

With the help of outstanding students at the Milano Finance Lab and Parsons School of Design of the New School and CCNY's School of Architecture, HCCI won $25,000 to be used for pre-development expenses. The rehabilitation and re-use of this site will bring several benefits to the community:

- Development of a blighted urban renewal site (site 23 in the Bradhurst Urban Redevelopment Authority).
- Elimination of a brownfield site.
- Provision of 15,000 square feet of office space to be leased by a financially secure corporation that will serve the frail elderly.
- Consolidation of HCCI offices will return ground-floor commercial to retail uses to service the growing residential community and generate income for HCCI.

In the meantime, HCCI explored additional ways to fund this development. Indeed, the organisation has successfully received support from a number of private sector organisations including Fleet/Bank of America, Chase, Citibank and M & T Bank. Greenpoint has agreed to provide 'fitness training' for local residents as well as homeownership counselling. Despite all this, funding for the capital project has been limited.

We are now incorporating housing above the proposed office space in order to meet a current government priority of affordable housing, which will help to subsidize the building of our central office. This sequence of events emphasizes how development often includes a greater number of compromises to meet current priorities of funders in order to meet a primary goal.

How HCCI and other non-profits can address the issues of changing demographics through housing and economic development

We will now address three of the current issues that are affecting the Harlem community. The issues are gentrification, housing affordability and access to home ownership.

Jane Jacobs notes, that in the 1970s the cost of a median-priced home escalated to a point where they were affordable to only 10% of families in

---

[11] DEC controls a pool of technical assistance funding from EPA that is available for site assessment. When an applicant is approved by DEC, the funds are used to pay an EPA team to do the assessment.

the US. 'Ninety percent could no longer afford to buy a "median" house' (Jacobs, 2004, p. 29). Renters also began to spend increasing proportions of their income for rent, and similar statistics showed up in Canada and England (Jacobs, 2004). She warns that this is one of the first indications of a declining culture and that, if we are to continue with life as we currently know it, we must solve the housing affordability problem or communities, as we now know them, will disappear.

Whenever we talk about housing affordability, the question arises: affordable for whom? When we talk about community redevelopment, in the face of gentrification, the question is who benefits from the redevelopment? Further we can ask 'Who benefits from local and regional planning and decision-making? Who benefits from public subsidies that attract investment into low-wealth communities? Who benefits from public investment in economic development and infrastructure projects?'

Equitable development seeks to ensure that the answers to these questions always include low-income and low-wealth community residents. Through work with PolicyLink, HCCI developed a commitment to building wealth for our residents as we rebuild the community.

HCCI staff and board members explored strategies for residents to gain an ownership stake in community business and real estate assets. With the assistance of PolicyLink, HCCI staff identified several models that could be adapted to Central Harlem, including a model of collective ownership in a community-serving business in Washington, DC (BIG Wash Laundromat); resident/cooperative ownership in a temporary employment agency in Baltimore, Maryland (SSC Employment); a cooperative home healthcare agency in the Bronx, New York (Cooperative Home Care Associates); and resident investment in community real estate in San Diego, California (Market Creek Plaza)[12].

In order for HCCI to fully explore these models, PolicyLink arranged a site visit to each project between July and September 2002. This gave board and staff members an opportunity to learn about the models directly from local experts and to build peer relationships that will help with future planning. Since the visits, PolicyLink has been helping HCCI to assess the feasibility of applying the different strategies to Central Harlem and to identify funding and technical assistance resources to move relevant models forward.

The site visit to BIG Wash enabled HCCI staff to learn about a successful business model that includes residents as investors and managers of a community-serving enterprise[13]. Since the site visit, we have been exploring

---

[12] Each of these models is described in detail in the PolicyLink *Sharing the Wealth: Resident Ownership Mechanisms* report, available at www.policylink.org.

[13] For more information on BIG Wash, see *Sharing the Wealth: Resident Ownership Mechanisms*, p. 63.

the relevance of the model to the development of business enterprises in Harlem.

Another model under consideration by HCCI was a cooperative home care model to serve the growing population of seniors in Harlem. The site visit to Cooperative Home Care Associates (CHCA) in the Bronx gave HCCI staff an opportunity to learn about one approach to meeting the growing demand for senior care in a way that enables workers to become owners of a for-profit enterprise. The worker-owned model advanced by CHCA is seen as a national model because it has produced higher-quality services, higher wages and lower turnover than other providers in the home care industry. Its success is largely attributed to the worker ownership structure.

A visit to the Empower Baltimore Management Corporation (EBMC) enabled HCCI staff to learn about another cooperative model: a worker-owned temporary agency. The agency, SSC Employment, was established in 1997 out of local organizing efforts in Baltimore's low-income communities. The founder of SSC Employment, Avis Ransom, is currently working with EBMC to help them develop cooperative businesses, owned by low-income/low-wealth workers, who will provide services to the biotech industry in Baltimore[14].

Finally, HCCI learned about ways to include residents as owners in commercial real estate development through a site visit to Market Creek Plaza in San Diego, California. The 20-acre commercial and cultural center is the nation's first commercial real estate project that will include low-income/low-wealth residents as shareholders, in partnership with the non-profit, philanthropic and private sectors.

These site visits enabled HCCI board and staff members to learn about the details of resident ownership models and to build peer relationships with board members and staff of organizations from several US cities that are advancing these models.

HCCI is currently exploring the adaptation of these models to commercial development projects in Harlem. For example, HCCI investigated the opportunity to invest in a garage on a block that includes 300 units of HCCI housing[15]. HCCI explored ways to include residents as stockholders in the purchase of the garage. Several options were explored, including the limited liability company (LLC) model, used by the Jacobs Center for Nonprofit Innovation in the development of Market Creek Plaza. This model would include residents as a class of stockholders within the LLC

[14] For more information on SSC Employment, see *Sharing the Wealth: Resident Ownership Mechanisms*, p. 63.

[15] The garage, currently owned by Uptown West Management LLC, is at 234 West 148th Street.

structure, along with other investors, including HCCI. Another approach would have HCCI purchase the property with investment from labor, private foundations and other investors and then develop the business with residents as owners.

Of the equitable development concepts, the expansion of ownership opportunities has been the most difficult to fund. The banks have supported financial fitness training and individual development accounts (IDAs), but to date, no one has provided funding for creating ownership opportunities.

When we look at the success of the most extensive housing redevelopment plan of any city in the nation, layered with the most well funded empowerment zone in the country, the unintended consequence of increased investment and rising property values is low-income families being priced out of the market.

America has spent billions on income support for the poor since the 1930s, but we have failed to empower these families to accumulate wealth in order to significantly change their economic futures. Today, 32 percent of all New Yorkers are asset poor, meaning they lack sufficient net worth to subsist at the poverty level for three months without other support. One in four households in New York have zero or negative net worth and New York ranks last in homeownership rates. Significant local ownership facilitates a community's ability to determine its future. Ownership creates stability, because residents have a long-term commitment where they have made a substantial investment. Our goal is to create a self-determinative community.

We envision that with steady progress, and under the leadership of HCCI, Central Harlem, and a once-disinvested community in New York City, will be widely viewed as a national model of equitable development. Harlem, in 2012, will be a community that is owned – literally and figuratively – by a broad cross-section of Harlem residents and community-based institutions. This achievement will be the result of an eight-year period of development and implementation of innovative wealth-building strategies as part of the Harlem Equitable Development Initiative. While the work of the Initiative will still be in progress, foreseen outcomes are tangible:

- Hundreds of families have transitioned from renters to owners of their homes. This includes purchasers of newly developed affordable cooperatives and former residents in buildings financed with low-income housing tax credits that have been converted to limited equity cooperatives at the end of the tax credit compliance period. HCCI led the conversion process and continues to manage the buildings on behalf of resident owners.
- Residents have purchased equity in a garage, a supermarket, a bakery and a laundromat in a revitalized neighborhood where many of them

live and work. These businesses were developed by residents, community-based non-profits and local churches, and are supported by a technical assistance team from HCCI. Private and public sector investors, labor pension funds, local churches and non-profit groups have joined residents as investors and business partners.

- Many local families own stock in a community-building real estate investment trust (REIT) that includes a diversified portfolio of properties. HCCI was the sponsor and developer of the community-building REIT concept.
- Residents have made these local investments – in home, business and real estate equity – beginning with savings accumulated in specialized individual development accounts called community-building IDAs. Matching funds from local banks, foundations, government and individual philanthropists who support innovative approaches to building community capital, supports the model.
- Residents participating in the initiative are vested stakeholders in their community. They are actively engaged in public policy and planning decisions because they have an ownership stake in sustained economic growth that will benefit their families, their communities and future generations.
- HCCI has an institutional home, having redeveloped a former brownfield site and created a permanent consolidated home office.

Local residents attend the Construction Trades Academy (CTA), a program established by HCCI with help from Local Initiative Support Corporation (LISC), where they learn valuable construction skills, financial literacy and gain employment on sites in Harlem and throughout New York City. CTA students form a watchdog group that advocates for environmentally safe construction practices and the hiring of local people for Harlem developments. Local Harlem schools and the CTA produce empowered citizens who contribute to and invest in Harlem's continued development.

How do we get from where we are today to our vision for the year 2012? Fulfilling the vision rests on using specific wealth-building strategies, including:

- Expanding the HCCI Homeownership office, a Central Harlem housing office that works with the public and private sectors to connect greater numbers of local residents to home ownership opportunities through aggressive marketing and continuous outreach.
- Providing pre- and post-home ownership counseling programs that help get people into their own home and then teach them to maintain them.
- Expanding comprehensive financial fitness training coupled with an IDA program.

- Preserving a portion of HCCI's tax credit properties through conversion to limited equity cooperative ownership (a strategy that will build home ownership opportunities for tenant families).
- Increasing the number of limited equity cooperatives that are financially stable and have well trained leadership.
- Expanding outreach and technical assistance to our membership and targeted audiences for HCCI home ownership and community building services.

The vision around home equity is to maximize opportunities for low-income/low-wealth residents in Harlem to own equity in their homes. Home equity will enable families to have stable housing; to build family assets as community real estate values appreciate; and to obtain other benefits that accompany home ownership. HCCI is committed to developing a range of home ownership opportunities for residents and in helping residents to access those opportunities. At the same time, the organization remains committed to developing and managing affordable rental units as an option for Harlem families, because we recognize that the responsibilities of home ownership are not for everyone.

Resident ownership mechanisms are one set of equitable development strategies that HCCI has viewed as particularly relevant to conditions in Harlem. As originally highlighted in the *Sharing the Wealth* report, resident ownership mechanisms (ROMs) are tools and strategies that enable residents to gain an equity stake in their communities. They link the community-development and asset-building fields by leveraging local economic opportunities to build assets among low-wealth/low-income individuals and families in the community. Finally, they create better sustainability because they increase the circulation of local dollars within the community.

HCCI is committed to building resident and institutional assets through ownership in community-serving businesses. In addition, while HCCI is not providing direct support for micro-enterprise development, it recognizes the importance of including this strategy as part of a broader resident ownership menu. The organization has entered into a collaborative effort with the Harlem Business Alliance to provide loans of up to $25,000 for physical improvement of small business spaces. HCCI is connecting with other Harlem-based institutions providing support for micro-entrepreneurs that successfully participate in our IDA program. In addition, HCCI has funded a loan fund that is administered by the Harlem Business Alliance.

These equitable development strategies will link people and place in the development process – to ensure that people benefit as our community improves. Equitable development will leverage private and public sector investment in communities in a way that produces financial returns for investors and social returns for community residents. Let us not only create affordable housing opportunities. We can rebuild communities while building a solid economic future for our residents.

References

Carle, R. and Decaro, L., Jr (eds) (1999) *Signs of Hope in the City*. Judson Press, Valley Forge.

Cruse, H. (1987) *Plural but Equal*. William Morrow and Company, New York.

Hickman, C. (2001) *Harlem Churches at the End of the 20th Century*. Dunbar Inc., New York.

Hooks, B. (1996) *Killing Rage*. Owl Books/Henry Holt and Company, New York.

Jacobs, J. (2004) *Dark Age Ahead*. Random House, Toronto.

Moore, K. (1988) A Harlem Coalition has Ideas. *New York Daily News* 2 November, p. MJ1.

Moore, K. (1989) locking horns over Harlem turf. *New York Daily News* 15 June, p. MJ1.

Scotland, R. (1999) Raising Lazarus, the resurrection and transformation of Harlem. In: Carle, R. and DeCaro, L.A., Jr (eds) *Signs of Hope in the City*. Judson Press, Valley Forge.

Smith, S. (2004) The church and affordable housing development (part 1). *Journal of Affordable Housing and Community Development Law* **13** (2): 166–170.

Walker, W.T. (1994) *The African American Church and Economic Development*. Martin Luther King Fellows Press, New York.

Washington, P. (1989) A Harlem Coalition has Ideas. *New York Daily News* 2 November, p. MJ1.

Other faith-based developments in the NYC area

Abyssinian Development Corporation
4 West 125th Street
New York, NY 10027
212-368-4471 Phone
646-442-6598 Fax
Ms Sheena Wright, President
1200 Housing units and retail development as well as a high school in partnership with the NYC Department of Education

BEC New Communities
67 Hanson Place
Brooklyn, NY 11217
718-858-8803 Phone
718-852-5597 Fax
Mr Dan Matthew, General Manager and CEO

Bridge Street Development Corporation
277 Stuyvesant Avenue
Brooklyn, NY 11221
718-573-6893 Phone
Rhonda Lewis, Executive Director

Church of the Master
86 Morningside Avenue
New York, NY 10027
212-666-8200 Phone
212-865-6431 Fax
law587@juno.com Email
Rev. Ronald Lawrence, Pastor
In the predevelopment phase of a 54-unit, market rate project

Concord Community Development Corporation
291 Madison Street, 3rd Floor
Brooklyn, NY 11216
718-638-7762 Phone
718-638-7804 Fax
CCDC@ureach.com Email
Ms Sylvia Kinard-Thompson, CEO/President
In the predevelopment phase of six-unit condominium project

Emmanuel Baptist Church
36 Saint James Place
Brooklyn, NY 11205
718-622-0908 Phone
718-622-3343 Fax
dthompson@ebc-ny.org Email
Deborah E. Thompson, Executive Director
Currently doing homeownership and financial literacy work

Fellowship Baptist Church/Fellowship Learning Center
3036 Richmond Terrace
Staten Island, NY 10303
718-981-9215 Phone
718-981-1959 Fax
Ms Cheryl Morrison, Edifice Committee
Developing a new child-care center adjacent to the church

First Church of Wyandanch
85 Parkway Blvd
Wyandanch, NY 11798
631-643-8777 Phone
631-653-6769 Fax
Rev. Michael Talbert, Pastor
In the predevelopment phase of developing nine, affordable, single-family homes for ownership

Greater Centennial Community Development Corporation
316 S. 8th Street
Mt. Vernon, NY 10550
914-664-1838 Phone
914-664-2067 Fax
Lauren Williams, Director of Real Estate Development
Developing a round of 16 affordable homes for ownership

Rockgate Community Development Corporation
770-784 Glenmore Avenue
Brooklyn, NY 11208
718-348-4839 Phone
718-348-7415 Fax
Rev. Jonathan Owhe, Pastor
Planning to expand current property which includes sanctuary and community space

Shiloh Community Development Corporation/Shiloh Baptist Church
185 Lincoln Road, Suite 101
New Rochelle, NY 10803
914-636-7505 Phone
718-207-3521 Cell
Mr Nathanial Adams, Chairperson
In the predevelopment phase of a 40-unit, affordable rental project

Upper Room Human Services Agency/Upper Room Baptist Church
180 Van Buren Street
Brooklyn, NY 11221
Bishop William G. Bennett, Pastor and Overseer
Trying to acquire property to house expanded services

9

The Transformation of UK Housing Associations: Responding to Change

David Mullins

Introduction

This chapter picks of the theme of institutional capacity within welfare regimes from Chapter 1 to explore the changing role played by housing associations within the UK. It uses a case study of a Birmingham-based housing association with its origins in social philanthropy movements of the 1920s to illustrate the interaction between housing institutions and the changing welfare regime over a period of nearly a century.

A series of critical incidents from the 1930s to 1990s illustrates the interaction between this association and the changing welfare regime. The case study shows an organization that has, throughout its history, operated in spaces defined by the regime but which has sought to expand those spaces through private experiment, individual and collective action. It has demonstrated the capacity to pursue independent social purposes and to influence (as well as responding to) changes in the wider welfare regime. The final section of the chapter discusses the implications of the case study for our understanding of welfare regimes and organizational transformation.

In Chapter 1 we discussed the ways in which concepts of welfare regime have been applied to international comparative research and the limitations of approaches that simply apply labels such as the 'Anglo-Saxon Liberal' regime to housing systems as different as those of the US and UK. We suggested that a more useful approach is to develop Esping-Andersen's (1990) notion of path dependency to explore the ways in which institutional capacity developed during one welfare era have influenced the shape of the regime in subsequent periods. The theory lends itself to longitudinal studies, and, as Esping-Andersen notes, it is the interaction over time between class mobilization, political coalitions and regime institutionalization that should be the focus of study. This chapter adopts such an approach, focusing on the interplay between certain institutions and the welfare regime in different eras.

The role of non-profit or limited-profit organizations is well established in the housing sectors of many developed societies. However, the precise institutional arrangements are extremely variable (Turner et al., 1987; Doling, 1997). In England in 2004 there were around 2000 housing associations, specializing in building, renovating and managing housing for lower-income groups in housing need, but sometimes providing a much wider range of services to neighbourhoods and communities. These organizations have progressively displaced local authorities as the main delivery agents for social housing, first through achieving a virtual monopoly on new building activity and, more recently, through wholesale transfer of existing stock to exploit their ability to harness private borrowing outside of the public sector accounts (Mullins et al., 1995). Housing associations now account for 35 percent of the social housing stock, the remainder being provided by the 200 or so local authorities who retain their landlord role (Pawson, 2004).

Welfare regime theory has been adapted and applied to a wide variety of aspects of welfare states including the origins and role of non-profit organizations. Salamon and Anheier (1998) suggested that the size and role of the non-profit sector reflects its embeddedness in social, economic and class relations, producing four main regime types (liberal, social democratic, corporatist and statist[1]) based on the level of state social welfare spending and scale of the non-profit sector. Like Esping-Andersen, they suggested that the non-profit sector in both the UK and the US could be best understood as a liberal model. This model depicts relatively low state welfare expenditure combined with a large and vibrant non-profit sector and dominant middle class.

Shifts in welfare regimes

However, it is clear that there have been significant variations over time and between different sectors and that it is important to focus on regime shifts rather than static classifications of regime types. The origins and relationship of English housing associations to the changing welfare regime in the UK have been examined in earlier work by the author (Mullins, 2000) to identify three apparent regime shifts during the twentieth century.

[1] A category added by Salamon and Anheier to Esping-Andersen's original three regime types to describe cases such as Japan with low state welfare spending and a small non-profit sector resulting from direct provision of benefits by employers.

Table 9.1 Shifts in regime types for non-profit housing in the UK.

Regime types	Regime shifts in role of non-profit housing organizations in UK
Liberal regime: low state spending, large role for not-for-profits (NFPs)	Nineteenth century Philanthropic Trusts and Model Dwelling Companies closely associated with a low state spending liberal regime. But role of philanthropy very limited in relation to need for decent housing
Social democratic regime: increasing state spending crowds out NFPs as service providers, new roles as campaigning, advocacy and recreation agencies	1919–79 rise of municipal housing in England associated with partial regime shift to universalist social democratic welfare state. Central role of NFPs declined, but variety of organizations grew: both campaign and advocacy based and directly stimulated by the state to fill gaps and complement municipal provision
Corporatist regime: high state welfare spending increases the service provision role for NFPs as they are incorporated into the state	Since 1979 liberal project to roll back the state created new role for NFPs. To some extent this has led to emergence of corporatist regime. NFPs increasingly interdependent with state, and moving beyond contracting to welfare pluralism. However, fragmentation between small number of large developers incorporated in state delivery and large number of others, some of whom retain civil society roles

Source: Salamon and Anheier (1998), Mullins (2000).

Table 9.1 shows these shifts in relation to Salamon and Anheier's account of regime types for the non-profit sector.

The period after World War I saw a gradual move away from a liberal regime in which a small philanthropic and limited profit (up to 6 percent return) sector had emerged in the absence of state housing provision to meet growing urban housing needs.

In the middle years of the twentieth century a social democratic welfare state embraced direct state provision of housing to an unusual degree (33 percent of all households by 1979). This to some extent 'crowded out' the non-profit sector, although it still continued to grow as old institutions survived and new ones were formed. By the 1960s there were new social movements in response to continued poor housing, homelessness and new immigration, leading to an increase in housing association formations. During this period part of the non-profit sector took on a more overt campaigning character, seeking to influence the welfare state, as well as providing complementary services to those provided directly by the state.

The final quarter of the century saw further regime change as some non-profit institutions were incorporated as part of a neo-liberal project to roll back the state. This was the most substantial period of growth of the sector. Much of this expansion involved new rather than existing institutions and during this period the extent of independence of the non-profit sector from the state was questioned. Nevertheless many organizations with social origins in earlier eras survived and accommodated themselves in a variety of ways to the new regime. There was some evidence, particularly after 1997, that some larger organizations were able to influence the regime by proposing and piloting new policies, suggesting a further regime change to something closer to corporatism. By 2004 a clear division was emerging between a core group of around 70 associations, identified as 'investment partners' selected for the procurement of new social housing on a large scale (Zitron, 2004), and the remainder of the sector which would be confined to housing management and neighbourhood roles. Some of the latter group can be expected to continue roles developed in earlier eras, maintaining links with civil society and campaigning for change while the former group has been more clearly incorporated within the state. This split suggests that not only can there be regime shifts over time, but different institutions will play different roles, making the labelling of regime types to depict the relationship between non-profit organizations and the state inherently problematic.

Transformations

The main purpose of this chapter is to examine evidence concerning the processes of transformation implied by the above account. Underlying the apparent stability of welfare regimes are constant negotiations and conflicts that may lead to transformations at organization, sector or regime level. Analysis of such transformations requires more detailed historical evidence relating changes in institutional behaviour, such as organizational expansion, repositioning or restructuring, to changes in the welfare regime and associated changes in values and beliefs about society. It is important to recognize the active agency role played by organizations in shaping their own environment (Mullins and Riseborough, 2001).

The approach taken in the chapter is to explore the changing role of one organization, drawing on data collected as part of a wider history project (Gulliver, 2000). This organization, a non-profit housing organization based in the English Midlands, is now known as Focus Housing Group. It has adapted to and sought to influence regime change for 75 years. The chapter proceeds by describing the social origins of Focus, then known as Copec, in the 1920s and identifying four significant transformations in its history. The chapter's concluding discussion considers what this case study tells us about the relations between welfare regime, organizational fields and individual organizations in the non-profit sector. Discussion of the social origins of Copec and four key transformations in the organization's history are arranged in the following sections:

- **The 'call to Christians to become responsible landlords'** – social origins of Copec in early twentieth century philanthropy.

- **Dropping distribution** – the move towards a non-profit charitable model in the 1930s and 1940s.
- **Campaigning for change** – developing a new relationship with the welfare state in the 1960s and 1970s.
- **Going for growth** – responding to the new liberal agenda in the 1980s and early 1990s.
- **'Turning around the supertanker'** – a social enterprise for the new millennium.

The 'call to Christians to become responsible landlords'

In 1926 Professor Frank Tillyard, of Birmingham University, wrote a pamphlet to accompany an appeal to churches and individuals to purchase shares and loanstocks in a new housing society. The title of the pamphlet encapsulates the social origins of Copec Housing Society of which he was to be the first chairman. This was a Christian-based initiative set in a local elite network; it was also part of a national movement of 'front line housing societies' (Garside, 2000) founded in the 1920s and 1930s to transform the models of nineteenth century philanthropy. It targeted the needs of people living in inner urban slum housing. In Birmingham this meant the 40,000 back-to-back terraces thrown up during nineteenth century industrialization.

Four factors were of particular significance in the social origins of Copec. First the changing nature of the state and the development of a direct housing role, second the transformation of models inherited from nineteenth century philanthropy, third the national stimulus of a new social movement, fourth its immediate environment in the local elite and religious networks of 1920s Birmingham.

Social origins theory points to the influence of the type of state; the extent of poverty and the adequacy of welfare provision made directly by the state in defining the spaces in which voluntary organizations operate. The scope of Copec's activity was influenced from the start by establishing a complementary relationship to the newly emerging state housing sector, while remaining largely independent of state funding. Prior to World War II, the city council concentrated its efforts on providing municipal housing in purpose-built estates on the outskirts of Birmingham. Copec, therefore, decided to focus on inner-city areas, particularly Aston, Hockley, Nechells, Sparkbrook and Small Heath, where courts of back-to-back housing with communal washhouses, ashpits and privies had been largely unimproved by either market, early philanthropy or state (Fenter, 1960).

National factors have tended to influence the legal and funding models, and housing management approaches of non-profit housing organizations as well as reactions to models developed in earlier periods. By the 1920s, a new style of limited profit organization was being developed. This was consciously different from the subscription charities, endowed charities and model dwelling companies of the previous century (Morris, 2000), but shared certain characteristics with the latter: raising capital from investors at less than the market rate of return (usually 2–6 percent per annum), and essentially 'top-down' in character, often established by people from various religious denominations and from local social elites. Usually termed housing societies, a distinctive feature was a determination to focus on improving slum housing. Fourteen such societies had been formed in London alone by 1934, and there were upwards of 50 nationally by the early 1930s (Garside, 2000). Many of these organizations have survived to the present day, some, such as St Pancras and William Sutton Trust and Focus itself, are amongst the largest modern housing associations.

The management model adopted by the new society was based on that developed by Octavia Hill in the late nineteenth century, a system of 'housing management and behaviour modification [bascd] on middle class women managers' (Malpass, 1998, p. 11). Initially, rent collection and other housing management and administrative work was undertaken by members of the Copec committee, especially women members, but a full-time manager, Margaret Fenter, was appointed in 1927. Trained in the Octavia Hill 'method', and gaining previous experience with the Commissioner for Crown Lands and St Pancras Housing Society, Fenter put into operation an efficient and caring, if somewhat paternalistic, approach to dealing with tenants and their problems.

The impact on Copec of a national movement of new housing societies is already clear. More specifically, Copec was one of four housing societies set up following a national Christian Conference On Politics, Economics and Citizenship in 1924; others were Altrincham, Bangor and Manchester (Garside, 2000). In Birmingham the new society was set up by a continuation committee drawing strength from involvement of a wide range of denominations. The Bishop of Birmingham, Hamilton Baynes, was a committee member from 1925 to 1932. As Frank Tillyard commented:

> 'Copec is a joint effort of the Christian community in the widest sense of the term. As the direct offspring of the Copec Conference, it was started with a representative committee, but it had not at first a direct link with either Unitarians or the Roman Catholics. In both cases this has been remedied. Attempts have been made to bring the various local churches into direct contact with the work of the Society by getting them to find money required for the acquisition and improvement of properties.' (Congregational Quarterly, 1932, cited by Gulliver, 2000)

Finally, the embeddedness of Copec in local elite and religious networks of 1920s Birmingham is apparent. There were especially close links with

Quakers who were well represented among business leaders. One such connection was through the Cadbury family and the Bournville Village Trust (BVT), an employer-sponsored housing organization, they had set up in 1895. Copec maintained very close connections with BVT, having some common committee members and using its architect's services. Both the Society's Chairman, Professor Frank Tillyard, and the Honorary Secretary, Miss Florence Barrow, were closely connected to the Birmingham Settlement, founded in 1899 and also part of a national philanthropic movement (Glasby, 1999). Other prominent families, such as the Barrows and the Sturges, were also involved in a network of the 'great and the good' that controlled the early Copec. The committee was also 'tuned-in' to local politics (two to three councillors sat on Copec's committee throughout the 1920s and 1930s, and the Chamberlains were supporters of Copec). The Society was managed by a committee of 12, all of whom were drawn from wealthy and influential families in Birmingham at the time. The relationship with local elites was longstanding. Committee members were known as Directors and there was very little turnover; by 1949, four members of the original twelve were still serving on the committee, with Florence Barrow having been Honorary Secretary for 25 years.

Having outlined the social origins of Copec we now turn to consider some of its main transformations over the following 75 years.

Dropping distribution

As the continuity of committee membership indicates, there was relatively little change in the organization before World War II. The society set about improving and managing inner area housing and raising public awareness about the plight of slum dwellers and housing needs amongst the local population. It employed only one member of staff until 1935 when a second was employed, increasing to five with a trainee scheme by 1940. It gradually increased its housing stock to around 300 by 1935, a level that would not be exceeded until the 1970s. There was however, one area in which there was a significant transformation in the 1930s: the move to a non-distributive charitable funding model.

One of the main features of the housing associations today is that they operate on a 'not for profit' basis. Strictly speaking this means that while surpluses may be accumulated they cannot be distributed to shareholders and board members. This was not the basis on which Copec was established.

As a Public Utility Society (established by the Housing and Town Planning Act 1909), Copec was empowered to issue shares and loanstock and to pay dividends of up to 6 percent. Copec used both shares and loanstock; these were not tradable on the open market but did involve distribution to holders. Unlike the £1 shares issued by Industrial and Provident Societies today, those issued by Copec in the 1920s and 1930s carried the possibility of the payment of dividends. The committee had considered it unwise to pay dividends until the end of 1927, but each year between 1927 and 1939 a 3 percent dividend was paid on all shares.

Loanstocks were issued to accumulate additional funds for the purchase and reconditioning of specific properties. For example, Copec's records indicate that loanstock was issued for properties such as those in Pope Street purchased in 1926. Stock holders were entitled to payment of dividends from the subsequent rental income. A fascinating debate took place in Copec's early years as to how these factors should be balanced. Frank Tillyard gives an insight into these discussions:

> 'In Birmingham there was much controversy as to the interest rate to be offered on the Loanstock. Some were for a businesslike 5 percent. Others were for a philanthropic 2.5 percent. The Committee made a shrewd guess that they were not likely to be able to work to a 5 percent basis so that on that footing what was overpaid to business lenders would have to be subtracted from the shareholders' dividends. On the other hand, the Committee hoped to make more than 2.5 percent, in which case the shareholders would profit at the expense of the philanthropic lenders. We finally settled on a 4 per cent Loanstock, and since the end of the first year (1925/26), the Society has been able to pay a 3 percent dividend on the share capital and carry a reasonable sum to a depreciation reserve.' (Congregational Quarterly, 1932, cited by Gulliver, 2000)

Perhaps this discussion already indicates unease about the principle of profit distribution that was to resurface in the late 1930s. In 1936 loanstock holders were invited to convert from 4 percent to 3 percent interest on the understanding that the savings in interest payments would be used for repairs and modernization, which were difficult to fund from rent income. By 1938, the society's capital included £13,979 in paid up shares and £48,300 in loanstock; £35,155 of the latter had been converted to the 3 percent rate.

By the late 1930s, Copec was undertaking a major review of its legal standing. This was precipitated by a debate in the society about the public perception of an organization making money out of slum housing. The 1938 Annual Report indicates that the issue was coming to a head:

> 'It has been increasingly felt that it is becoming more and more difficult to describe the work of the Society as a commercial venture, since the adequate housing of the lowest paid workers is a work of social importance from which it is not possible or desirable to expect more than a very modest return on capital.'

Copec converted to charitable status in 1939, thereby also gaining tax advantages in the form of relief from Schedule A tax liability and excess profit tax during the war. No further dividends were paid on shares but interest was paid on loanstock. Stockholders were encouraged to offer their holdings to the association for charitable projects or to redeem their holdings. As late as 1952 the interests of stockholders were still an important consideration for Copec. Compensation payments were received from the City Council following acquisition of some of Copec's properties. Stockholders were consulted and invited to gift their loanstock holdings to support a new scheme, Bryony House, for older people or to redeem their loanstock holdings at an agreed price.

Thus Copec had gradually shifted its funding basis from providing a limited return on investment to shareholders and stockholders towards a charity model in which income came either from gifts or from trading activities, in the form of rental income. It had also begun to dip its toe in the water of state funding, first receiving public subsidy through the City Council for rehousing associated with demolitions, but the society's records indicate very clearly that it still retained a sense of independence from the state.

This transformation into a new kind of charity was not influenced by internal debates alone. When Copec was set up the decision to use a legal model available at the time was largely the result of external influences. Subsequent experience, working with this model, had convinced the Copec committee that it was not entirely appropriate to the purposes or indeed the image of Copec, and when the opportunity arose, a change was made. It is important, however, not to discount the impact of external influences on this decision too. The formation of the National Federation of Housing Societies in 1935, of which Copec was a founder member, had provided the opportunity to discuss alternative models. As Miss Fenter put it:

> 'Much useful information has been obtained and exchanges of views arranged as to the work of other societies in addition to which the Federation acts as the mouthpiece of housing societies in negotiation with the government.' (Fenter, 1960)

Campaigning for change

The next transformation in Copec's role took even longer to mature than had dropping distribution. This involved the organization re-inventing itself to engage in new ways with the post-war welfare state. The expansion of the welfare state into activities that had previously been Copec's core business posed challenges that took time to respond to.

The period between World War II and the mid 1960s was relatively quiet for Copec, an era in which its initial mission had largely been achieved and long-standing committee members and staff, including Miss Fenter the first paid staff member, had left the organization. The very success of the earlier campaign for improvements to Birmingham's inner areas had caused a reduction in Copec's own housing stock (114 of its properties were demolished between 1945 and 1950). Instead it was allocated a temporary 'bit part' in managing acquired properties on behalf of the City Council which now assumed the leading role in redeveloping the inner areas. The old 'settlement', in which Copec had worked in the inner areas with existing properties, albeit with resources totally inadequate for the task, while the City Council built new homes in the suburbs, was now over. Copec had long recognized the need for action on a scale that only the municipal authority could provide. Nevertheless, Copec's records indicate that the organization experienced a sense of loss of purpose associated with these campaign achievements.

Copec eventually found a new role as a result of two main influences. First, the emergence of new social movements and the re-invigoration of the voluntary sector and with it voluntary housing and a more overt campaigning approach in the 1960s. Second, the full realization of this new role was dependent on a transformation in the frameworks of the welfare state itself, leading to significant funding for voluntary housing in the 1970s.

The 1960s is commonly seen as a time of dramatic social change in which new attitudes began to transform the conformity of the 1950s. In the UK, this was an era in which poverty was rediscovered and campaigns were established to achieve social change. In the housing world this was symbolized by the birth of Shelter in 1966 within a few days of the transmission of the groundbreaking drama documentary 'Cathy Come Home' by the BBC. Both the documentary and the campaign demonstrated to the public and policy makers alike that the problems faced by homeless people and those living in substandard housing were not eradicated. Shelter was bound up with the housing association movement from the start.

Three new housing associations were formed in the West Midlands as part of this movement, all of which were eventually to merge with Copec. These new associations were managed by voluntary committees drawn from the original campaigning groups, supplemented by people drawn from other voluntary organizations, the church, academic institutions, and to a lesser extent the business community. In Birmingham, Copec and Birmingham Housing Trust (BHT) received funding from Shelter and were most closely involved in the local Shelter campaign. Despite their different social origins and traditions (BHT was a 'brash newcomer' according to its founding director David Mumford), Shelter managed to persuade the two organizations to merge in 1970. Although BHT had been much smaller than Copec, its influence on the merged organization was probably the stronger. This was apparent as a renewed focus on voluntary sector purposes of housing people no-one else would, at the same time as a more aggressive approach to securing development opportunities to expand the housing stock.

The Shelter campaign went much wider than simply supporting organizations like BHT and Copec. The principal aim was to achieve changes to the welfare state to provide a right of access to housing for homeless people. There was also an agenda to move housing associations on from depending primarily on charitable fund raising to being involved in state-funded programmes to rehabilitate poor housing and to house homeless people. These aims were to be substantially achieved over the next decade.

While public funding had been available since 1919, it was not until the 1960s that housing societies were employed on a significant scale as instruments of central government housing policy. It was a Conservative administration that first decided to use non-profit housing organizations to pursue public policies for new programmes in 1961 and 1964 for better off, mobile workers who needed to rent until they could afford to buy. However, it was the Housing Act 1974 that brought them within the ambit of the public sector by introducing generous financial support and associated regulation for housing associations targeting low-income households and housing needs. This soon led to a majority of housing associations becoming dependent upon the Exchequer for their subsequent growth and deficit funding of their running costs. As Malpass (1998) has commented:

> 'It has been recognised that since the mid-1970s, as housing associations moved towards a closer relationship with government and to a more central role in housing policy, the price of increased financial support was a certain loss of independence.'

At the time associations appear to have seen the Act as an opportunity to deliver their mission rather than as a threat to autonomy. The Birmingham Copec Housing Trust (the result of the Copec/BHT merger) registered with the Housing Corporation as soon as it could and Chairman John Duncan wrote that:

> 'we are regarding this time of development as the most significant stage in the Trust's life since Copec and BHT merged in 1970. The Trust now has the opportunity to respond to the housing needs on a wider front than previously.' (Copec, 1975)

The impact on the association was dramatic. Following another merger with one of the Shelter generation associations in Wolverhampton in 1975,

Copec Housing Trust was formed to manage 2000 homes and to develop 300 new homes each year. Management was decentralized and branch offices opened in Wolverhampton and Sandwell. New services were set up, including a Special Projects department to coordinate housing with support to groups referred to at the time as 'battered wives' and 'the disabled', and a housing advice service was set up in partnership with Birmingham City Council.

This long period of transformation began and ended with some important partial successes of campaigns initiated by Copec to influence the shape of the welfare state. The long-standing aim of the early Copec had been to secure significant improvements in the back-to-back housing which made up much of the inner areas of Birmingham. This campaign saw some of its aims achieved in the post-war period when the local state brought significant resources to bear on slum clearance and rebuilding. A long period of search for a new role followed, culminating in involvement in the Shelter campaigns of the 1960s. Copec, together with newly formed associations including BHT and Wolverhampton Housing Association, was involved in campaigns whose aims were partially realized by a new regime of public funding for housing associations and later (in 1977) by legislation to provide local authorities with a duty to assist certain groups of homeless people. Ironically when the new role was found, the focus was again on rehabilitating older housing in the inner areas, albeit on a larger scale than before.

The expansion achieved by Copec in the decade in which public funding was introduced was out of all proportion to what had been achieved in the previous 50 years of largely voluntary effort. Between 1970 and 1980 the number of homes owned increased by a factor of 10 to 3500, stock value increased by nearly 20 times to £25 million and the number of staff employed increased more than fivefold to 95.

Going for growth

This transformation in the scale of Copec's activities set the scene for the next period in which growth appeared to take over as the main raison d'etre of the organization. Between 1985 and 1995 the organization grew to own nearly 10,000 homes with a capital value of £385 million and increased its staff to over 600. This was achieved not only through taking advantage of the public funding programme to develop new homes for rent, but also through expanding into home ownership programmes and other activities and through a number of further mergers with other housing associations.

It is possible to rationalize this focus on growth with realizing the aims of the earlier period; as David Mumford puts it:

'we had a tunnel vision, there were 40,000 on the city's housing list and now there was money to expand. I wanted to do things in a big way.'

However, it is also possible to see the increasing expansion in the 1980s and early 1990s as a symbol of the new competitive mood in the housing association sector. This mood was associated with the election in 1979 of a new Conservative government with a radical reform programme for the welfare state. For housing this meant the end of half a century of growth of the municipal sector and gradual replacement by the non-profit sector.

The role allocated to the non-profit sector by the Conservative government in this period was very different to the agenda of social action, the rediscovery of poverty and campaigning for the extension of the welfare state in the 1960s. Now the emphasis was on contracting for the state, efficient performance and, from 1988 onwards, harnessing private finance and competition to produce the largest number of homes for the lowest amount of public expenditure. The 1988 Housing Act is usually seen as a major turning point for housing associations through their 're-privatization'. Randolph (1993) comments:

> 'The Conservative government intended the Housing Act 1988 to stimulate a major shift in the orientation of housing associations. The objective was to move them firmly away from the quasi-public housing sector which associations had become as a result of the Housing Act 1974 and back towards the private, or more precisely, the "independent" rental sector.'

The effect of the 1988 Housing Act on the development of Focus was profound. Within a year of the Act's introduction, Copec and three other Midlands associations had formed the Focus group as a vehicle for joint bidding for development funding and raising private finance. By 1991, all but one of these had permanently merged to form Focus. Another association, Shape, joined the group in 1994.

So, there is evidence that there was structural transformation in response to funding and other opportunities. What additional evidence is there that the organization was transformed by the neo-liberal regime associated with the Conservative government of the 1980s and early 1990s in ways other than a simple increase in size? Three examples illustrate the ways in which Copec/Focus embraced the new competitive ethos to 'go for growth':

- The first is geographical expansion; David Mumford comments that in this period: 'we saw our core area as the Midlands – from Stoke to the M4', a huge expansion of earlier definitions. This was viewed with suspicion by other associations and by some local authorities. David Mumford recalls that this caused problems in his dual role as a Housing

Corporation Board Member. 'We were never the favourite. The Corporation was worried that we were getting too much of the cake. I found out that the Midlands office had set a maximum limit of 10% of the programme to go to us. I took this up with the Board. There was no such policy.'

- The second example of expansionism is more specific, and came in the very last year of growth of the Corporation's programme in 1992. The Chancellor's autumn statement brought forward £580 million from the next year's programme to address problems associated with a prolonged slump in the owner-occupied market in the run-up to a general election. Focus was one of 27 successful applicants nationally who were identified as capable of spending this sum in record time to purchase existing properties from individual owners or from private developers. Focus bought 578 homes using £14 million of public funding matched with £11 million private finance. Undertaking a programme on such as scale over such a short period of time, much of it in areas where Focus did not already have a presence, brought both benefits and costs to the organization. While this programme had provided another opportunity to demonstrate the ability to deliver government objectives it also contributed to a loss of community focus and sense of overall purpose in the organization.
- The third example of a willingness to go for growth with only limited reference to the social purposes and values of the organization was extensive involvement in low-cost home ownership activities through two special-purpose subsidiaries. While facilitating home ownership is a legitimate objective for housing associations, it was never seen as an organizational priority by Copec. The justification for involvement was an opportunistic one. These activities were undertaken because there was funding available, as a result of the top slicing of the Housing Corporation programme nationally for home ownership initiatives. David Mumford recalled that 'my view was if they're going to do it, we're going to be there'. A third subsidiary undertook home ownership development without public subsidy. Here the aspiration was to generate surpluses that would the subsidize the core purposes of the organization.

During the 1980s and early 1990s Focus had become one of the most successful 'contractors of the state' in the housing association sector. By demonstrating the ability to deliver, the association was able to play a role in housing provision in the Midlands that would have been beyond the wildest dreams of Copec's founders. However, when the association came to take stock of its purpose and direction in advance of the 1997 general election, there was a clear view that a further transformation was necessary to again bring Focus in to line with changing times. This was summed up by one commentator who said that everyone had known what the old Copec had stood for, but the purpose of Focus was becoming unclear.

'Turning around the supertanker'

When a Labour government came to power in 1997, Focus, like other associations that had been heavily involved with the previous government's agenda, was already engaged in a major repositioning exercise. In Focus' case this involved a high-profile relaunch of the association soon after the election as a Social Investment Agency 'which provides social goods and services, usually in partnership with others, to communities which experience significant and multiple disadvantage and lack of opportunity and investment'. The association's mission was restated accordingly as 'to relieve poverty and provide opportunities in priority communities'.

At the same time they set about trying to influence the policy agenda and to demonstrate relevance to the new policy priorities (Mullins and Riseborough, 2001). Initiatives, such as the New Deal for Work, the New Deal for Communities, Health Improvement Zones and the Social Exclusion Unit, set agendas that Focus began to engage in at once through concrete projects.

The rethink had actually started 2 years before the election with a corporate planning process in which Focus determined to go back to it roots. It was decided to place the regeneration of communities at the centre of its organizational strategy, to tackle poverty and deprivation in the communities where it worked and to offer a wider regeneration service to partners. A new emphasis on 'responsible growth' led to disposal and swapping of some stock outside of these core communities. This approach was influenced by another merger in 1994 with a smaller and financially overstretched but visionary association, Shape, which had extensive experience of community regeneration activities. It is no coincidence that the rethink also coincided with the appointment of a new chief executive and followed a significant reduction in availability of public funding for continued development-led expansion.

The social investment approach included a number of elements. An anti-poverty strategy aimed to improve the quality of life of communities and those on welfare benefits and provide incentives for those seeking work. Economic regeneration initiatives facilitated access to pre-vocational training guidance, work experience and employment. Local employment was stimulated through local procurement policies, support for existing businesses, encouragement of new enterprises and direct investment in the housing stock. Community health was targeted with the health needs of tenants and other residents promoted through better use of existing resources and levering in new resources. Lastly, Focus began to improve the quality of the physical environment through its Home Improvement Programme.

Summing up this transformation, which is still in progress at the time of writing, Chief Executive Richard Clark commented:

'many people have compared what we are trying to do to changing the course of a supertanker – and I'd say we're half way there. In the last twelve months we've pulled the organisation around from being "bricks and mortar" led to face in the right direction, towards social investment. Now the task is to move in that direction.'

Discussion

This case study has provided an opportunity to develop our earlier analysis of social origins and transformations (Mullins, 2000) through detailed application to an organization that has operated through several welfare regime eras. It has provided evidence to support the view that it is necessary to examine very closely the relationship between welfare regime, organizational fields and individual organizations.

Welfare regimes

First it is clear that the role played by non-profit organizations can only be understood in the context of the broad shape of the welfare regime, and the spaces provided for such organizations to develop. The history of Focus Housing Group has followed rather closely the transformations in the welfare regime and opportunities provided by housing policies.

Copec was formed after the liberal era of nineteenth century industrialization and urbanization when the absence of state intervention beyond the most basic public health legislation had provided a potentially large space in which philanthropy could develop. Its origins were also after the questioning of the limited scale and performance of nineteenth century models had already led the state to accept greater responsibility for housing and had identified local authorities as the main agencies to deliver. This left a smaller space for organizations such as Copec to develop in, complementing municipal provision by improving and managing existing slum dwellings rather than building new homes in the suburbs. Nevertheless Copec was not an isolated example of a new voluntary organization set up to occupy this space, but was one of over 50 new-style 'front-line' housing societies established in this period.

When the state finally responded to campaigns by Copec and others to undertake more radical improvements to inner areas after World War II it again chose to allocate the main role to municipal authorities. Copec was left with a declining housing stock and a marginal role, assisting the municipal authority in short-term management of properties to be demolished. The resolution to the long search for a new role that followed was eventually inspired by engagement with new social movements in the 1960s, but the new role was not fully realized until the state expanded the space available by generous funding of a complementary voluntary sector public housing programme in the 1970s.

The one key respect in which regime change has had a significant impact on this organization has been in the opportunities provided for organizational growth. For most of the social democratic regime era the organization played a very limited complementary role to the municipal housing sector, and for the first 40 years of operations, the organization had no more than 350 properties in management and employed no more than seven staff. However, during the final stages of the social democratic period, the state provided greater opportunities for the voluntary sector to expand to deliver some of the services that they had campaigned for. In just 10 years the organization's housing stock and staffing had increased fivefold to 1400 and 40 respectively. The emergence of a neo-liberal regime, in which the non-profit sector was encouraged while municipal housing declined, led to much greater expansion. In the next 20 years total stock increased steadily to nearly 10,000 and staffing to over 600 by 1995.

Incorporation by the state in the 1970s produced an unprecedented increase in the scale and complexity of operations and in the asset base of Copec. However, this period of expansion as part of the public sector in a social democratic welfare regime was not seen as challenging the values base of the organization. As David Mumford was later to comment:

> 'The 1974 Act did not change what we were trying to do, but it enabled us to do it more quickly. For the first time we were solidly backed. If we could put up good enough ideas for schemes the funding was there.'

In contrast, the next period of expansion was associated with a transformation of the welfare regime as the new right began to roll back the state, with a particular emphasis on reducing the municipal housing role. In many ways the housing association sector provided a test bed for the new role to be played by the wider voluntary sector in the emerging 'mixed economy of welfare'. The introduction of private finance and the use of competition to allocate funding to associations were all aimed at a transformation of the ethos of associations away from those of the public sector. The extent to which Focus went for growth in this period, often into activities and localities that had not previously been seen as core to its mission, may be seen as evidence of the success of this privatization project.

Re-evaluation of the growth-hungry 'bricks and mortar' based vision of housing associations in the most recent period of transformation may also be interpreted as a response to changes in the welfare regime. After 1992 funding for new development by housing associations went into rapid decline. Housing policy moved away from a 'numbers game' and was rapidly becoming a sub-set of wider regeneration policies. Organizations like Focus were forced by these shifts in resources and policy aims to question their own role. After the 1997 election the launch of the Social Investment Agency concept was a clear attempt to engage with and influence a new government's agenda.

Organizational fields

The case study has also indicated the importance of considering the wider networks of relationships, or organizational fields (DiMaggio, 1986) within which non-profit organizations operate. Each of the main transformations of Focus has seen interaction with other organizations, and sometimes by collective action that has influenced elements of the welfare regime.

The formation of Copec can only be fully understood in relation to social movements of the time. Two movements of direct influence were the Christian Conference on Politics, Economics and Citizenship in 1924, which led to the foundation of at least three associations, and the wider 'front-line' housing society movement focusing on slum housing. More locally in Birmingham the foundation was based in an elite network of religious, business and civic leaders.

The transformation to charitable status and away from profit distribution in the late 1930s was partly influenced by discussion of structures through the newly formed National Federation of Housing Societies.

The changes of the 1960s and 70s were again strongly influenced by national and local social movements, this time based on the rediscovery of poverty and the formation of Shelter, the national housing campaign. This campaign played a significant role in developing the infrastructure of a new 'housing association movement' and, in the case of Copec, stimulating an alliance between old and new actors. It also, in part, led to changes in the regime itself that were to produce unprecedented opportunities for Copec to expand through public funding.

The recent transformation into a Social Investment Agency has also been strongly influenced by wider networks of relationships, including those with partner agencies who have enabled Focus to develop a broader vision of its role. A key influence was that of Shape which brought into Focus its considerable experience and track-record of non-housing activities, particularly its neighbourhood regeneration work, development of employment and training initiatives and its experienced staff and committee. The Shape philosophy prefigured the approach to social investment activities that characterized Focus in the late 1990s.

The extent of organizational autonomy

Given the previous discussion it would clearly not be possible to argue that Focus has enjoyed the independence to develop and transform itself entirely in line with internal choices. However, the case study provides considerable evidence that its transformations have reflected internal as well as external factors and that on occasions it has been able to influence its organizational field and indeed the wider welfare regime to a degree. There is also a case to be made that throughout most of its history Focus has been consistent in the values and social purposes that it has addressed itself to.

Taking the question of values and purpose first, there has been a remarkable consistency in the association's involvement in improving and managing older inner-city housing stock throughout its history. This was a particularly striking feature of the long search for a new purpose in the post-war period. When the new role was eventually found after 20 years, it turned out to be remarkably similar to the old role, rehabilitating older housing. Only in the competitive growth phase of the late 1980s and early 1990s could it be argued that Focus strayed significantly from this core role, when it became involved in building new homes on greenfield sites over a wide geographical area and for sale as well as for rent.

Examples of the ability to reject externally imposed options include the association's early decision not to take advantage of subsidy available to Public Utility Societies from 1919 onwards, preferring the greater independence afforded by charitable and loanstock funding. It is believed that the first public subsidy drawn by Copec was payments from the City Council under the Slum Clearance Act 1930 for rehousing people displaced from demolished properties.

The first significant transformation of Copec was as a consequence of internal unease with the profit distribution model that had been adopted at the time of establishment. The decision to switch to charitable status and to encourage stockholders to convert to donations represented a conscious decision to sacrifice growth for integrity of ethos.

Examples of the ability to influence the field and regime have included the strong focus on campaigning and demonstration projects throughout the organization's history. As Professor Tillyard noted in 1926, 'public reforms have always come in the wake of private experiment'. Copec's long campaign for radical action in the inner-city slums had its answer in post-war municipal programmes. Involvement in the Shelter Campaign eventually led to the 1974 Housing Act (that provided public funding for housing associations) and the 1977 homelessness legislation, which led to local authorities having a responsibility to house homeless persons.

On a smaller scale staff and committee members of Focus have enjoyed a substantial influence on the development of the housing field at various points in its history. Over 40 Copec students were trained in the Octavia Hill method in the 1930s and early 1940s, many of whom went on to manage housing throughout the country and further afield. Miss Fenter herself was elected President of the Society of Housing Managers the year after she retired from Copec in 1954. A later director, David Mumford, combined his role of director with membership of the Board of Housing Corporation for a number of years after 1974. In this role he was able to influence the level of resources for the Midlands and to take up restrictions such as zoning and informal limits to the share of the regional programme going to individual associations.

In the modern era Focus' early championing of the social investment agenda has clearly influenced the government's policy agenda and the strategies of many other organizations in the field. A distinctive feature of this period has been the ability of the organization to access funding from a much wider range of sources than hitherto. This has produced a very different pattern of resource dependency (Aldrich, 1976) to the 1970s when expansion was largely based on a monopoly purchaser, the Housing Corporation. In the recent era public funding from a range of programmes, including Single Regeneration Budget, New Deal and European Social Fund, had been the backbone of the Social Investment Agency. Private finance has also been very important with a private sector loan book of some £160 million mainly related to traditional housing development activities and plans to develop a special purpose lending vehicle to borrow £150 million and on-lend for social investment activities. Meanwhile, the earlier emphasis on charitable funding had, like the initial dependence on loan stock and shareholding, now been displaced by these larger scale public and private funding streams.

New challenges were emerging in the early 2000s as government priorities for housing were more clearly articulated through the Communities Plan (ODPM, 2003), the re-inventing of the Housing Corporation's investment role (Housing Corporation, 2003) and the closer interest of the Treasury in the delivery of housing outcomes through the concept of public service agreements. In this period the stakes were raised for organizations that perceived tensions between their community accountability and civil society roles on the one hand and their role as procurers of housing for the state on the other. Focus Housing Association was one of the 70 lead associations selected for the first year of the new investment partnering regime, receiving an allocation of £21 million to develop around 600 new homes over a 2-year period. While in itself this activity stream might not be expected to shift the priorities of a large and diverse community business with a stock holding of over 12,000 rented homes, the associated compliance regime places the association under closer governmental scrutiny and within a select group of associations that may increasingly be perceived as having a distinct agenda to non-developing associations.

Conclusion

The chapter has taken the example of one organization's history to provide a more detailed picture of the processes of negotiation and conflict associated with transformations of the welfare regime for non-profit housing in England in the twentieth century. In this way it has contributed to debates about the social origins and transformations of non-profit institutions which have so far been informed mainly by cross-sectional comparative research.

Key transformation events have been identified in the chapter. These provide concrete examples that have enabled greater understanding of the processes of transformation and the relationships between not-for-profit organizations, organizational fields and welfare regime. In every case the transformations described have involved an interaction between regime influences, the organization's own strategies and intermediary relationships with other bodies within the organizational field.

The micro approach adopted has enabled us to track the relationships between the organization and a succession of welfare regimes. It can be seen that this organization emerged from a particular welfare regime and that it initially relied strongly both on the institutional forms available for philanthropic organizations at the time and on its ability to attract elite supporters to the cause. However, the organization subsequently adapted and transformed itself in relation to both internal and external factors. External factors included changes in the welfare regime and funding opportunities to carry out its work. Internal factors included the influence of committee members and staff, for example in early debates about the validity of profit distribution, later in the encouragement first of an expansionary momentum, and most recently in 'turning around the supertanker' to embrace a new agenda of social investment and responsible growth.

Intermediary relationships with other institutions were also significant. In particular, influences arising from changing values in society had a major impact on this organization's history, for example the emergence of reform movements in the 1960s and 1970s re-invigorated the early campaigning zeal and re-established a clear role after a period of uncertainty. In many respects the mid-1990s can be seen as a similar period when a new vision was crafted from contemporary social influences to restore a sense of purpose as well as re-aligning the organization to the latest shifts in the welfare regime.

Overall the examples of organizational transformations discussed in this chapter indicate the capacity of Focus throughout its history to act as a social entrepreneurial organization, spotting opportunities in the spaces provided by the welfare regime, seeking to expand these spaces through individual influence and 'private experiment' and through collective action as part of wider social movements that have, at times, played their part in transforming the welfare regime itself.

Acknowledgements

I would like to acknowledge the research undertaken by Kevin Gulliver for his book *Social Concern and Social Enterprise*; this chapter draws mainly on the arguments in Chapter 7 of that book drafted by myself. Supervising Kevin's research and editing his book provided an excellent opportunity to develop some of the ideas originally set out in my paper *Social Origins and Transformations, The Changing Role of English Housing Associations* published by Voluntas in 2000. I would also like to thank Prime Focus Housing Group for the opportunity to study archive material on its history and to the many people who agreed to be interviewed for this project, including in particular David Mumford, who was Director for a quarter of a century, and to Richard Clark, the current Chief Executive. Finally, I would like to thank Jeremy Kendall, Peter Malpass and Moyra Riseborough for comments on various drafts of this chapter.

References

Aldrich, H. (1976) Resource dependence and interorganizational relations. *Administration and Society* 7 (4): 419–54.

Copec (1975) *Annual Report of Birmingham Copec Housing Trust*. Copec, Birmingham.

DiMaggio, P. (1986) Structural analysis of organizational fields: a blockmodel approach. *Research in Organizational Behaviour* **8**: 335–70.

Doling, J. (1997) *Comparative Housing Policy. Government and Housing in Advanced Industrialised Societies*. Macmillan, Basingstoke.

Esping-Andersen (1990) *The Three Worlds of Welfare Capitalism*. Princeton University Press, Princeton.

Fenter, F.M. (1960) *COPEC Adventure. The Story of the Birmingham COPEC House Improvement Society*. Buckler and Webb, Birmingham.

Garside, P.L. (2000) *The Conduct of Philanthropy. William Sutton Trust 1900–2000*. Athlone Press, London.

Glasby, J. (1999) *Poverty and Opportunity. A Hundred Years of the Birmingham Settlement*. Brewin Books, Warwick.

Gulliver, K. (2000) *From Social Concern to Social Enterprise. The Origins and History of Focus Housing Group*. Focus Housing Group, Birmingham.

Housing Corporation (2003) *Re-inventing Investment*. Housing Corporation, London.

Malpass, P. (1998) *The Work of the Century. The Origins and Growth of the Octavia Hill Housing Trust in Notting Hill*. Octavia Hill Housing Trust, London.

Malpass, P. (2000a) The discontinuous history of housing associations in England. *Housing Studies* **15** (2): 195–212.

Malpass, P. (2000b) *Housing Associations and Housing Policy.* Macmillan, Basingstoke.
Morris, S. (2000) Defining the non-profit sector: some lessons from history. *Voluntas* **11** (1): 25–43.
Mullins, D. (2000) Social origins and transformations, the changing role of English housing associations. *Voluntas* **11** (3): 255–75.
Mullins, D. and Riseborough, M. (2000) *What are Housing Associations Becoming?* Final Report of Changing with the Times Project. University of Birmingham, Housing Research at CURS No. 7, Birmingham.
Mullins, D. and Riseborough, M. (2001) Non-profit housing agencies: 'reading' and shaping the policy agenda. In Harris, M. and Rochester, C. (eds) *Voluntary Organisations and Social Policy*, Chapter 11. Macmillan, Basingstoke.
Mullins, D., Niner, P. and Riseborough, M. (1995) *Evaluating Large Scale Voluntary Transfers of Local Authority Housing*. HMSO, London.
Office of the Deputy Prime Minister (2003) *Sustainable Communities; Building for the Future.* (The Communities Plan.) Office of the Deputy Prime Minister, London.
Pawson, H. (2004) Reviewing stock transfer. *Housing Studies Association Spring Conference, Sheffield Hallam University*.
Ragin, C. (1998) Comments on 'Social Origins of Civil Society'. *Voluntas* **9** (3): 261–70.
Randolph, B. (1993) The re-privatisation of housing associations. In: Malpass, P. and Means, R. (eds) *Implementing Housing Policy*. Open University Press, Buckingham.
Salamon, L. and Anheier, H.K. (1997) *Defining the Nonprofit Sector. A Cross-National Analysis.* Manchester University Press, Manchester.
Salamon, L. and Anheier, H.K. (1998) Social origins of civil society: explaining the nonprofit sector cross-nationally. *Voluntas* **9** (3): 213–48.
Turner, B., Kemeny, J. and Lundqvist, L.J. (1987) *Between State and Market: Housing in the Post-Industrial Era*. National Swedish Institute for Building Research, Gavle.
Zitron, J. (2004) *Towards Investment Partnering.* Chartered Institute of Housing, Coventry.

10

The Rise and Fall (and Rise) of Community Development Corporations and Black and Minority Ethnic Housing Associations

Harris Beider

Introduction

Recent years have seen much interest in social capital, community development and the role of housing organizations in the management of increasingly diverse communities. Much influential work on these themes has been undertaken in the US with focus on the specific contribution of community development corporations (CDCs) as a conduit for social and community renewal. This chapter will compare the role and performance of CDCs with black and minority ethnic (BME) housing organizations in the UK as agents for the provision of housing to low-income communities and community mobilization. Both types of organization have emerged in the recent past with a tradition of grassroots campaigning, working with increasingly diverse communities, actively supported at first by macro policy drivers but now operating within a less helpful environment. They need to meet the expectations of government, private sector sponsors and local communities within a changing demographic base and fluid policy framework. We will use sectoral reviews, key issues and criticisms to assess the relative independence of CDCs and BME housing associations from the external environment and explore a new role which may be somewhat removed from their origins. Comparative analysis of organizations based in differing housing systems requires us to use a welfare regimes approach to make sense of their past and help chart the future.

Welfare regimes and housing organizations

The influential work of Esping-Andersen (1990) helps us to understand the role of organizations operating within different housing systems and the role and remit of community development corporations (CDCs) in the US and BME housing associations in the UK. The emergence of each sector could be seen as a response of the state to grassroots mobilization drawn from black and minority communities to meet the crisis of capital withdrawal, housing decay and general disinvestment of inner urban areas during the 1960s (US) and 1980s (UK). In this way, both organizations can be seen to have shaped their external environment but also interacted with policy and political drivers to maximize housing-based investment. Welfare regimes theory enables us to make sense of the emergence and interaction of CDCs and BME housing associations with the state, as well as supporting comparative analysis across different systems.

There is an alternative view to convergence of US and UK welfare regimes. This suggests that the UK has much more in common with European countries rather than the US. Historic support and development of the welfare state had survived attacks by the Conservative government on public spending during the 1980s. More recently, the New Labour government has committed increased spending on health, education and housing as well as radical proposals to cut child poverty. Public spending has risen to 41 percent of national income for the period 2003/04 which is four percent higher than 1999/2000 (Emerson et al., 2003, p. 42). It could be argued that continuing public concern on issues such as health and education may be the result of residual popular support for welfare provision and an active rather than passive state. In short, the UK cannot be simply seen as following a US-type liberal regime as discussed below. However, the long-term trend does appear to be towards convergence between the US and UK but further research is required to make the case convincingly.

Using cross-country data sets to review variation in social rights and the welfare state in terms of the relationship between the state, market and family, led Esping-Andersen to identify the clustering of societies around three types of welfare regimes. The first, known as the liberal welfare state, tends to be characterized by robust support from the state for market allocation of goods and services and stringent, stigmatized and means-tested benefits. The second regime is described as being corporatist. That is, it is viewed as having embedded social rights which are sometimes provided by the state but an important role is also played by the church and family. The final regime, social democratic, was underpinned by state intervention in the provision of rights and services and the avoidance of the dualism between the state and the market and between working class and middle class interests (Esping-Andersen, 1990; Mullins, 2000). These are not meant to be pure demarcated models but Nordic countries tend to be social democratic, continental Europe countries espouse corporatist principles whilst the US and UK seem to be framed within a liberal regime type.

> 'The consequences of this type of regime minimises de-commodification-effects, effectively contains the realm of social rights, and erects an order of stratification that is a blend of relative equality of poverty among state-welfare recipients, market differentiated welfare among the majorities and a class dualism between the two.' (Esping-Andersen, 1990, p. 27)

This view is consolidated by empirical understanding of poverty and disadvantage within the US and UK. In the US, poverty and disadvantage are shaped by ethnicity and spatial concentration within cities and neighbourhoods. The African-American poverty rate is 22.7 percent and that of Hispanics is 21.4 percent. Taken together, this represents 8 million African-Americans and Hispanics living below the official poverty threshold of $17,960 (for a family of four). The poverty rate for these two groups is twice the national average. More than 40 percent of the poor Americans are concentrated in cities (US Census Bureau, 2001). In the UK, research conducted for the Social Exclusion Unit revealed that more than 50 percent of the black and minority ethnic population lived in the 44 most deprived local authorities across England. These areas contain four times as many people from black and minority ethnic groups than other areas (Social Exclusion Unit, 2000). This racialized poverty in both the US and UK reduces access to decent housing, health and education. It can lead to growing polarization and dualism on class and ethnicity (Home Office, 2001), the emergence of neighbourhoods known as 'cities within cities' (Venkatesh, 2000) and a growing black underclass (Wilson, 1980). Poverty concentration within neighbourhoods and amongst black and minority ethnic groups provides a challenging set of circumstances for both CDCs and BME housing associations, many of whom are based in these locations.

Discussion on welfare regimes seems to suggest similarities between the UK and US. The importance of the market, low levels of state intervention and commodification of housing should result, other things being equal, in a large number of active not-for-profit organisations such as BME housing associations and CDCs. However this view does not sufficiently take into account the importance of changing governmental policies, community action shaping housing policy and short-term responses to stifle urban unrest. The emergence of BME housing associations and CDCs should be placed in this wider context. During this chapter, we will describe the key characteristics of each sector and consider how these organisations have been shaped by macro policy as well as influenced change.

Interestingly, CDCs and BME housing associations developed in politically conservative times. The emergence of CDCs can be traced back to the 1960s and the failure of the Great Society Reforms. This led to the Nixon administration. Similarly, the first BME housing associations were registered by the Housing Corporation during the 1980s in the second and third Thatcher-led Conservative governments. It could be argued that the birth, growth and consolidation of both types of organisation demonstrate the importance of sector lobbying to win resources for poor communities. This is an aspect which has been ignored by welfare regimes literature. Indeed, BME housing associations and CDCs appear to demonstrate a level of independence from welfare regimes in both the UK and US. Their emergence, rise and problems will now be discussed in turn.

The emergence, role and problems of community development corporations

There are no clear definitions of what constitutes a CDC or an accurate account of their size and scope. The most comprehensive national surveys of the sector are now nearly a decade old and this makes it difficult to develop a baseline understanding of these organizations. They have been described as 'non-profit community-based organizations that, as part or all of their programmes, are engaged in housing development, commercial facilities development, or business enterprise development' (Vidal, 1992); 'bodies that are partly voluntary organisations, partly private developers and partly arms of government' (Twelvetrees, 1989); 'small community based organisations having non-profit tax exempt status, a volunteer board and an emphasis on physical redevelopment . . . who have taken on the heroic task of trying to rebuild communities devastated by capitalist disinvestment' (Stoecker, 1996). Drawing together these strands suggests that CDCs are small, voluntary neighbourhood organizations operating within distressed economic areas and providing low-income housing and social and economic services to communities.

The birth of CDCs was set against the economic and political failures of the 1960s. Economic growth and the civil rights struggles had failed to heal a society which was becoming increasingly fragmented along class, race and gender lines and polarized within cities. The CDC was at once attractive to politicians of the left and right. To the former, they were seen as organizations that were rooted in community mobilization and empowerment of the most disadvantaged sections of the population. For the latter, they were viewed as a perfect example of 'bootstrap' capitalism, self-help and entrepreneurism (Bratt et al., 1994).

The growth of CDCs has taken place in three distinct and sequential clusters. First generation CDCs, which developed during the 1960s, came about as a result of growing public concern on race and poverty. Community mobilization together with support from leading politicians and private financiers led to the implementation of the 1966 Special Impact Programme Amendment to the 1964 Economic Opportunity Act. This measure led to the direct involvement of the private sector, in collaboration with government and communities themselves, on the deep-seated problems of urban poverty. Critically for some, the growing corporate interest in poverty, and soon after financial support, compromised the principle of radical and grassroots mobilization as capital and community are unlikely bedfellows (Stoecker, 1996; Filner, 2001). The Act led to the direct financing of the first 100 CDCs, including the first such organization, Bedford-Stuyvesant Restoration Corporation, in New York City. The focus was on employment creation rather than meeting the housing needs of low-income communities.

Second generation CDCs during the 1970s mobilized on the community displacement impact of urban renewal and focused on building low-income housing. It is estimated that up to 1000 CDCs were developed during this period, helped in part by the 1974 Community Services Act which launched dedicated support for CDCs in the form of the Community Development Block Grant (CDBG) and specified the importance of each CDC involving local people in the process (Vidal, 1992). The 1977 Community Reinvestment Act stated that banks and other financial lenders should invest in local communities. This made it easier for CDCs to access finance but further removed them from their community roots (Stoecker, 1996).

Third generation CDCs developed during the drastic federal budget cutbacks during the Reagan administration of the 1980s, and were premised on public–private partnerships and the emergence of low-cost finance intermediaries such as LISC and the Enterprise Foundation. The 1990 Affordable Housing Act led to new steps to address poverty through initiatives such as the Home Investment Partnership Programme which provided block grants for states and localities to support low-income housing projects (National Affordable Housing Management Association, 2004).

The growth of CDCs to the current estimate of between 2000 and 2500 organizations is a result of a mix of complex factors that include mobilization by communities, the search for solutions to pressing urban problems by politicians and, increasingly, the role of private capital. Growth has been maintained despite an increasingly difficult financial climate at federal and state level. The average CDC is one that is likely to be based in the urban centres of the north-east and mid-west US. Organizations themselves are unlikely to be large agencies and will probably have fewer than seven full-time employees and an annual budget of $700,000. Although involved in a range of supply-side initiatives, the provision of low-income housing is the key business activity for more than 75 percent of CDCs, with, on average, 21 housing units being developed and managed each year by a typical organization (Vidal, 1992). It is estimated that the CDC-owned and -managed housing stock amounted to 400,000 in the mid 1990s (National Congress of Community Economic Development, 1995). This low-cost housing had been supported by significant federal, state and local funding which during the first two phases of CDC growth (1996–1980) was over $400 million (Stoecker, 1996).

Seen in this context, CDCs appear to represent an innovative and radical attempt at grassroots capitalism, at once encouraging financiers to see the logic in investing in economically devastated communities and enlisting the support of communities to be involved in local decision making. They have been showcased as an effective conduit to support job creation, retail expansion and small business development as well as low-income housing (Ford Foundation, 1998). However, even their most ardent admirers realize that the CDC success story has not been completely without fault. To critics, their performance has been patchy and they are indicted as responsible for the corporate takeover of local communities. It is these criticisms that we will now address.

There is wide agreement that the problems of urban poverty, segregation and lack of access to affordable housing are worsening (Briggs, 2005). At best, CDCs are small-scale responses to housing and economic problems that have national and global roots. The housing output of CDCs in the 1990s was 0.07 percent of housing need (Twelvetrees, 1989). Stoecker states that the pace of capital disinvestment together with federal cutbacks make it impossible for CDCs to respond effectively (Stoecker, 1996). The increasing array of tasks undertaken by CDCs together with the need to access funding from a wide variety of funders leaves them stretched and exposed. Macro policy shifts and short-term budget increases, such as those emanating from the 1990 Affordable Housing Act, place an unfair burden on CDCs to do even more from a weak base. The view is that organizations are set up to fail.

The second problem with CDCs goes back to their messy conception and birth during the 1960s. Access to private sector funds, together with corporate representation on CDC boards, results in a compromise of radical community mobilization roots. The dependency on external funding is not unfamiliar to most voluntary sector organizations. However, given the specific anti-racist and tenant-controlled history of many community organizations before they became CDCs, the shift leads to many becoming simply sub-contractors of government. Thus, CDCs are 'based in communities but not community based' (Stoecker, 1996, p. 7). Some critics have argued the growth of private sector influence and funding of CDCs has led to the focus on supply-side solutions, such as job training, business creation and retail development, rather than the demand-led solutions, such as capacity building and community mobilization (Twelvetrees, 1989, p. 170; Filner, 2001, p. 7). Representation on boards of management of CDCs often consists of stakeholder interests from foundations and local authorities. This helps to satisfy financiers that loans and investments are secure but can lead to perception that the CDC represents outside interests ahead of the community in which it is based.

The third problem to highlight is the danger of community displacement and market disorganization. Competitive funding regimes may result in CDCs competing with each other to develop housing opportunities and create jobs. In this situation larger CDCs are much more likely to succeed and squeeze out smaller operators. Creating opportunities for low-income shared ownership and job programmes may result in the economically mobile leaving communities and further reducing social capital. Increased dependence on external funders reduces the scope and inclination to participate in empowerment and historical advocacy roles (Stoecker, 1996) as CDCs are unlikely to be supported to carry out such initiatives. In addition, these activities may be viewed as too radical and critical of the political and financial establishment to deem support.

Increased dependence on private sector and state funders, professionalization required to demonstrate ability to manage programmes and staff and boards that are out of touch with local people are some of the criticisms that have been made about the role and remit of CDCs in the US. As we will now see, these have an echo when we consider the evolution and performance of BME housing associations in the UK.

Black and minority ethnic housing associations: growth, consolidation and the struggle for survival

The origins of the black and minority ethnic housing movement may be traced to the sharpening anti-racist struggles of the 1970s and 1980s (Sivanandan, 1982). Increasing controls and restrictions on immigration, commencing with the 1962 Commonwealth Immigration Act, combined with discriminatory practices within housing and related markets, led to black and minority ethnic communities being concentrated in some of the poorest neighbourhoods and suffering multiple disadvantage (Social Exclusion Unit, 2000; Hall, 1987). Local authorities were slow to respond to the deepening disadvantage of local black and minority ethnic communities until the 1980s. Solomos suggests two events which helped to lead to local interventions of race equality. First, the Section 71 Amendment to the 1976 Race Relations Act placed a duty on local authorities to eliminate unlawful discrimination and promote equality of opportunity (Solomos, 1993, p. 104). Second was the catalytic impact of the 1981 riots in Liverpool, Birmingham and London between mainly groups of black and minority ethnic youth and the police (Benyon and Solomos, 1987). These national events shaped local policy responses and led black political caucus groups to campaign for increased race equality and radical left-wing local authorities to widen equal opportunities to include the active support of black and minority self-help organizations.

Lobbying and campaigning on the need to establish BME housing associations should be placed in this context of legislative changes, urban disturbances and public mobilization. The decision by the Housing Corporation in 1986 to establish a Black and Minority Ethnic Housing Policy to 'encourage, run and create separate black-run organizations as a channel for providing rented housing' (Harrison, 1995, p. 88) was a result and response to some of these external pressures. In addition, we should also note that the housing research highlighted the failure of white-led housing associations to meet the needs of black and minority ethnic tenants in terms of providing access to housing and providing a culturally sensitive service (Harrison, 1995, p. 89). That the new policy was initiated during the second Thatcher government, when public expenditure was being substantially reduced and local authorities attacked by a number of policies (such as the right of tenants to buy their council houses), shows the effectiveness of campaigning and the pressing need of government to intervene in the deepening crisis within cities.

The definition of a BME housing association was first used in 1992 and has not changed since. The Housing Corporation stated that these are housing associations designed to meet the needs of black and minority ethnic communities and who have a management committee of which 'at least 80 percent are from ethnic minority communities . . . African, Asian, South-East Asian ethnic or racial origins' (Housing Corporation, 1992, p. 1). Difficulties with adopting a representative view of participation and control has led to problems for BME housing associations working with increasingly diverse communities.

It should also be noted that recent research in the UK on black and minority ethnic involvement in housing stock investment decisions shows that accountability and involvement of communities should not be viewed simply through the lens of representation on a board or management committee. Capacity building, formal and informal methods of involvement as well as mapping communities were essential to build trust and participation (Mullins et al., 2004).

Housing Corporation support to BME housing associations took place in three phases. We have already commented on the first 5-year Policy between 1986 and 1991. This is marked by direct grants and registration of 44 BME housing associations which helped these organizations to develop, build and manage housing and increase access for black and minority ethnic tenants to good-quality, low-cost housing. Moreover, organizational development created employment and management opportunities for black and minority ethnic practitioners to lead new bodies and become committee members. This phase may be characterized as focused on empowerment, meeting needs and increasing the representation of black and minority ethnic people in the voluntary housing sector (Harrison, 1995, p. 91).

The Housing Corporation continued its support to BME housing associations in its second 5-year plan from 1991–1996. The theme – *An Independent Future* – emphasized the importance of working with BME housing associations to secure a financially viable future. Additional monies from Housing Corporation capital and revenue budgets were invested in BME housing associations but were within a context of declining housing budgets to support housing association development. Importantly the 1988 Housing Act led to housing associations needing to secure increasing amounts of development funds from the private sector. This meant higher rents for tenants to pay for housing association projects and less focus on resource-intensive initiatives, such as urban renewal and rehabilitation of housing stock.

The problem for BME housing associations with private funding is that private lenders need reassurance from housing associations that loans will be repaid. Asset value, management capacity and robust governance are indicators that are taken into consideration when agreeing to a loan. Most BME associations were established after 1988 and were small both in terms of size and management skills (Royce et al., 1996). Consequently they were seen as a high risk.

The Housing Corporation began to review the extent to which BME housing associations could have a viable future (Harrison, 1995, p. 93). This took the form of grouping associations, which could become viable and receive Housing Corporation support to build houses. Inevitably these were the larger BME associations which began expanding their operations to cover wider parts of the country. This led to smaller, grassroots BME housing associations being 'muscled out' from communities where they had worked for many years. Apart from increasing conflict within the BME housing sector, the new funding regimes showed that the need to access support from private financiers meant shedding the radical roots of BME housing associations. Community mobilization, in a pattern not dissimilar to CDCs, was sacrificed at the altar of supply-side solutions to rescue disadvantaged urban communities.

The final Housing Corporation Policy – *An Enabling Framework* – from 1998–2003 is embedded within a consumerist approach. The focus is not on meeting the needs of BME housing associations but helping to empower black and minority ethnic housing tenants of all housing associations 'towards ensuring that the expectations of black and minority ethnic communities are achieved, whether through the provision of adequate social housing by the full-range of landlords (only some of which will be black-led) or through enabling members of the community to effectively participate in the delivery of services, through equality in the workplace, in management and in board membership' (Whitehead et al., 1998, pp. 5–6).

The approach taken emphasizes the important role that white-led housing associations should play in having a workforce and board that is representative of the diverse communities which they serve. In this respect, the Housing Corporation can use its regulatory powers to intervene, suspend staff and withdraw financial support to ensure that policies are being followed. This has not being implemented explicitly, but the profile of race equality has nevertheless been raised. Though the funding regime during the final policy does not seek actively to support the registration of new BME associations to the same extent as in 1986–1991, the race equality agenda has a high profile within the sector.

The catalytic impact of the Macpherson Report (1999) into the murder of the black teenager, Stephen Lawrence, led to changes in the landmark 1976 Race Relations Act which emphasized the positive duty of all publicly funded bodies to promote race equality. The diversity agenda was further supported by the National Housing Federation led Race and Housing Inquiry (2001), which demonstrated that there was much work to be done by the social housing sector in meeting the needs of black and minority ethnic tenants as well as providing opportunities for black and minority employees to reach their potential in housing associations. It is important to reiterate that the discussions were focused on governance and business performance rather than actively supporting a programme that led to BME associations. These are 'softer' areas which ask for organizational reconfiguration as opposed to resources to empower black and minority ethnic organizations and communities.

Housing Corporation policy shows a declining interest in supporting BME housing associations and a shift to focusing on the needs of black and minority ethnic housing consumers. The Housing Corporation would argue that, whilst registering 62 BME housing associations was a considerable achievement with support of over £500 million capital budget, it was equally important to address the role of over 2000 registered housing associations in meeting race equality targets.

Reviewing the performance and future role of BME housing associations has been the focus of recent research (Lupton et al., 2004). There is little doubt that the BME housing sector has helped to create opportunities for black and minority ethnic housing practitioners in new jobs and at management committee levels, created a growing asset base in terms of housing stock owned and managed by associations and levered in resources for economic and social development. As a sub-sector, these achievements mark it out as being one of the most successful examples of BME enterprise development in the UK (Housing Corporation, 1998). Against this is a raft of problems and issues that now need to be considered.

BME housing associations are young, small organizations that are vulnerable to changes in policy and economic climate. Most notably these have been rent regulations imposed by the Housing Corporation to counter the high rents charged by housing associations. The policy of keeping rent increases to 1 percent above the rate of inflation hit much harder those associations that did not have generous financial reserves. BME associations do not have the ability to cross-subsidize rents from surpluses generated from previous development. Rents are higher than average and this makes BME housing associations vulnerable to the charges of creating barriers to tenants who may wish to work but cannot afford high rents (Whitehead et al., 1998). The combined drive of the Housing Corporation (reduce rents) and the government (address social exclusion and benefit dependency) sharpens the scrutiny on BME associations.

The development role for housing associations is becoming increasingly complex and expensive. The capacity of BME housing associations to undertake development work is under strain. Research has shown that partnership arrangements with larger white-led associations often become unequal with less role and remit given to BME associations (Whitehead et al., 1998, p. 4). Sometimes the need to buy in skills and expertise to support these projects leads to capital and resources leaking from the communities in which associations are supposed to be investing. There is a danger that the development trend for most BME associations will reduce and the role of building houses for low-income black and minority ethnic communities will be lost.

The reforming race equality drive of BME housing associations can also be seen conversely as a casualty of the higher profile of the housing association sector as a whole. Legislation and inspection have reduced the scope for housing associations to opt out of race equality initiatives. Some actively promote programmes to meet the needs of black and minority ethnic staff and tenants. Increasing opportunities for black and minority housing tenants was a key part of the most recent Housing Corporation Policy and white-led associations house far more people from these groups than their BME peers. Commitment to race equality and changing organizational culture leads to questions on the need, identity and sustainability of BME housing associations (Lupton et al., 2004, p. 20).

Comparison between community development corporations and black and minority ethnic housing organizations

Having discussed the CDC and BME housing sector, their growth and problems, we are in a better position to discuss key issues and a new agenda for action. In the US, and perhaps to a greater extent the UK, the policy shift has been to focus on communities *per se* rather than race equality. This should be seen as part of the wider reconfiguration of national, state and local governments. Social exclusion, residualization and social capital are concepts that use race equality as a proxy. The changes in Housing Corporation policy from 1986 to 2003 can be contextualized as a retreat from empowerment and focus on management and performance (Harrison, 1995).

It could be argued that both CDCs and BME housing associations have had to 'buckle down' within an operating environment that has seen significant flow of support from private sector sources and increasing inspection by regulatory bodies. The need to develop skills and capacity leads to changes in organizational make-up and culture. Grassroots mobilization on community politics becomes less important than charting financial projections and meeting growth targets. Radicalism is necessarily diluted.

Both organizations tend to operate in weak markets where there is growing residualization, polarization and reduced social capital. Building successful businesses in such an environment would be difficult, but may be more so for organizations that have to meet targets of job creation, housing lettings and tenant participation. This may be further compromised as they are being seen less as a community advocate and simply a provider of contracted services (Stoecker, 1996). In the UK, many BME housing associations are located in areas of low and declining demand (Lupton et al., 2004, p. 20) where priorities are not so much on increasing the housing stock but on renewal and regeneration. For small organizations, taking on the results of national and global restructuring (and finding answers) may prove difficult.

The transformation of organizations from grassroots bodies responding to the needs of communities to professional institutions meeting the demands of the private finance sector leads to a debate on representation and involvement. Both CDCs and BME housing associations enjoy stating that they enable black and minority ethnic communities to be involved in decisions about their neighbourhood. The pioneering CDC in Bedford-Stuyvesant included representatives from large foundations, finance houses and national politicians. Some in the CDC sector saw their organizations as based on tenant and community participation, community builders and advocates for the people (Twelvetrees, 1989, p. 179). However, working within a government-defined policy framework and securing increasing amounts of funding from the private sector leads to these principles being compromised.

BME housing associations appear to be following the same route as CDCs in terms of funding and responding to an externally defined policy framework. Representation on management committees does include black and minority ethnic communities but typically these are neither tenants nor local community activists. These organizations are advised by the Housing Corporation to develop robust management systems which tend to favour the inclusion of legal and finance experts as much as tenants and community activists. Nor are BME housing associations always led by members of the black and minority ethnic community. Three out of 60 registered BME housing associations recruited a white chief executive and many more have white people at senior levels within management teams. Forty percent of tenants of BME housing associations are white (Housing Corporation, 1998).

Despite this, it is clear that some BME housing associations still cherish their activist roots. For example one chief executive states, 'As long as there continues to be social exclusion, racism, discrimination and institutional racism, the role of BMEs is paramount, representing the grassroots of BME communities and meeting unmet needs' (Lupton et al., 2004, p. 22). This rhetoric has not kept pace with the reality of reforms in housing and finance which we have discussed above. The argument also confuses discussion on the sensitive areas of representation, accountability and involvement. Housing association board members have legal responsibility for the due care of their organization rather than representing any specific interest group.

Taking this further, it is unrealistic for board members to represent BME communities in areas of increased diversity. Recent migration from Eastern Europe, Somalia and Iraq leads to conceptually 'new' black and minority ethnic communities who will have very little in common with 'old' black and minority communities from the Caribbean, India and Pakistan. This diversity has been further deepened by labour mobility and asylum dispersal. Language barriers, increased awareness on 'new' communities and accommodating the jostling for power, access and general community politics present barriers to representation. BME housing associations were formed through lobbying from activists drawn from 'old' communities. Tenants are consumers. A BME association board needs to reflect and not simply represent their interests, which may in itself become an unattainable goal within an increasingly fragmented community (Mullins et al., 2004, pp. 16–18).

Some members of the BME housing sector acknowledge that BME associations cannot fully represent the needs of tenants and communities and are working with 'first force' organizations. These are embedded within communities, display high levels of participation and may not be linked to housing. Accountability lines are short, they are in touch with grassroots opinion and less beholden to external funders (Mullins et al., 2004, p. 27). Alliances between BME housing associations and radical 'first force' organizations to provide housing and meet community needs help to defragment the confusion on representation and accountability. The same issues may also be applicable to CDCs (Stoecker, 1996).

Towards a conclusion and agenda for action – the rise of CDCs and BME associations?

We have attempted to compare the role, performance and problems of CDCs in the US with BME housing associations in the UK. This is located within a welfare regimes approach popularized by Esping-Andersen. There appears to be a level of convergence between housing systems. Policy is focused on increasing social capital within residualized neighbourhoods, privatization of municipal housing stock and reduced role of the state in the provision of goods and services but sharper roles and responsibilities in terms of citizenship. Organisationally CDCs and BME housing associations have been encouraged to use their community skills and entrepreneurism to strengthen weak markets by providing good-quality housing, rebuilding social capital and participating in wider neighbourhood renewal strategies. As the state continues to withdraw from service provision and increases the role of private capital as both funders and partners of low-income housing development, black and minority ethnic communities may be asked to become active citizens in return for narrowly prescribed welfare benefits. However, government interventions such as the establishment of the Social Exclusion Unit in 1997 and the publication of the Communities Plan in 2003 demonstrate that the UK has the capacity to revert back on occasions to a generous and expansive welfare regime. The expectation is that government will continue to respond to urban crisis such as the 1981 riots or the 1999 Macpherson Report but the trend will be towards organizations operating within a residualized welfare state.

Have CDCs and BME housing associations been shaped by their external policy environment? The emergence of both organizations took place within liberal welfare regimes. Indeed, some have claimed their relative success as an entrepreneurial triumph which has been facilitated by a retreating state. They have not been 'crowded out'. To an extent both CDCs and BME housing associations owe their growth and consolidation to macro policy drivers by the federal government and the Housing Corporation respectively. They have responded to a favourable set of policy and funding mechanisms which coincided with public policy interest in race equality and urban disadvantage. However, CDCs continued to grow despite severe cutbacks in federal aid during the 1980s whereas the number of BME housing associations registered slowed down during the second Housing Corporation Policy of 1996–2001, and especially after the 1998 Housing Act which emphasized the role of private finance in supporting housing development. More research is required to come to a conclusion about this growth disparity but we could suggest that the multi-faceted and entrepreneurial nature of CDCs allowed them to absorb financial cutbacks better than BME housing associations that were largely dependent on Housing Corporation funding. This, together with the fact that CDCs have been based in communities for nearly 40 years, as opposed to the relative youth of BME housing associations of 20 years, help to explain the organizational independence of the former to external policy drivers.

An agenda for action could be put forward for CDCs and BME housing associations. This should recognize that the goals of community representation and advocacy may be unattainable within increasingly diverse and fragmented communities. Accountability and communication with communities may be far more realistic than representation at board level and help to clarify roles and responsibilities. Both organizations could consider splitting community mobilization and developmental roles, with the focus on the latter where skills and expertise can be used to support low-income housing. Intra-community alliances with grassroots organizations as well as inter-community partnerships with other organizations may help to meet the housing and related needs of consumers more effectively. It may be appropriate for CDCs and BME housing associations to consider that their role as agents of community mobilization, advocates on behalf of poor communities and champions of race equality needs to adapt to the demands of entirely different circumstances, communities and policy drivers now, and perhaps in the future.

References

Benyon, J. and Solomos, J. (eds) (1987) *The Roots of Urban Unrest*. Pergamon Press, Oxford.

Bratt, R. et al. (1994) *Confronting the Management Challenge: Affordable Housing in the Nonprofit Sector*. Community Development Research Center, New School for Social Research, New York.

Briggs, X. (2005) More Pluribus, Less Unum? Changing the Geography of Race and Opportunity. In: Briggs, X. (ed.) *Metro Dilemma: Race, Housing Choice and Opportunity in America*. Brookings, Washington DC.

Emerson, C. et al. (2003) *A Survey of Public Spending in the UK*. Institute of Fiscal Studies, London.

Esping-Andersen, G. (1990) *The Three Worlds of Welfare Capitalism*. Polity Press, Cambridge.

Filner, M. (2001) *Community Development Corporations: An Historical Overview*. Center for Democracy and Citizenship. Research Case Study Series, University of Minnesota.

Ford Foundation (1998) *Seizing Opportunities: The Role of CDCs in Urban Economic Development*. Ford Foundation, New York.

Hall, S. (1987) Urban Unrest in Britain. In: Benyon, J. and Solomos, J. (eds) *The Roots of Urban Unrest*. Pergamon Press, Oxford.

Harrison, M. (1995) *Housing, 'Race', Social Policy and Empowerment*. Avebury Press, Aldershot.

Home Office (2001) *Community Cohesion: A Report of the Independent Reviewing Team*. HMSO, London.

Housing Corporation (1998) *Black and Minority Ethnic Housing Policy*. Housing Corporation, London.

Housing Corporation (1992) *An Independent Future: Black and Minority Ethnic Housing Association Strategy 1992–1996*. Housing Corporation, London.

Lupton, M. et al. (2004) Serious players. *Inside Housing* January: 20–22.

Macpherson, W. (1999) *The Stephen Lawrence Inquiry: Report of an Inquiry by Sir William Macpherson of Cluny, Cm 4262-1*. HMSO, London.

Mullins, D. (2000) Social origins and transformations: the changing role of English housing associations. *Voluntas* **11** (3): 255–76.

Mullins, D., Beider, H. and Rowlands, R. (2004) *Empowering Communities, Improving Housing: Involving Black and Minority Ethnic Tenants and Communities*. Office of the Deputy Prime Minister, London.

National Affordable Housing Management Association (2004) *Resource Information*. NAHA www.nahma.org

National Congress for Community Economic Development (1995) *Tying it All Together: The Comprehensive Achievements of Community-Based Development Organizations*. National Congress for Community Economic Development, Washington, DC.

Race and Housing Inquiry (2001) *Challenge Report*. National Housing Federation, London.

Royce, C. et al. (1996) Financing black and minority ethnic housing associations. In: *Findings*. Joseph Rowntree Foundation, York.

Salamon, L. and Anheier, H. (1998) Social origins of civil society: explaining the non-profit sector cross-nationally. *Voluntas* **9** (3): 213–48.

Sivanandan, A. (1982) *A Different Hunger*. Pluto Press, London.

Social Exclusion Unit (2000) *Minority ethnic issues in social exclusion and neighbourhood renewal*. HMSO, London.

Solomos, J. (1993) *Race and Racism in Britain*, 2nd edn. Macmillan, London.

Stoecker, R. (1996) *The Community Development Corporation Model of Urban Redevelopment: a Political Economy Critique and an Alternative*. Revised paper first presented to 1995 American Sociological Association Annual Meeting.

Twelvetrees, A. (1989) *Organising for Neighbourhood Development: A Comparative Study of Community Development Corporations and Citizen Power Organisations*. Avebury Press, Aldershot.

US Census Bureau (2001) *2000 Statistical Abstract of the United States*. US Census Bureau.
Van Vliet, W. (ed.) (1997) *Affordable Housing and Urban Development in the United States*. Urban Affairs Annual Reviews 46. Sage, London.
Venkatesh, S. (2000) *American Project: The Rise and Fall of the Modern Ghetto*. Harvard University Press, Cambridge.
Vidal, A. (1992) *Rebuilding Communities: A National Study of Urban Community Development Corporations*. Community Development Research Center, New School for Social Research, New York.
Whitehead, C. et al. (1998) Rents, viability and value in black and minority ethnic housing associations. In: *Findings*. Joseph Rowntree Foundation, York.
Wilson, W. (1980) *The Declining Significance of Race*. University of Chicago Press, Chicago.

Conflict, Participation and Power

11

Housing, Regeneration and Change in the UK: Estate Regeneration in Tower Hamlets, East London

Stephen Hall

Introduction

This chapter presents and contextualizes empirical case studies of two differing approaches to estate regeneration in the London Borough of Tower Hamlets: the Tower Hamlets Housing Action Trust (HAT), one of six HATs in England; and Poplar HARCA, a Local Housing Company established with Estate Renewal Challenge Fund (ERCF) resources. In presenting these case studies, the chapter will provide a brief history of estate regeneration in England, and attempt to link the case studies to on-going debates in this domain. The important and continuing role of physical regeneration (despite the rhetoric of 'holistic' renewal), community involvement and stock transfer will be noted. It will be argued that a consistent characteristic of estate regeneration policy during the past two decades has been an ambiguity concerning the objectives and intended beneficiaries of intervention. This ambiguity can result in unsustainable or even perverse outcomes for regeneration programmes. It will also be argued that estate-focused initiatives overlook the crucial fact that socio-economic peripherality is more than simply a function of geography. In respect of all these key questions, the Labour government of Tony Blair, elected in 1997, represents a strong continuity with, indeed consolidation of, past trends.

Estate regeneration policy in England

As in most European countries, large social housing estates were constructed in England during the post-war period in response to housing shortages prompted by war damage and demographic change (Van Kempen et al., 2005). These estates proved initially to be popular, state-of-the-art, mass housing for working-class households. During the late twentieth century, they became the subject of a variety of problems: physical decay plus social and economic crisis. These problems have been compounded, in the UK, by the 'residualization' of social housing (Van Kempen et al., 2005). The social profile of the client base of social housing has narrowed dramatically. Private home ownership has become increasingly attractive and drawn middle-income groups away from social housing. The resident base of large estates has, thus, become characterized by a concentration of low-income households, experiencing multiple social and economic disadvantage (Van Kempen et al., 2005). In response, the estates have become the focus of significant state-led regeneration activity.

The aims and activities of the regeneration of social housing estates in England have undergone numerous changes in the past two decades. Initially, the problems of estates were not interpreted in a structural social and economic context. Improved local housing management was the key priority, as in the Priority Estates Project, launched in 1981.

In the 1980s, the mainstream urban regeneration policy of the Thatcher government was dominated by a physically deterministic approach, designed to create profitable investment opportunities for the private sector on former industrial sites, as part of a broader strategy to engender an 'enterprise culture' in the inner city through initiatives such as Urban Development Corporations and Enterprise Zones (Deakin and Edwards, 1993; Thornley, 1993). Parallel programmes for social housing estates were developed, including Estate Action or 'EA' (Pinto, 1991, 1993) and Design Improved Control Experiment or 'DICE' (Coleman, 1985), which prioritized physical 'bricks and mortar' style housing regeneration.

In the 1990s, a more 'holistic' approach to regeneration, 'Challenge Funding', that sought to reconcile social and economic questions and to promote multi-sector partnership working and community involvement, was advocated by the Major administration (Oatley, 1998; Hall and Mawson, 1999). Estate regeneration became incorporated into generic, area-based policies such as City Challenge (Atkinson and Moon, 1994; Oatley, 1995) and the Single Regeneration Budget or SRB (Hall, 2000; Rhodes et al., 2003), which addressed economic development, employment, ethnic minority needs, crime, environmental improvements and community infrastructure, as well as housing problems. Large-scale estate-based physical regeneration was the exception rather than the norm within these programmes (Nevin et al., 1997).

From the mid 1990s, a parallel estate-based regeneration focus was introduced, orientated around the transfer of local authority housing stock to housing associations, which permitted significant levels of private finance to be secured for estate regeneration, in addition to measures to tackle broader economic and social issues such as unemployment and crime (HACAS Chapman Hendy, 2002). The Conservative government was an active supporter of stock transfer as a vehicle for the de-municipalization of social housing as well as achieving regeneration objectives. The 1985 and 1988 Housing Acts permitted local authorities to dispose of their stock and to use the capital receipt to invest in housing improvement (NAO, 2003). Stock transfers were encouraged through Housing Action Trusts (HATs) since 1992, Large Scale Voluntary Transfer (LSVT) since 1988, and the Estate Renewal Challenge Fund (ERCF) since 1995.

HATs were introduced in the Housing Act 1988 and evolved into a more inclusive, holistic regeneration scheme during the years of the Major administration (Evans and Long, 2000; Tiesdell, 2001). HATs are 'non-departmental public bodies' (NDPBs) which assume responsibility for the regeneration and management of disadvantaged estates deemed beyond the modest resources and competencies of local authorities. The initial transfer and choice of post-HAT landlord are sanctioned by tenant ballot (Tiesdell, 2001). Six HATs were designated in 1992/3 in Birmingham, Hull, Liverpool and London (Stonebridge, Tower Hamlets, Waltham Forest). The HATs received government grants of up to £260 million (Tiesdell, 2001), far larger than is typical for regeneration schemes in England. Ostensibly HATs were agencies to develop estates in a 'holistic' fashion. However, the bulk of their efforts have been channelled into physical housing regeneration, involving very substantial demolition and rebuilding programmes (Evans and Long, 2000; Tiesdell, 2001). HATs were also given powers to facilitate the provision of local amenities and provide support to local business (Evans and Long, 2000).

By 2003, local authorities had entered into transfer agreements resulting in the transfer of over 740,000 dwellings, or 18 percent of the total local authority stock (NAO, 2003). The majority (i.e. 90 percent) of early transfers were self-financing, located in southern England, and pursued through LSVT (NAO, 2003). The ERCF provided a government dowry to cover the negative value of the housing stock in disadvantaged areas. The National Audit Office estimates that stock transfers had raised £11.6 billion of private finance (NAO, 2003). These figures must be interpreted in the context of an estimated repair backlog of £19 billion nationally (NAO, 2003).

In the mid 1990s, much media and government debate focused on the problems of the 'worst estates' (e.g. 'joy-riding', concentration of single parents). Indeed, in 1997, newly elected Prime Minister, Tony Blair, chose to make his inaugural speech from the Aylesbury Estate in London. The approach of the Labour government to estate regeneration is strongly influenced by its broader political project, the so-called 'Third Way' (Tiesdell and Allmendinger, 2001; Hall, 2003). This project has resulted in surprisingly little change to mainstream housing policy (Marsh and Cowan, 2001) but a renewed emphasis on social and spatial inequalities. It thus reversed the decline of England's cities through measures introduced in the first 'urban' White Paper for a quarter of a century (DETR, 2000), tackling the problem of social exclusion locally, through the National Strategy for Neighbourhood Renewal (SEU, 2001), and tackling the contrasting demand of low housing demand in parts of the north and Midlands and housing shortages in London and the south, plus promoting social mix, through the Sustainable Communities Plan (ODPM, 2003). The defining features of the Labour government's approach may be summarized as follows: 1) an emphasis on supply-side measures, such as improving employability, housing, neighbourhood management and public service provision at the neighbourhood level; 2) local intervention in which responsibility for enhancing social inclusion is devolved to a district and neighbourhood level, albeit in the context of national guidance and targets; 3) the pursuit of a co-ordination dividend through 'joined up' working by local partnerships; and, 4) a 'bootstraps' approach to renewal, inspired by a renewed interest in ideas of local community (Etzioni, 1993), self-help and mutuality as a vehicle to build and exploit 'social capital' to tackle area deprivation (Forrest and Kearns, 2001).

Stock transfer is also an important element of the Labour government's objective of ensuring that all social housing is of a suitable standard by 2010 (NAO, 2003). The government have disbanded ERCF and transferred finance to the Housing Investment Programme (HIP), which will support the transfer of up to 200,000 dwellings annually (Murie and Nevin, 2001).

Tower Hamlets Housing Action Trusts and Poplar Housing and Regeneration Community Association

Introduction

The Borough of Tower Hamlets is located immediately to the east of central London. Tower Hamlets represents clearly the paradoxes of London's 'world city' economy with its concentrations of social problems adjacent to the world-class real estate of the City of London and Docklands (Fainstein et al., 1994; Buck et al., 2002). Tower Hamlets was recognized in the government's Index of Multiple Deprivation (2000) as the most disadvantaged local authority area in England (average of ward level averages). All Tower Hamlets' 19 wards were in the most deprived 10 percent of wards nationally and 17 in the most deprived 5 percent. However, the borough has also been characterized by extreme levels of housing demand. The average ratio of house prices to household income, at 5.27 (Wilcox, 2003), is the

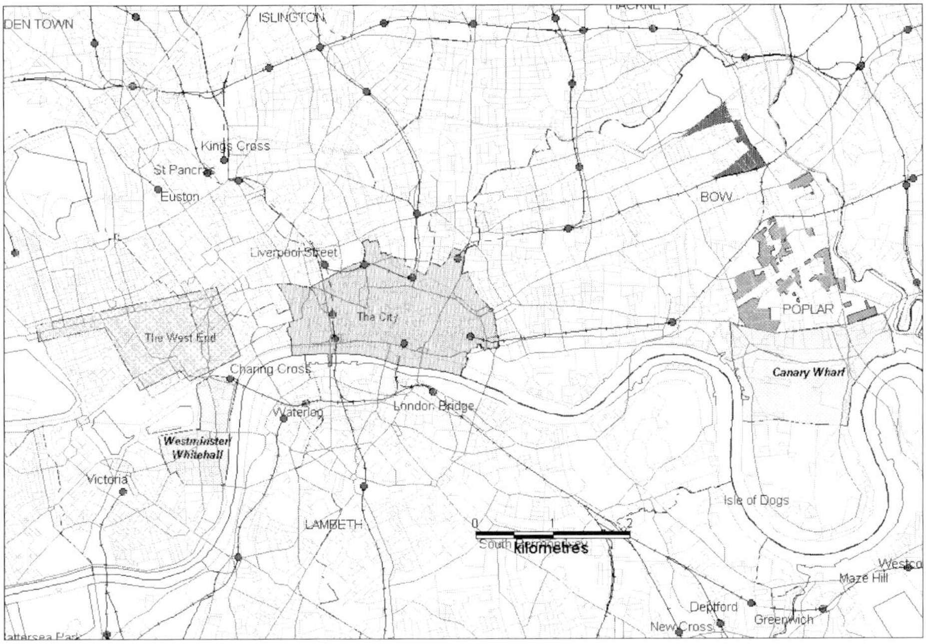

Figure 11.1 Location of Bow and Poplar.

15th highest of any local authority area in England. This compares to an average ratio in London of 4.8 and a national average of 3.4 (Wilcox, 2003). Local residents are effectively disenfranchised from the borough's housing market.

The case study areas, Bow and Poplar in the east of Tower Hamlets, are traditional working-class neighbourhoods on the southern periphery of the declining industrial sub-region known as the Lea Valley. However, the adjacent sub-region is employment rich; to the south is Docklands; to the east, in the neighbouring borough of Newham, is development associated with the Channel Tunnel Rail Interchange and 2012 Olympic Stadium and Village; to the west is the City of London. The location of the case study estates is indicated in Figure 11.1.

A basic statistical profile of London, Tower Hamlets, Bow and Poplar is provided in Table 11.1. The borough has a younger age profile, more significant black and minority ethnic population, lower employment and higher unemployment rates and a far higher proportion of social housing and flatted accommodation than is the norm for London (itself atypical in these respects) or England and Wales. These traits are even more marked in the case of Poplar. Bow has a slightly older age profile and less significant black and minority ethnic population, but a higher proportion of social housing and flatted accommodation than is the norm for London and Tower Hamlets.

Table 11.1 Profile of case study estates.

Area	Population less than 16 years old (%)	Population more than 75 years old (%)	Population of white origin (%)	Population of BME origin (%)	Population employed (%)
Bow	19.2	6.4	68.4	31.6	50.3
Poplar	30.4	3.9	41.5	58.5	39.0
Tower Hamlets	22.9	4.0	51.4	48.6	49.1
London	20.2	5.9	71.2	28.8	60.2
England/Wales	20.2	7.6	90.9	9.1	60.6

Area	Population unemployed (%)	Housing: owner-occupied (%)	Housing: social rented (%)	Housing stock: houses (%)	Housing stock: flats (%)
Bow	7.3	22.9	64.6	14.4	84.7
Poplar	8.1	16.8	74.1	10.5	88.5
Tower Hamlets	6.6	29.0	52.5	16.3	82.1
London	4.4	56.5	26.2	51.1	46.9
England/Wales	3.4	68.9	19.2	80.4	18.0

Source: Census 2001.

Bow is located in the north-east of Tower Hamlets. Much of the housing stock outside the large post-war social housing estates was built during the Victorian period and is heavily gentrified. The Bow estates are inhabited predominantly by white working-class residents. This is a product of past housing allocation policies in which white families tended to be allocated to the most modern housing, in the east of the borough.

Three estates (Lefevre, Monteith, Tredegar), comprising 1585 dwellings, mostly high-rise units built in the 1960s and 1970s and formerly belonging to the London Borough of Tower Hamlets (LBTH), were transferred to Tower Hamlets HAT ('THHAT') in 1993 (following a ballot of tenants). An earlier proposal to designate a HAT for the Ocean Estate in Stepney (now the site of a New Deal for Communities programme) was rejected by tenants.

The objectives of THHAT were to bring about lasting regeneration of the area, empower residents to make informed decisions about their housing choices and give them a stake in the community by: building new homes or improving existing homes for all residents; ensuring effective housing management; providing a wider choice of tenure and forms of home ownership; and improving the social, living and environmental conditions in its area through improved job opportunities and youth and community facilities.

The majority of the HAT's efforts, in practice, have been devoted to housing renewal. This is reflected in the proportion (80 percent) devoted to housing development and management. Almost all the pre-existing stock has now been demolished, 1000 new homes built, in a low-rise traditional 'neighbourhood' format, and a further 150 improved.

In 1995, the HAT estimated the lifetime cost of its development programme to be £175 million. However, in 1996, the government fixed its lifetime grant at £123 million. The HAT was not permitted to borrow additional resources to make good the shortfall. It therefore entered into partnership with Circle 33 Housing Association and jointly established Old Ford Housing to complete the remainder of the physical regeneration programme and manage the new stock.

In 1998, the HAT established a Community Development Trust, Bow People's Trust (BPT), to undertake community and economic regeneration. The HAT committed to provide £4.4 million in grant in aid to BPT, over a 7-year period, to be supplemented by other sources (e.g. National Lottery, European Social Fund). The HAT has more recently become engaged in a wider spectrum of activity in partnership with the local authority and other bodies, in the context of the National Strategy for Neighbourhood Renewal.

Poplar is located in the east of Tower Hamlets, immediately south of Bow and north of Docklands. Its diverse land use consists of light industrial premises interspersed by numerous small social housing estates and intersected by trunk roads, canals and railways. The neighbourhood, due to its proximity to the Port of London, has historically been associated with immigrant communities.

The area has been unique in east London for its 'monolithic' social housing stock. There has been very little private housing in the Poplar neighbourhood during the post-war period. This position has been changing since the opening of the Docklands Light Railway (DLR), which has prompted considerable pressure for new private housing developments adjacent to DLR stations in Poplar.

Poplar Housing and Regeneration Community Association (HARCA) was the first urban housing company to be established through the ERCF. Seven diverse and widely dispersed estates, comprising 6360 units formerly the property of the London Borough of Tower Hamlets, were transferred to Poplar HARCA (PH) from 1997 following tenant ballots. The new organization has seven key objectives: to encourage and develop investment, develop partnerships, develop effective neighbourhood and housing services, make Poplar safe, encourage local participation, fulfil residents' potential through education and training, and maximize resources.

The physical regeneration programme (now complete) involved the demolition of some 10 percent of the stock and the refurbishment of the remainder using a £53 million dowry from the ERCF, plus £91 million of private finance. PH's economic and social regeneration budget has been £1 million per annum, raised from rent income and other sources. Each estate has a HARCA centre which is a multi-purpose building used for a variety of social and economic functions.

PH is an active networker and has sought to develop a role in managing neighbourhoods in the Poplar area outside the estates for which it is the principal landlord. Its key partners are Leaside Regeneration, an urban regeneration company, funded via SRB Round 4 resources and set up in 1999 to promote the physical and economic regeneration of the north-east of Tower Hamlets and west of the adjacent borough, Newham, and the Bromley by Bow Centre (BBBC), a community organization which runs a variety of health, education and training, enterprise and family support projects. In partnership with Leaside Regeneration and BBBC, PH has established the Poplar Area Network (PAN), which subdivides the Poplar area into five discrete neighbourhoods. Partners include the principal local statutory, business, community and voluntary sector actors.

Housing and physical regeneration

The basic physical characteristics of the Bow estates at the time of HAT designation may be described as follows:

- Monteith, comprising three 22-storey tower blocks, four smaller blocks, plus sheltered accommodation and an old people's home, housed mainly an elderly population. The estate is conveniently located for Victoria Park, immediately to the north.
- Lefevre, comprising four 10-storey blocks forming a physical 'wall' between the adjacent urban motorway and the rest of Bow plus five smaller blocks and a community centre, had a mix of ages among its tenants and was in a reasonable state of repair at HAT designation.
- Tredegar, the newest of the three estates, comprising 16 medium- and four low-rise blocks plus a shopping and community centre block, had a younger population with a reputation for anti-social behaviour. The smaller units were of a traditional, rather than systems built, construction and have remained in good condition.

The designation of the HAT was essentially driven by the condition of the housing stock in the early 1990s. The estates had not been significantly improved since construction and had been the subject of failed Estate Action bids. Many blocks suffered from classic symptoms of obsolescence and decline. In design terms, the long corridors inherent to the deck access design of some blocks made convenient escape routes for muggers and burglars.

The HAT Masterplan described the housing component of the programme as a 'new start', the objectives of which were:

- The demolition of *all* the existing housing with the exception of a tower block containing 130 flats, a sheltered housing scheme of 30 flats and nine existing homes.
- The development of 1097 new homes, of which 49 percent were to be houses with gardens.
- Improvement of the remaining housing.
- The development of a further 60 homes as a mixed-tenure scheme in partnership with the private sector.

The HAT's housing renewal programme started in spring 1995 and was due to be completed in October 2003. However, a significant amount of work remained, to be completed by Old Ford, at the time of the HAT's exit in summer 2004.

The HAT's first new homes were completed in September 1996. These were built to a higher specification than could be achieved through Housing Corporation grants elsewhere. Allocation policy has also been more generous than elsewhere locally. All adults of 16 years and above are entitled to a double bedroom. HAT tenants are, therefore, privileged in a borough which suffers from some of the highest levels of overcrowding within the UK.

Not surprisingly, the HAT's most recent quality-of-life survey (THHAT, 2002) indicated a high degree of satisfaction with the newly built housing stock, especially size and number of rooms, the residents' choice process and the quality of the workmanship.

The HAT operated a generous allocation policy towards 'hidden households', especially tenants' children, within its existing resident population. The HAT has sought to keep families and communities together, and has seen re-housing hidden households as one way of achieving this. This is a contentious issue in the context of the acute levels of housing need that exist in Tower Hamlets. However, these new tenancies are a very small proportion of overall lettings, the majority of which have been direct applications to Old Ford for Assured Short Hold Tenancies (ASHTs), often to key workers, in stock earmarked for demolition.

The designation of the HAT essentially 'ring-fenced' the estates. There were no new allocations to the area following designation. This means that the demography of the estate has been 'frozen' during the past decade. It has therefore seen considerable physical renewal but very little social renewal. This is in contrast to Tower Hamlets generally, which has witnessed rapid social change, with gentrification in the private sector, and increased proportions of minority groups, especially the Bengali community, within the social housing sector.

The Poplar estates were built over a long period of time: during the 1930s (London County Council), 1950s (Poplar Metropolitan Council) and 1960s (Greater London Council). The estates are very diverse in terms of their age of construction, predominant dwelling type, design and location. Therefore, residents experienced a range of housing-related problems.

The basic physical characteristics of the estates may be described as follows:

- Aberfeldy comprises mainly three-storey flats plus some new houses. The estate suffers from extreme physical isolation and is tightly bounded by major trunk roads plus the River Lea.
- Bow Bridge and Coventry Cross comprise four- and five-storey flats built during the inter-war period. The estates have a large minority ethnic population.
- Burdett comprises mostly flats with some maisonettes and houses. It has a large Bangladeshi population.
- Lansbury comprises four- and eight-storey flats and maisonettes. A significant minority of the residents are pensioners. Part of the estate was built for the 1951 Festival of Britain. It has been described as one of the most popular and successful post-war housing schemes in London.
- Lincoln comprises flats and cottages. The resident population is the most diverse of the HARCA estates.
- Teviot comprises three-storey flats and maisonettes. It has a multi-ethnic population. The estate is extremely isolated and is cut off from the rest of Poplar by major railway lines.

LBTH had previously succeeded in securing EA, DICE and SRB funds to improve the estates but need outstripped local resources. The original housing stock had a negative value of £53 million. This was underwritten by a £54 million ERCF dowry, in addition to which, PH secured loans from Abbey National (£69 million) and BNP Paribas (£22 million). Private finance was easy to secure given high level of demand for housing in East London.

The housing development programme was implemented in partnership with East Thames Housing Group, which did most of the physical work so PH could build up its competence in economic and social regeneration. The physical regeneration programme was completed in 2002 and all housing is categorized as 'decent'. Some 10 percent of the original stock has been demolished, to be replaced with 500 new homes. The balance has been refurbished to give it an extended life of 30 to 50 years. New-build housing (237 units) has replaced large blocks on the Aberfeldy estate.

The key issue at the time of HARCA designation was the 'manageability' of the housing stock. That is, problems such as benefit dependency and anti-social behaviour, rather than issues related to the housing stock itself. Indeed, the decision to designate the HARCA in those estates where it operates was related to maximizing the number of housing units that could be improved using the ERCF grant. Estates in the area with more profound structural problems were specifically excluded from the HARCA's stock.

The key problem for some estates is accessibility. This is a particular problem in Aberfeldy and Teviot, both of which are isolated from the rest of the area by major arterial road routes, railways and the River Lea.

PH has remained part of the LBTH Common Housing Register. As such, it accepts nominations from the local authority. Nevertheless, turnover has remained low – a function of the extreme levels of demand for housing in the area, despite Poplar's reputation as one of the less popular neighbourhoods within the borough's Choice-Based Letting system.

The major physical regeneration programme locally is the Limehouse Cut Regeneration Programme. This is a £1.4 million project, funded through Leaside Regeneration's SRB6 programme, to regenerate and improve the 1.5-mile Limehouse Cut Canal. The scheme is led by British Waterways London and supported by Leaside Regeneration and Lea Rivers Trust. It aims to make the Cut a safer, more accessible and exciting amenity for members of the local community.

Economic and social regeneration

There are no significant employers within the Bow area. The local manufacturing employment base has declined dramatically and many of smaller workshops have been converted into flats.

Educational attainment levels locally are low. There is a lack of choice of good schools and many parents place their children on waiting lists for schools in nearby Bethnal Green. There is a lack of aspiration to want to go to school, let alone college or university. For many local people, therefore, work is low-skilled occupations: childminders, cleaners, school assistants and dinner ladies.

Physical and social mobility is low, particularly for older residents and minority communities. The problem is not necessarily one of physical access to external job opportunities, as public transport connections are good, but that local people are very parochial. Generally speaking, younger residents travel out of the neighbourhood to work. Most of these are employed in low-pay, low-status occupations in Docklands, the City of London and the West End.

There is a high level of entrepreneurial expertise within the East End. However, the process of social betterment invariably means successful residents leave the neighbourhood, often moving further out of London to Essex and the north-eastern suburbs. It is difficult to retain aspirational households within east London.

Tower Hamlets has high concentrations of chronic long-term illness, single parents and mental and physical health problems among its residents. The most important local health issues are high levels of under-age pregnancy – the conception rates for females aged between 15 and 17 in Bow's Park Ward are the highest in Tower Hamlets and twice the national average (EDAW, 2003) – drug addiction and asthma.

Robbery (mugging) is a significant problem in the Bow area. The robbery rate in Park Ward is nine times the national average (EDAW, 2003). There exists, within the East End, a very materialist culture of 'public show'. Local people spend a high proportion of their disposal income on their appearance and this culture contributes to the vicious circle of crime (e.g. teenagers are mugged for their expensive footwear).

The East End of London has a long-standing association with gangs and organized crime within the white working-class community. More recently, local black youths and younger elements of the Bangladeshi community have organized into gangs and have a keenly developed territorial rivalry with gangs from neighbouring Hackney.

The HAT's four strategic objectives in economic development are:

- To increase job choice and provide training for residents.
- To enable residents to have more money in their pockets in the short term.
- To enable residents to access money that can be of collective benefit.
- To meet needs and maximize economic opportunities for those not seeking or able to gain full-time work.

Bow People's Trust (BPT) has been the *de facto* delivery agent for economic and social regeneration projects. BPT's employment and training provides assistance with preparation of CVs and application forms, interview technique, plus specific vocational needs including: computing; basic skills (literacy and numeracy); and English for Speakers of Other Languages (ESOL). BPT also runs a free job brokerage service for local companies that seeks to match job seekers with employment opportunities and provides support for community businesses.

Bow schools are the target of a number of Neighbourhood Renewal Fund sponsored projects with an aggregate budget of £730,000. These aim to address the causes of under-achievement and barriers to progress in targeted schools through investment in the provision of full-time home–school liaison workers, evening ICT classes for families and parenting classes, literacy support projects, academic mentoring, additional staff, plus revision classes.

A Neighbourhood Renewal Fund supported project, *St Paul's – A New Heart for Bow*, aims to create a 'Healthy Living Centre' for Bow. This includes an exercise centre, community hall, healthy eating café and community art space. The project has a budget of £250,000 and targets disabled people and the elderly. It is led by the Church of St Paul with St Mark.

The police are involved in community issues and hold a forum for residents as well as attending the Neighbourhood Renewal Local Area Partnership meetings. In addition, knife workshops are offered to local children about the dangers and implications of people carrying knives and guns.

In Poplar, the local industrial estates provide employment for many local white working-class and black households plus commuters from neighbouring Newham and Essex but there are limited opportunities for the local Bengali community.

The local skills base is very limited. A PH-commissioned survey (MbA, 1998) reported that two thirds of respondents had no UK-sourced academic qualifications and four fifths had no vocational qualifications. Moreover, the job search horizons of local residents are very parochial. This parochialism is reinforced by the fragmented built environment of Poplar with its numerous physical boundaries, roads, canals and railways.

The Poplar estates are characterized by high levels of poverty and benefit dependency. Two thirds of households include an individual in receipt of some form of benefit whereas one tenth have no use of *any* of the following: phone, washing machine, drying facilities, study space, cable or satellite TV, a lift, a private garden or a balcony.

Unsurprisingly, Poplar is characterized by significant health problems. The area has the highest number of amputees in London (due to diabetes and circulatory problems); heart disease, strokes and cancer also rank highly. There are many teenage pregnancies among young married Bengali girls, sometimes only 16–17 years of age, and there is a high incidence of mental health problems.

The survey indicated considerable anxiety among local residents over crime-related issues. Burglary is a significant problem in the Poplar area. The burglary rate in Poplar's East India Ward is nearly twice the national average (EDAW, 2003).

PH has a dedicated community and economic regeneration department. Its mission statement is 'To work with local people to define and achieve their own financial and social aims and to create an environment where they want to live'. Each estate has a Neighbourhood Centre, in which a Community Area Director (CAD), responsible for regeneration, is based. In partnership with Bromley by Bow centre, the CADs have formulated 'Community Plans' for each estate.

There is a small Excellence in Cities (EiC) Action Zone in Poplar. Education Action Zones were established in the Schools and Framework Act

1998 with the objective of raising standards in schools. EiC Action Zones (small zones) are designed to address the educational problems of the major cities where standards have been consistently low. They typically focus on a single secondary school and its associated primary schools. There are currently 109 EiC Action Zones in England. Each receives a grant of £250,000 a year for a period of 3–5 years. Tower Hamlets-Poplar EiC Action Zone began in January 2001, and targets Lansbury Lawrence School and seven associated primary schools. The zone's objectives are: to secure a steady increase in parental attendance at school events; to enable teachers to identify pupils' progress and areas of development; and to ensure vulnerable pupils show no signs of deterioration on transfer, with 50 percent continuing any previous improvement. The key external partners involved are Tower Hamlets College, PH, Sure Start, LBTH Local Education Authority and Citigroup.

In 1998 the BBBC building was named one of the first leading examples of a Healthy Living Centre. The approach to health and regeneration involves promoting health in its most holistic sense, resulting in a close working relationship with general practitioner partners and primary care teams. The Bromley by Bow Health Team includes: complementary therapists, artists, community nurses, general practitioners, gardeners, multi-cultural youth, family and community care teams.

PH works closely on crime prevention with LBTH Youth Service, police (including PH subsidies for police overtime), Victim Support, Community Safety Unit, Social Landlords Crime and Nuisance Group and Crime Concern. A private contractor provides an estate ranger service designed to counteract crime and anti-social behaviour. Crime Concern has been commissioned to undertake research into a community safety strategy for Poplar. Various Neighbourhood Renewal Fund supported projects, with an aggregate budget of £560,000, are deployed to counter crime. These aim to: extend PH's neighbourhood 'ranger' programme to crime 'hot spots' elsewhere; install closed circuit television systems in Poplar High Street and in the local college; identify and engage 50 young people at risk of committing crime and providing them with alternative leisure pursuits such as football and cricket workshops; provide outreach work to support prevention and greater up-take of rehabilitation programmes; engage young people in positive community activities, e.g. local football teams, weekend activities, art, video and film workshops, weight training, summer youth groups and outreach work; remove abandoned cars promptly from estates, and engage 'vulnerable' young people in community filmmaking, local history courses, the arts, local football leagues.

Partnership and community involvement

For the past decade, Bow HAT has had formal responsibility for the regeneration of the three estates. It has sought to achieve some of its objectives in partnership with its successor organizations: Old Ford and Bow People's Trust. The HAT bequeathed a trust fund endowment on exit, the resources of which were generated through the sale of surplus land. BPT, Old Ford and other local groups are eligible to apply to the trust for funding. Local residents have been involved in the regeneration programme in a variety of ways.

The designation of the HAT was subject to ratification through a ballot of tenants. In spring 1993, tenants were balloted on the HAT proposals. The turnout was 76 percent, with 65 percent of tenants voting for HAT designation. Residents also have the right to choose their landlord after the HAT has ceased to exist. Overwhelmingly, they have chosen Old Ford Housing.

The key strategic decision-making body is the HAT board. This comprises 12 appointed members, of whom three are residents. There are four sub-committees that cover the work of the HAT's directorates: housing management; development; community and economic development; finance and administration.

The HAT Masterplan was formulated by consultants, under HAT supervision, with close involvement of residents. Tenants were involved particularly in the design stage of the programme. Consultation methods included open meetings, resident design group meetings, surgeries for consultations with architects, open days, 'Planning for Real' sessions, coach trips to existing housing schemes, newsletters, comment books in estate offices, preference indicators 'stick a dot' against options and a questionnaire on the Masterplan. Residents were involved in choosing the internal layout of their homes (including fixtures and fittings and colour schemes in kitchens and bathrooms) and the external appearance of the home, including street layout.

The HAT programme has, thus, prioritized the needs and aspirations of *existing* tenants. Many of these were resident in the Bow area at the time of the construction of the estates in the 1960s. This was a period when public consultation and engagement were uncommon.

The East End of London has a well developed community and voluntary sector infrastructure, particularly that related to the relief of poverty. Much of this is partly a product of the philanthropy of the Victorian age and is concentrated, in particular, in the west of Tower Hamlets. There is a general paucity of faith, community and voluntary sector organizations in Bow. Much of the community activity within the HAT estates has been housing-orientated. This is, perhaps, not surprising, as poor housing has been the driver for the regeneration of the estates, and local organizations, e.g. tenants' associations, have prioritized this single issue. In April 2000, a new forum, TML, which represents Old Ford tenants on the three estates, was established. However, it has reduced its general level of activity, perhaps because its immediate housing needs are in the process of being satisfied.

There is a problem with integration on the estates. It is difficult to secure the participation of some minority ethnic groups (e.g. within TML). This is especially true of the small Somali communities on the estates.

In Poplar, the housing development programme was implemented in partnership with East Thames Housing Group. Other key partners include, as noted, Leaside Regeneration (an Urban Regeneration Company) and Bromley by Bow Centre.

Tenants have been involved in the regeneration programme and neighbourhood management in a number of ways. All ERCF programmes are subject to ratification by a ballot of tenants. Three tenant ballots were held prior to transfer. The first, (Lansbury North, Lincoln North, Burdett) had a 66 percent turnout and 72 percent yes vote. The second took place the following year. The third ballot failed. A small group of tenants joined PH independently in 2001.

Thc main PH board has 18 members, comprising five LBTH nominees, seven residents (six tenants and one leaseholder) and six independents (including one from ETHG).

PH developed a comprehensive framework for community involvement in November 1998. A Community Involvement Support Team has been established and is funded, partly, by the Housing Corporation's Community Training and Enabling grants programme.

Each estate has an Estate or Area Board, each of which nominates representatives to the Joint Estates Panel, which itself nominates seven representatives to the PH board. Estate Boards are responsible for dealing with immediate local issues, e.g. housing management and community regeneration, and have sub-committees and working groups for this purpose. Each Estate Board has 20 full members, including a proportionate number of tenants and leaseholders and these have a 2-year term of office. All residents are eligible for election. The chair and vice chair of the boards also reflect the tenure split locally. Some 27 percent of the membership of estate management boards is from BME groups, who comprise more than half the overall population. The Joint Estates Panel is the main strategic body for local residents. Its purpose is to consider issues relating to the services provided by PH. The Panel comprises 12 tenants and six leaseholders

Tenants' satisfaction surveys indicate increasing participation rates in estate management: from 7 percent in 1999 to 50 percent in 2001. Tenants were less satisfied with opportunities to get involved in 2001 (52 percent) than in 1999 (72 percent) but more satisfied with the extent to which their views were taken into account by PH: 52 percent in 1999 compared to 60 percent in 2001. PH performs better on this indicator than LBTH but less well than the HC comparator group, although most members of this are not located in disadvantaged areas (Housing Corporation, 2001).

As noted above, East London has a well developed community and voluntary sector infrastructure but this has been under-developed in the eastern neighbourhoods of Tower Hamlets. Poplar is no exception. At the time of HARCA designation, there were no pre-existing tenants' associations. The only community and voluntary sector organizations were philanthropic (Bromley by Bow Centre, Salvation Army and churches). There was no indigenous 'grassroots' activity.

Conclusions

The case studies of Tower Hamlets HAT and Poplar HARCA illustrate a number of important and unresolved questions concerning estate regeneration policy in England.

First, the two case studies show – notwithstanding the rhetoric of 'holistic' regeneration – the continuing importance afforded to *physical* housing regeneration. It is now widely acknowledged that 'bricks and mortar' type investment alone (e.g. Estate Action) has not led to sustainable regeneration. There has, it is commonly argued, been a shift in investment from physical to human resources during the past decade (Forrest and Kearns, 1999). City Challenge and the Single Regeneration Budget witnessed a diminution of estate-based physical housing renewal (Hall and Nevin, 1999). In this respect, the two case studies are both atypical, as they are fundamentally housing-led, and typical, as they have pursued parallel economic and social regeneration efforts. It has also been argued the importance of housing has been neglected in the thinking of the Labour government and, particularly, in the National Strategy for Neighbourhood Renewal (Forrest and Kearns, 1999; Maclennan, 2000), which considers employment to be the primary success factor (Hall, 2003). Nevertheless, the strategy explicitly acknowledges physical improvement as government priority and introduces a *Decent Homes* standard: a 'commitment' to bring all 'social housing up to a decent standard by 2010, with a third of this improvement taking place by 2004, over half of which will be in the most deprived areas' (SEU, 2001, p. 40). More recently, the government's Sustainable Communities Plan (ODPM, 2003), which seeks to address the dual problems of the lack of affordable housing in London and the south, on the one hand, and low demand in parts of the north and Midlands, on the other, provides for substantial new build in the former and demolition and re-building in the latter.

Second, the two case studies illustrate the continuing importance afforded to *stock transfer*. This has been justified by governments on a number of grounds. For the Conservative government, the de-municipalization of social housing was a self-justifying ideological objective. In addition, it has been argued that transfer brings added value to the regeneration process in that: it levers in private resources (and, thus, limits public borrowing); it offers improved organizational competence (as politicized, bureaucratic, multi-purpose local authorities are replaced by single-purpose bodies, e.g. HATS, or entrepreneurial housing associations); and, it provides choice for tenants. However, each of these arguments can be countered. Local authorities could be empowered by change in public borrowing regulations. The *a priori* argument that they are organizationally inferior is prejudicial rather than empirical. Tenant choice is limited if transfer is conducted *en masse*. Nevertheless, the process has continued under a Labour government. Local authorities can seek additional investment in their housing stock through three options but these all require a separation of housing strategy and management responsibilities: *stock transfer*; *arm's length management organizations* or ALMOs (the transfer of management responsibilities to independent ALMOs – although local authorities remain the legal landlords, high-performing ALMOs are eligible for additional government housing investment funds); and the *private finance initiative* or PFI (the signing of contractual arrangements between local authorities and private consortia which refurbish and manage the stock for a specified period in return for a fee).

Third, the case studies illustrate a crucial question that has historically remained unresolved and continues to be so in the latest package of strategies implemented by the Labour government. That is, *who are the intended beneficiaries of investment*: existing residents, future residents, statutory agencies with fixed investments (e.g. housing, schools) and inflated managements costs (e.g. policing) in declining neighbourhoods, or a combination of these? In the case of THHAT, the answer is atypical and unambiguous. It can be argued that the *existing* residents of the estates have led a privileged existence during the past decade. They have benefited from a uniquely well resourced regeneration programme, resulting in the construction of new homes built (within limits) to their own (high) specification. In addition, HAT designation effectively excluded the possibility of more difficult tenants being housed locally and permitted a generous policy of housing the children of existing tenants. However, there are important questions arising about the sustainability of such an approach, not simply because of cost. The re-absorption of the estates into the broader housing market of Tower Hamlets in 2004, following HAT exit, will create significant challenges in terms of an influx of new residents who may differ from the existing population in respect of age, race, socio-economic circumstances and lifestyle.

The case of Poplar is more typical. The regeneration programme in Poplar has been less well funded, although the estates have still been fortunate to be in receipt of a significant renewal programme, and the estates have remained within the broader Tower Hamlets housing market. The entire housing stock has been refurbished during the past 5 years or so and is now classified as being of a statutorily 'decent' standard. However, it can be argued that physical housing conditions have not been fundamental problems in the Poplar estates. The key challenge has been 'manageability' (e.g. anti-social behaviour). The area is characterized by profound social and economic problems characteristic of many inner-urban neighbourhoods. These affect existing residents, deter future residents and impose disproportionate costs on the public services. Moreover, as in Bow, these problems have proved far more intangible and illusive than physical regeneration.

Fourth, the case studies illustrate the significance of community involvement. The participation of local residents in housing management has been a priority of estate regeneration programmes since the 1980s. More recently, local residents have become involved to a greater degree in the formal decision-making process of regeneration programmes through participation in representative forums and other activities such as 'Planning for Real'. The process is central to the regeneration agenda of the Labour government. Attempts to involve local people in estate regeneration have met with uneven success (Hastings et al., 1996; Taylor, 2000) as a result of the lack of experience of partnership working and limited financial and personnel resources of many tenant groups. More importantly, given the ambiguous objectives of regeneration cited above, the relative importance of the needs and aspirations of existing residents, let alone future ones, are rarely defined. Are they, of primary importance, *primus inter pares*, or merely those of one group of stakeholders among many? The case of Bow HAT is, again, unambiguous and atypical. The existing (and 'ring fenced') tenant-based organization has been intensively involved in the design stage of the physical regeneration programme. This has reinforced the outcomes of the programme as cited above. Questions of sustainability are again pertinent. As the original objectives of improving housing conditions have been achieved, tenants have become less interested in participating in decision making locally. The renewal of social capital locally has been limited. The case of Poplar is, once again, more typical and illustrates the difficulties of involving and reconciling the aspirations of a diverse, vulnerable and fluid resident population who represent one group of stakeholders in a multi-stakeholder regeneration programme, the objectives of which are indeterminate.

Finally, the case studies demonstrate that urban communities may be *peripheral* in an economic and social sense, even when they benefit from a highly central geographical location and that locally focused intervention may thus be thwarted. The Labour government's approach to economic and social regeneration, embodied in the National Strategy for

Neighbourhood Renewal, advocates improving the employability of residents of disadvantaged neighbourhoods through supply-side measures (Hall, 2003) such as those deployed by the HAT and Poplar HARCA (e.g. training, counselling, work incentives, childcare). This thesis has been criticized for neglecting demand-side considerations, especially the long-term decline of the employment base of Britain's cities (Turok and Edge, 1999). However, whilst the level of employment opportunities and amenities accessible in the immediate vicinity is poor, both Bow and Poplar are located in a jobs-rich sub-region due to their proximity to Docklands and Canary Wharf to the south, the City of London and West End to the west, plus the Newham development area to the east. Despite the strong aggregate level of employment demand in East London, almost insurmountable supply-side barriers exist in neighbourhoods like Bow and Poplar. A number have been cited in this chapter: the low levels of educational attainment locally; adequate transport links that are not exploited because of concerns about security and access; and a built environment that functions as a psychological barrier, reinforcing parochial attitudes and limiting social mobility, as well as a physical one. Estate regeneration in the UK has historically been handicapped by an 'inward' looking approach to regeneration (Hall, 1997) that assumes that problems and solutions are co-terminus (i.e. found at the level of the estate). Labour's emphasis on community, self-help and mutuality are merely likely to reinforce this particular bias.

References

Atkinson, R. and Moon, G. (1994) The City Challenge Initiative: An overview and preliminary assessment. *Regional Studies* **28** (1): 94–7.

Buck, N. et al. (2002) *Working Capital: Life and Labour in Contemporary London*. Routledge, London.

Coleman, A. (1985) *Utopia on Trial: Vision and Reality in Planned Housing*. Shipman, London.

Deakin, N. and Edwards, J. (1993) *The Enterprise Culture and the Inner City*. Routledge, London.

Department for Environment, Transport and the Regions (2000) *Our Towns and Cities: The Future – Delivering an Urban Renaissance*. DETR, London.

EDAW (2003) *Tower Hamlets Ward Data Report*. London Borough of Tower Hamlets, London.

Etzioni, A. (1993) *Spirit of Community: the Reinvention of American Society*. Simon and Schuster, London.

Evans, R. and Long, D. (2000) Estate based regeneration in England: lessons from Housing Action Trusts. *Housing Studies* **15** (2): 301–7.

Fainstein, S. et al. (1994) *Divided Cities: New York and London in the Contemporary World*. Blackwell, Oxford.

Forrest, R. and Kearns, A. (1999) *Joined-up Places: Social Cohesion and Neighbourhood Regeneration*. York Publishing Services, York.

Forrest, R. and Kearns, A. (2001) Social cohesion, social capital and the neighbourhood. *Urban Studies* **38** (12): 2125–43.

HACAS Chapman Hendy (2002) *Beyond Bricks and Mortar: Bringing Regeneration into Stock Transfer*. Chartered Institute of Housing, London.

Hall, P. (1997) Regeneration policies for peripheral housing estates: inward and outward looking perspectives. *Urban Studies* **34** (5–6): 873–90.

Hall, S. (2000) The way forward for regeneration: New Labour and the single regeneration budget. *Local Government Studies* **26** (1): 1–14.

Hall, S. (2003) The 'Third Way' revisited: 'New' Labour, spatial policy and the National Strategy for Neighbourhood Renewal. *Planning Practice and Research* **18** (4): 265–77.

Hall, S. and Mawson, J. (1999) *Challenge Funding, Contracts and Area Regeneration: A Decade of Innovation in Policy Management and Co-ordination*. Policy Press, Bristol.

Hall, S. and Nevin, B. (1999) Continuity and change: a review of English regeneration policy in the 1990s. *Regional Studies* **33** (5): 477–82.

Hastings, A., McArthur, A. and McGregor, A. (1996) *Less than Equal: Community Organisations and Estate Regeneration Partnerships*. Policy Press, Bristol.

Housing Corporation (2001) *Inspection of Poplar HARCA*. Housing Corporation, London.

Maclennan, D. (2000) *Changing Places, Engaging People*. York Publishing Services, York.

Marsh, A. and Cowan, D. (2001) New Labour, same old Tory housing policy? *Modern Law Review* **64** (2): 260–79.

MbA (1998) *Dishing the Dosh Household Survey*. Poplar HARCA, London.

Murie, A. and Nevin, B. (2001) New Labour transfers. In: Cowan, D. and Marsh, A. (eds) *Two Steps Forward – Housing Policy into the New Millennium*. Policy Press, Bristol.

National Audit Office (NAO) (2003) *Improving Social Housing Through Transfer*. The Stationery Office, London.

Nevin, B., Loftman, P. and Beasley, M. (1997) Cities in crisis: is growth the answer? An analysis of the first and second round of the Single Regeneration Budget Challenge Fund. *Town Planning Review* **68** (2): 145–64.

Oatley, N. (1995) Competitive urban policy and the regeneration game. *Town Planning Review* **66** (1): 1–14.

Oatley, N. (ed.) (1998) *Cities, Economic Competition and Urban Policy*. Paul Chapman, London.

Office for the Deputy Prime Minister (OPDM) (2003) *Sustainable Communities: Building for the Future*. Office for the Deputy Prime Minister, London.

Pinto, R. (1991) Central–local interaction in renovating run-down estates: the view of housing authorities on the Estate Action initiative. *Local Government Studies* **17** (1): 45–62.

Pinto, R. (1993) An Analysis of the Impact of Estate Action Schemes. *Local Government Studies* **19** (1): 37–55.

Rhodes, J., Tyler, P. and Brennan, A. (2003) New developments in area-based initiatives in England: the experience of the single regeneration budget. *Urban Studies* **40** (8): 1399–426.

Social Exclusion Unit (SEU) (2001) *A New Commitment to Neighbourhood Renewal – National Strategy Action Plan*. Cabinet Office, London.

Taylor, M. (2000) *Top-down Meets Bottom-up: Neighbourhood Management*. Joseph Rowntree Foundation, York.

Thornley, A. (1993) *Urban Planning Under Thatcherism – The Challenge of the Market*, 2nd edn. Routledge, London.

Tiesdell, S. (2001) A forgotten policy. A perspective on the evolution and transformation of Housing Action Trust policy. *European Journal of Housing Policy* **1** (3): 357–83.

Tiesdell, S. and Allmendinger, P. (2001) Neighbourhood renewal and New Labour's Third Way. *Environment and Planning C* **19** (6): 903–26.

Tower Hamlets Housing Action Trust (THHAT) (2002) *Quality of Life Survey 2002*. Tower Hamlets Housing Action Trust, London.

Turok, I. and Edge, N. (1999) *The Jobs Gap in Britain's Cities – Employment Loss and Labour Market Consequences*. Policy Press, Bristol.

Van Kempen, R., Dekker, K., Hall, S. and Tosics, I. (eds) (2005) *Restructuring Large Housing Estates in Europe*. Policy Press, Bristol.

Wilcox, S. (2003) *Can Work Can't Buy: Local Measures of the Ability of Working Households to become Homeowners*. Joseph Rowntree Foundation, York.

12

Erasing the Community in Order to Save It? Reconstructing Community and Property in Community Development

James DeFilippis

Introduction

The issue of conflict in community politics is an important one for this book. This is simply because conflict has been increasingly assumed away in contemporary theory and practice of urban community development. The retreat from conflict, and therefore politics in its most meaningful sense as an idea and concept, must be understood and addressed intellectually and politically, if we are going to move towards more equitable and just urban communities for us to live in. This chapter presents a critique of the dominant theories and practices of community development, and a more applied discussion of alternatives to the dominant perspectives, in light of this critique.

This chapter will begin with a discussion of this retreat from conflict and politics in contemporary theories of community and community development. It will progress to talk about how these theories are embedded in the larger context of urban neo-liberalism, state retrenchment and gentrification. The chapter will then critique this dominant framework, referred to here as *neo-liberal*[1] *communitarianism*. Central to the discussion

[1] Liberal is being used here not in its normal American sense, but in its classic meaning in political philosophy (and the non-American world). That is, to refer to a belief in small government and individual rights.

is the question of property, and the role of property relations in connecting, *or not connecting*, people to the places in which they live. Current theory and practice, in other words, are creating a framework where the displacement of the communities we are allegedly trying to develop is the logical and unsurprising outcome of community development. The chapter then briefly discusses theories of property and ownership, before concluding with descriptions of forms of housing development that offer a set of ways forward in community development.

'Neo-liberal communitarianism' and the retreat from politics

There are several different strands of thought in community development in the Anglo-American world. These strands are varied, and in some ways quite distinct from each other. But while they may appear to be rather disparate, they all share a belief that communities are defined by relationships that are, in the last instance, 'win–win' relationships – whether the relationships in question are those within a community, or between people in a community and the larger world. What they do with this understanding, however, varies and it is important for us to examine these different schools of thought and practice.

Market-based community development

Community development has increasingly become entrepreneurial and market-based in its understanding of urban problems and poverty, and logically, in its programmatic responses to that poverty. This embracing of the market has, in many ways, been developing over time and actually has its roots in the ambivalence of the earliest American community development corporations (CDCs) in terms of their understandings of the roots of 'the urban crisis' in the 1960s (see DeFilippis, 2004). But this ambivalence has rather unambiguously moved towards a clear, overarching perspective that the market is *the* way to develop inner cities for the benefit of those in the inner city. As Michael Porter, perhaps the most visible proponent of this perspective, put it, 'a sustainable economic base can be created in inner cities only as it has been elsewhere: through private, for-profit initiatives, and investments based on economic self-interest and genuine competitive advantage' (Porter, 1997, p. 12). But Porter, and his *Initiative for a Competitive Inner City*, are far from alone, and this is a dominant direction in community development in the US (Pierce and Steinbach, 1987; Bendick and Egan, 1991; Berger and Steinbach, 1992; Nowak, 1998; Carr, 1999; Grogan and Proscio, 2000; Community Development Partnership Network, 2003;

Weissbourd and Bodini, 2005). It is perhaps a recent Brookings Institute paper which states the logic most nakedly when it argues:

- 'To address poverty, create wealth.
- Wealth is created by investing in assets.
- The economic mechanism for asset investment is the market.
- Therefore, to increase wealth in poor communities, expand market activity to the assets of those communities.' (Weissbourd and Bodini, 2005, p. 5)

Along with the turn towards the market has come a tendency to see community development as about individuals, rather than *communities* per se. This is not surprising since mainstream economics has little place for, and attached little importance to, the idea of community. Individualizing community development through the market plays out in a variety of ways, and can be seen from the increasing pressure being put on community development credit unions to function like other credit unions (for example, through credit scoring, risk-based lending), to the increasing emphasis on workforce training and small business development in community development, or the Ford Foundation's decision that the only kind of housing that it will fund is individual home ownership (Pitcoff, 2003).

Shiny, happy communitarians[2]

While it might seem a bit paradoxical, given the individualized, market orientation described above, the second dominant thread in community development is a powerful re-assertion of the idea of *community*, and a particular version of *communitarianism* (Sites, 1998). This communitarian framework is one that posits a belief that there are shared interests among individuals in a community, and thus community development should be about creating the social relationships which allow those mutual goals to be realized. There are two principal figures in this understanding of community. The first is John McKnight, who has argued for a framework of community development centered on 'community-based assets' (Kretzmann and McKnight, 1993; McKnight, 1995; see also Shragge and Fisher, 2001). The second is Robert Putnam, whose work on social capital (Putnam, 1995, 1996, 2000) has become almost axiomatic in community development theory and practice (see DeFilippis, 2001). Social capital, for Putnam and his many followers, 'refers to features of social organization such as networks, norms, and social trust that facilitate

[2] With apologies to REM.

coordination and mutual benefit' (Putnam, 1995, p. 67). Voluntary associations, therefore, are not based on interests, and therefore do not contain a kernel of conflict (since interests in society might be understood to not coincide) but rather are 'features of social life – networks, norms, and trust – that enable participants to come together to pursue shared objectives' (Putnam, 1996, p. 34). McKnight's work on community-based assets is a close sibling to Putnam's. The basic goal is to mobilize assets to build community involving 'virtually the entire community in the complex process of regeneration' (Kretzmann and McKnight, 1993, p. 345). Thus, not only do these frameworks argue that relations within communities tend to be largely 'win–win' relations, but both take that framework one step farther to assume that *individual* gains and interests in the community are synonymous with *collective*, or community, gains and interests. This particular form of communitarianism therefore fits rather comfortably with the individualistic, market-based perspectives and policies that govern community development activity.

Moving beyond community and 'Shaming the Inside Game'[3]

Finally, the expansion of the idea that social relationships are 'win–win' relationships is, in the third dominant stream of thought, extended beyond the community and to society at large. There are two different ways in which this has occurred. The first is the growth in the last decade of non-confrontational forms of engagement and organizing. Community development is now a collaborative process, and the original, unapologetically conflictual, ideal of black power, or the neo-Alinsky style organizing have been rejected (Epstein, 1996). Michael Eichler, the President of the 'Consensus Organizing Institute,' which has received a great deal of attention in the field (see Gittell and Vidal, 1998), described a project that the Institute had worked on as having, 'demonstrated the essential attribute of consensus organizing: instead of taking power from those who have it, consensus organizers build relationships in which power is shared for mutual benefit . . . Cooperation, rather than confrontation became the modus operandi for solving a neighborhood problem' (Eichler, 1998). Within the current understanding of cooperation, there is almost contempt for past organizing efforts, and Grogan and Proscio state, 'The community organizing and planning of that period [the 1960s] was soon squandered on divisive or extremist political tactics, including the in-your-face style of protest that Tom Wolfe famously dubbed, "mau-mauing"' (Grogan and Proscio, 2000, p. 66). The dominant view is that low-income inner-city residents

[3] This label is borrowed from Imbroscio, 2005.

have a shared set of interests with the larger society they exist within, and organizing and development should be structured accordingly. Power relations are therefore downplayed or completely ignored in this framework.

The second way social relations between people in low-income communities and the larger society are assumed to be harmonious and conflict-free is through the increasing emphasis on regionalization as the solution to communities' problems. This perspective begins with the assumption that low-income inner-city communities are failing because they are too isolated from the social, economic and political mainstream – and in this way shares much with the framework of *social exclusion/inclusion* which has been so significant in the UK and western Europe. The solutions, therefore, are not to look within communities, but rather to build regional frameworks (governmental, not-for-profit, etc.) and de-concentrate low-income people (through policies that encourage or require their relocation), to address the problems in low-income neighborhoods. Combining the work of William Julius Wilson (Wilson, 1987), and the critiques of community development as inefficient and ineffectual (see, for instance, Lemann's influential piece in the *New York Times Magazine*, Lemann, 1994), the closely related frameworks of regionalization and poverty de-concentration have become the received wisdom in American public policy circles.

Together these three sets of perspectives, which dominate the theory and practice of community development, can best be described as a form of *neo-liberal communitarianism*. There are significant problems with these approaches, which I will outline shortly, but first, a detour must be taken through the politics and policies of urban neo-liberalism to see how these frameworks fit into the larger political economy in which they are being conceived, received and acted upon.

Community and the state in the contemporary Anglo-American political economy

The frameworks described above fit very well with the larger trajectories of the state in both the US and UK, which have been increasingly towards systems of governance that are neo-liberal in their logic, institutional arrangements and power relations. Neo-liberalism has been thoroughly studied and analyzed in Anglo-American social sciences (see, for instance, Brenner and Theodore, 2002), so I don't want to devote a significant amount of time and space to describing or analyzing neo-liberalism. For the purposes of this discussion, there are three primary components of urban neo-liberalism in the UK and US. First, there has been a re-organization of the state/capital/labor relationships, with the state becoming increasingly unwilling (and, to a lesser degree, unable) to intervene in the economy in ways that limit the ability of capital to maximize profit. This has been particularly true in the realm of private property and the capitalist development thereof. Second, and emerging from the first, is a shift that has occurred in urban politics as local governments have increasingly moved from being concerned with managerial activities (schools, hospitals, fire, sanitation, local transportation, etc.) towards being centrally involved in economic development activities – often as risk-taking entrepreneurs in the process. The fact that policing – very much a traditional role of local government – has become increasingly emphasized in local politics in the last decade is a good indication that the trend towards urban neo-liberalism is not about the weakening of the local state per se. Instead, it's a re-articulation of the roles and goals of the local state; a withdrawal from social service provision mixed with a re-assertion of the state's roles in promoting capital accumulation and 'social control'.

The third characteristic of urban neo-liberalism, which follows from the first two, is the privatization of formerly social or public functions and activities. By privatization I mean three different things. The first is the simple shift to the market – which is what is most commonly meant by privatization. The second is privatization as the shift of a social or public function to a private – that is, within a household – function or activity. The third form of privatization is non-profitization – that is, shifting public functions or activities to a not-for-profit entity, which is usually located in communities with high densities of people who need the services provided socially because they cannot purchase them individually within their households. This, of course, is a reminder that in both households and not-for-profits, the market remains a defining arbiter of decisions and capacities. This shift to market is evident across the board in the US (although it is slightly less so in the UK). Communities are the realm in which the reproduction of society, or social reproduction, occurs, and in virtually every component of social reproduction – housing, health care, childcare, senior citizen care, and education – there is an unambiguous move towards the privatization of those functions.

Privatization as a shift to not-for-profits brings with it a recognition of the importance of the idea and *ideal* of community to Anglo-American political thought and public policy in the last 15 years. 'Community' plays a central role in the philosophies of New Labour, the 'New' Democrats and Christian Conservatives. And so community development theory and practice have evolved in concert with state policies in both the UK and the US that have systematically withdrawn the public sector from social services. This embrace of 'community' has been a central part of the thinking of both the 'New' Democrats under Clinton (and Hillary Clinton's treatise on her political philosophy was appropriately called 'It Takes A Village') and the Bush administration, with its promotion of faith-based organizations in social service provision. But it is probably New Labour that has been the most explicit in its embrace of community. As Tony Blair stated:

> 'At the heart of my beliefs is the idea of community. I don't just mean the local villages, towns and cities in which we live. I mean that our fulfillment as individuals lies in a decent society of others. My argument is that the renewal of community is the answer to the challenges of a changing world.'[4]

The presumptive next Prime Minister, Gordon Brown, also certainly appears set to continue this emphasis on community (Brown, 2004; Fyfe, 2005). While it could be argued that the embrace of community by policy makers is simply a cynical effort to justify the retreat of the public sector from its obligations, that would be unfair, and there is every indication that policy makers in both the US and UK believe that the community is the most effective arena to address a myriad of social issues.

There is, however, an exception to this belief, which has already been described above. This exception is important to recognize and understand because it points us towards some (but not all) of the fundamental problems with neo-liberal communitarianism in its community development and public policy forms. The exception to the idea that community and social capital are important to development, is, ironically enough, poor communities. Thus, despite the nominal understanding of 'social capital' as being pivotal to individuals' life chances, and despite the embrace of community-based service providers, there is also the simultaneous goal of de-concentrating poverty. And American urban and housing policies have had the de-concentration of poor people as a central goal for the last 15–20 years. This fundamental contradiction – between, on the one hand, embracing social capital and community, and, on the other, promoting (or requiring) mobility and de-concentration – can only be explained by an erasure of the social ties that already exist within low-income communities. This is predicated on the belief that poor communities lack 'social capital' and even lack *community* itself.

Community development, place and the problems of neo-liberal communitarianism

This removal of community from community development is intimately related to the assuming away of conflict; and it is often the same steps in logic that perform both erasures simultaneously. The problems with the

[4] Quoted in Levitas, 2000.

dominant perspectives, their assumptions and blind spots are manifold and fully enumerating them would extend beyond the scope of this chapter. However, a few key issues need to be raised here, and the critique will proceed through the three sets of perspectives in turn.

Assuming that the market should be the organizing mechanism for community development ignores the histories of capital disinvestment and decline in most inner-city communities. More often than not any investment capital in them will need to come from the outside. The potential for inner-city residents to have some say over what kind of investment and what kind of development gets undercut by a framework which relies on the free market investments into the inner city. Embracing the market by itself leaves control over capital and economic development firmly and squarely beyond the reach of people in inner-city communities.

Also, why should it be assumed that capital investment in a place would necessarily benefit the people in a place – particularly given the lack of control over that investment noted above. Or, put another way, why should we assume that the interests of capital are the same as the community? Capital is interested in the extraction of profit, which is needed to reproduce itself and to generate wealth for the *individuals* that own it. This is inherently true in capitalism. We should no more assume the shared interests of, for instance, black inner-city business owners and their workers than we would the shared interests of the shareholders of Nike and its sweatshop workers in Indonesia. And while both the local business owners and their workers might share an interest in a community (because of their dependence on it), that in no way eliminates their inherent respective class interests. One need only think of Detroit as a classic case of how capital investment need not lead to benefit by the people. For most of the last 100 years Detroit has been the home of three of the largest industrial corporations on the planet, existing, in all of their corporate affluence, side-by-side with one of the poorest urban populations in the developed world.

The problems with communitarians are similar to those of the market-based theories of community development. First, the individual is *not* the same as the group. We cannot assume that individual gains and interests are the same as those of the larger community. This point is especially valid in low-income urban neighborhoods. As Blakely and Small astutely observe in their critique of Porter, 'there is little connection between [individual] work and better outcomes for ghetto dwellers or for their communities. The signals are very clear. To move up economically requires moving out. As a result, the link between job and social betterment is lost collectively as achievers move out of the ghetto' (Blakely and Small, 1995/1996, p. 166).

Second, the communitarian perspective assumes that people in a community share common interests simply by virtue of living in the same area.

But clearly communities are diverse, conflict-laden and contested places and spaces. Landlords and renters, business owners and workers, and service providers and recipients are all fairly obvious examples of how interests are not shared by people in a community. They demonstrate that individuals' self-interest cannot be 'added up' to yield a community's interests – if they could, then whose interests would be 'added up' and whose would be left out of the equation? For interests to be shared requires the construction of shared interests – not the *a priori* assumption of their existence. The assumption of shared interests, again, becomes a vital step in the erasure of the real and tangible social relations that constitute communities.

The problems of moving the assumption of shared interests from within the community to the larger society involve similar sets of issues as the market-based and communitarian theories discussed above, but make explicit what was only implicit up to this point. That is, if the best ways to assist low-income communities are to fold them in with the larger metropolitan area or disperse them outright, then where does that leave low-income communities? In the first case, it leaves them in a position of political dependence on a larger suburban population that has demonstrated that it largely does not care about them – evidenced by the fact that they have left urban neighborhoods and interact with cities only in terms of going to downtowns for work or leisure, if at all. This population, we are now told, has shared interests with the low-income residents, and the fault lies in community development efforts that have historically not recognized that fact. This is a recipe for the drowning of the voices of low-income, inner-city people, and their further erasure from the public discourse. The de-concentration of the poor, of course, brings with it the assumption that poor communities are empty of social networks and social ties – and in this way, mirrors how Putnam's framework of social capital has been understood by policy makers. That is, if we assume that social networks generate capital, then we have confused the level of social networks with the ability of those social networks to generate capital. We therefore render invisible social networks unable to generate capital because of the poverty of those in those networks. We therefore will not be able to even see that which we are destroying, in the efforts to de-concentrate the poor and spread them around metropolitan areas.

When the problems with the three frameworks come together, and are taken to their logical conclusion, low-income communities can be viewed as simply empty places waiting to get filled. Community development therefore easily becomes place development – with little regard for the people in those places. The clearest expression of this was demonstrated in an extended memo in *The Neighborworks Journal*, which is the journal of the Neighborhood Reinvestment Corporation (NRC), by James Carr, senior vice-president of the Fannie Mae Foundation[5]. In it he issued a call for 'A New Paradigm for Community Reinvestment'. The new paradigm called for greater collaboration between community developers and outside investors and businesses. It included a promotion of the idea of place marketing in which community development projects could take on names such as 'The Woodlands,' 'Celebration' and 'Redwood Shores'. He even stated, 'some of these places could be treated as urban blank slates, where the development takes on an image the investors choose' (Carr, 1999, p. 21). In this new paradigm, the first role for government is to, 'assist private firms to extract value from community assets' (Carr, 1999, p. 22).

This willingness to treat low-income communities as blank slates by those who claim to speak for them has powerful echoes with the history of gentrification in urban areas. Gentrification and the displacement that usually accompanies it are powerful processes in the transformation of space. And in so doing, they disrupt and even destroy the social relations in the community prior to the period of gentrification. Perhaps the most powerful expression of this process was the 1992 film, *'Delivered Vacant'*, which chronicled this erasure in Hoboken, NJ, as the residential population of virtually an entire city was displaced within the space of only 8 years. While there has been debate about the extent to which gentrification leads directly to displacement, even those who have argued that gentrification does not lead to displacement often explain the lack of direct displacement as being a result of non-market-based interventions into the housing market (such as rent controls and public housing – see Freeman and Braconi, 2004). Such public sector interventions are increasingly rare and shrinking, and thus the threats of displacement from gentrification increasingly real.

Since the emergence of gentrification it has become untenable to argue that reinvestment is a desirable end in-and-of-itself for people in poor communities. Instead, rightfully conceived, reinvestment needs to be understood through the lens of questions such as: what kind of reinvestment? For whom? Controlled by whom? Such rudimentary questions are inherently and explicitly political, and they recognize that issues of urban development inevitably involve winners and losers. Unfortunately, they are too often not asked and the politics of community change are impoverished as a result.

Gentrification has left residents of low-income neighborhoods in a situation where, since they exert little control over either investment capital or their homes, they are facing the 'choices' of either continued disinvestments and decline in the quality of their neighborhood, or reinvestment that results in their displacement. Too many community groups

[5] The NRC and the Fannie Mae Foundation are two of the primary funders of community development in the US.

have bought into this false set of choices, and either oppose reinvestment to preserve their social ties and networks (in short, their community), or, conversely, become active agents in neighborhood gentrification and displacement. In New York, for example, not-for-profit community organizations, most notably Hope CDC in East Harlem (Dávila, 2004; Pacenza, 2002), have been openly promoting gentrification and have been doing market rate owner-occupier housing construction. Such a stance from community development organizations mirrors that of neo-liberal urban governance as local governments in the US and the national government in the UK have played an increasingly active role in promoting gentrification in Anglo-American cities (Smith, 2002).

Property and community development

But we must have more nuanced understandings of both community and development. We must recognize social conflict in normal community relations. We must also recognize that people in places build and construct social relations with other people who live near them. Those social relations may, or may not, yield wealth for those involved, but the lack of wealth by no means demonstrates a lack of relations. In the context of the preceeding discussion, community development should create structures that link people within places to the physical space of those places. We need, in short, a different set of relations to property – because it is only through a process of connecting people to place, institutionally and structurally, and in ways that empower people within communities, that we can get beyond the emptiness of communities in the context of development. In order to do this, we need to take a step back and reconstruct property. This, in turn, requires an understanding of what property is, and what it can be.

Reconstructing property

Property ownership, as it is understood in liberal ideology, refers to 'any property structure where full (or nearly full) rights to possess, use, manage, alienate, transfer, and gain income from property are granted to individuals' (Christman, 1994, p. 15). But this view of property, which has its distant roots in Locke's 'labor theory of property' (Krueckeberg, 1995), conflates very different kinds of rights. Property ownership, instead of being a coherent set of rights and interests (or a 'bundle of rights,' as it is often referred to), can be divided into two distinct sets: those of possession and those of income. This statement is, of course, not a new one, and actually stems from the nineteenth century arguments of Proudhon and Marx (who referred to them as use values and exchange values, respectively). But while this dual set of interests is often understood as the interests of the community vs. the interests of capital (Logan and Molotch, 1987; Mollenkopf, 1981), that dichotomy is much too simple. For while the interests of possession and income are not synonymous, and need to be disentangled, homeowners clearly possess both simultaneously.

Davis (1991, pp. 43–61) has argued that instead of two (use and exchange), there are actually six sets of interests that come with home ownership (three for each of the larger interests). He disaggregates exchange value interests into: equity (the property value); liquidity (the cash available through either rents or use as collateral); and legacy (the inheritable nature of property wealth). He also separates use interests into: security (the ability to stay in a home); amenity (the quality of the home); and autonomy (the level of personal control over the home). Homeowners, by virtue of their tenure, are able to transform those interests into rights. Renters, by virtue of their tenure, cannot.

With this typology we can construct a framework for affordable housing construction and community development that is theorized and enacted in ways that both allow for wealth creation and development without displacement, and build upon the social networks already in place. The use interests of home ownership need to be protected and extended universally. No one should live in fear of displacement, or live in a substandard house or apartment, or feel powerless within their homes. But given the limited amount of capital in low-income neighborhoods, forms of community development that privilege the individual and place development will almost inevitably displace the people in place. We must also recognize that most poor households simply do not have the income or wealth to become individual homeowners[6]. We thus need to extend the use interests of property, while transforming the income interests, to allow communities, and their already existing social relations, to benefit from them. Rather than assume that people's interests are shared, and that property development equals community development, we need to construct shared interests and allow property development to yield *community* development.

The question then becomes the pragmatic one of what this understanding of property looks like in practice. Does any of this make sense on the ground? There are various different forms of community development that have embraced alternative notions of property development, and which link people both to each other in their community, and to the physical space of the community. In general, these take three different forms:

[6] According to the 2001 Survey of Consumer Finances, the most recent one available, the median total family income in American renter-households was $24,700 per year. The median total family net worth was only $4800 (Aizcorbe et al., 2003, pp. 5 and 7).

limited-equity cooperatives, mutual housing associations (MHAs), and community land trusts (CLTs). This chapter will now discuss each in turn.

Limited equity housing cooperatives

Housing cooperatives (co-ops) have been part of the American housing stock for well over 100 years. Some feminists and socialists in the nineteenth century argued that collectively owning housing was just as important in a transformed society as was workers owning the means of production (see Stone, 1993). The early co-ops, however, were luxury developments called 'home clubs' which prided themselves on their exclusivity (Siegler and Levy, 1986). It wasn't until the twentieth century that co-ops became more politically and socially progressive in their motivations. In the inter-war period many union-built co-ops were constructed in the US, the bulk of them being in New York City. With the support of public policies, including tax breaks on property value appreciations and direct subsidies, co-ops have become a significant, if still rather peripheral, part of American housing. There are more than 1.5 million cooperative housing units in the US, and co-ops are by far the largest form of collectively owned and controlled housing in the US. Most of these, however, are market-rate or luxury cooperatives with little social or political goal or mission.

But this is certainly not true for all cooperatives, and there are about 400,000 housing units in limited equity co-ops nationwide (PolicyLink, 2002). This figure is slightly misleading since they are highly geographically concentrated and the absolute majority of these are located in New York City (Rae, 1997). The number of co-ops grew dramatically and very rapidly through the 1970s, as part of the self-help housing that emerged in this period as a result of the disinvestment and abandonment in many inner-city areas (Kolodny, 1987). This was particularly true in New York City, where abandonment led to a substantial stock of city-owned housing that was often converted by the tenants into limited equity co-ops (Leavitt and Saegert, 1990; Rae, 1997). Co-ops have also played a vital role in New York City in maintaining some form of affordable housing in several gentrifying areas – most notably in Hell's Kitchen or the rather famous creation of limited equity co-ops from squatters' buildings in the Lower East Side.

A housing cooperative is similar to other forms of home ownership in that the resident is not just a tenant but also an owner, and therefore possesses a set of rights and responsibilities. That, however, is where the similarities end. And what residents own is not their housing unit, but a share in the corporation that owns the housing complex. The corporation owns and controls the housing development, and the residents are the shareholders of the corporation. The key factor that distinguishes limited-equity from other, more commonly found, co-ops is that the price of the owners' shares is not determined by the real estate market (as it is for most co-ops), but by a set formula determined by the particular cooperative's by-laws and subscription agreements. Those by-laws restrict the resale price of the cooperative shares, and thus limit the households' equity, in order to guarantee the permanent affordability of the housing units in the cooperative. The extent to which equity is limited ranges from zero-equity agreements, in which resident-shareholders receive back what they initially put in when they decide to leave the cooperative, to agreements that allow for some price inflation and a return on any additional investments that might have been made. While such equity restrictions are legally enforceable if written into the governing by-laws of the cooperative, there is always the real possibility of the cooperative members dissolving the co-op or lifting the equity restrictions. In order to guard against this possibility, and ensure that the interests of the current resident-shareholders do not displace the long-term interests of the community, third-party agreements are often part of the conversion of a rental building to a limited equity co-op. This can either be with a government agency that is subsidizing the co-op, a community organization, or sometimes a community land trust.

Community land trusts

The community land trust (CLT) 'model' was created in the late 1960s by civil rights activists who were concerned about the decline in black property ownership in the South. These concerns led to the creation of the Institute for Community Economics in Massachusetts (Institute for Community Economics, 1982). As its origins would suggest, the model was originally constructed as a rural preservation model designed to protect farmers from creeping and speculative development, or foreclosure of family farms by hostile financial institutions. Since the early 1980s, hostile landlords and financial institutions have imported it into urban areas to protect affordable housing from creeping and speculative development, or the abandonment of affordable housing. There are 86 CLTs already operating in the US, and there are an additional 32 in the process of being created. Collectively these contain about 5500 units of housing, in which almost 75 percent of residents earn less than half of their area's median income (DeFilippis, 2002). A disproportionate number of these are in the north-east, and New England in particular. It should also be stated that the CLT model's recent increase has largely been in areas of rapid economic growth in the last 15 years – the San Francisco Bay area and the Pacific north-west, where anti-sprawl measures and large in-migrations of high-paid labor have made land values and housing affordability substantial policy and political issues.

In a CLT, a community organization owns and manages the land, while the residents 'own' (cooperatively through either an MHA or a limited-equity cooperative, or sometimes individually or even as affordable rental units) only the housing units located on the land. The central feature of land trusts is the separation of buildings from the land they are upon. The individuals 'own' the housing by virtue of long-term (usually 99 years, with inheritance rights) ground leases that they sign with the organization keeping the land in trust. The housing costs paid by residents vary from CLT to CLT, but in all cases there are strict limitations on housing costs and the resale price that can be sought by resident 'owners' when they choose to leave. Additionally, CLTs always possess the right of first refusal when a unit is to be sold. The argument is simply that people should collect on any investments they make in the units, but any further appreciation in the housing value is socially created and therefore does not belong to the individual.

The governing boards of CLTs are usually comprised as follows: one third lease-holders (residents); one third non-resident community members; one third public officials. CLTs thus limit resident control (relative to MHAs or limited-equity co-ops) and emphasize larger community and public control. This reflects the understanding of the communal nature of land that is a constant theme within the remit of land trusts.

Mutual housing associations

Mutual housing associations (MHAs) emerged in the early 1980s after a Congressional delegation went to Germany to study the German mutual housing model. The first MHA in the US was established in Baltimore in 1982, and in the years since then their numbers have grown to include over 30 different MHAs operating in the country. That growth, however, primarily took place through the 1980s and early 1990s. Numbers have stagnated in the years since then. Nationwide, almost half of all MHA residents earn less than 50 percent of their area's median income (AMI), with another 40 percent earning between 50 percent and 80 percent of AMI. So they are serving a distinctly low-income population, but not the poorest of the poor (DeFilippis, 2002).

An MHA is similar to a large-scale limited-equity housing cooperative, although it differs from most co-ops in that the residents do not own shares in the cooperative of their residence (which could then be resold when they move). Instead, the housing is off the market entirely and ownership of it is permanently in the hands of the association. Residents rent their units from the association but, at the same time, the residents constitute the association. The model is therefore substantially different from housing produced in either the for-profit real estate market, or the not-for-profit

CDC sector. While MHAs are similar in some respects to CDCs (primarily in that they build affordable housing for low-income people), they differ in two basic ways. First, while most CDCs advocate traditional forms of land and housing tenure, albeit more affordably, MHAs reject these, and instead promote and implement a form of mixed ownership, with individuals and communities possessing different components of the rights, interests, and responsibilities of ownership. Second, while many CDCs are participatory in character, and understand organizing to be an integral part of what they do, many others do not. Conversely, community and resident participation is written into the governance structures of MHAs, which, accordingly, stress this component of their development work.

MHAs are expansionist organizations, having the explicit goal of taking more and more housing out of the private market and moving it into collective ownership (which further distinguishes them from most limited-equity housing cooperatives, since they are largely stand-alone entities). There are two reasons for this. First, since they believe in the model of mixed collective/individual ownership, they want to bring more units into this framework. Second, the greater the number of units owned and controlled by the groups, the bigger the economies of scale realized by the organizations in the maintenance and management of the properties. Prospects for realizing their goal of fiscal self-sufficiency are substantially increased. Costs for MHAs are front-ended in the acquisition process, making the units more affordable and the organizations more fiscally viable over the long term. Finally, the collective ownership is structured so that the units are permanently taken off the private housing market.

In MHAs the residents pay monthly housing charges to the association, which can either be fixed, with periodic adjustments just to keep pace with inflation, or are paid by the residents as a percentage of the residents' incomes (between 26 and 29 percent). In either payment system, the logic that determines the cost of the housing is not the market, but the agreed policies of the MHA. Residents also contribute to the financing of their units and buildings by performing mandatory maintenance work, often as much as ten hours a month. Along with their monthly housing charges and maintenance work, in most MHAs residents must also pay a one-time membership fee ranging from around one thousand to several thousand dollars, which is used to help capitalize the associations. This fee earns interest, is refundable when the resident leaves the association, and is often viewed by residents and staff alike as an initial, if limited, form of equity. Residents who can't afford this initial investment are able to borrow the money from their association, and pay it back over an extended period of time.

Conclusion: the contours of change in community and property

This chapter began with a discussion of contemporary theories of community development. It critiqued such theories for both assuming away all differences within communities and emptying those communities of social relations. It argued that current practices of community development are such that they often facilitate and support gentrification and displacement. An alternative viewpoint was then presented that made a case for an understanding of community that recognizes social relations in place, and therefore advocated for forms of development that build on those relations. These forms of property development not only connect poor people to places, but they also do so in ways that allow those people to realize some individual wealth. They have community and/or resident control built into their governance structure, and thus they allow low-income residents to capitalize (in the literal economic sense of that word) on their social relations. In short, they are worthy of a great deal more attention and support than they have received thus far. But given the trajectory of the state philosophies and policies in both the UK and the US, this support is unlikely to occur.

References

Aizcorbe, A., Kennickell, A. and Moore, K. (2003) Recent changes in US family finances: evidence from the 1998 and 2001 survey of consumer finances. *Federal Reserve Bulletin* January: 1–32.

Bendick, M. and Egan, M.L. (1991) *Business Development in the Inner-City: Enterprise with Community Links*. Community Development Research Center. New School for Social Research, New York.

Berger, R. and Steinbach, C. (1992) *A Place in the Marketplace: Making Capitalism Work in Poor Communities*. National Congress for Community Economic Development, Washington, DC.

Blakely, E. and Small, L. (1995/1996) Michael Porter: new gilder of ghettos. *The Review of Black Political Economy* **24**: 161–83.

Brenner, N. and Theodore, N. (eds) (2002) *Spaces of Neoliberalism: Urban Restructuring in North America and Western Europe*. Blackwell Publishing, Oxford.

Brown, G. (2004) Keynote Speech to the National Council for Voluntary Organizations Annual Conference. http://www.ncvo-vol.org.uk. (Accessed 7 October 2005.)

Carr, J. (1999) Community, capital and markets: a new paradigm for community reinvestment. *The Neighborworks Journal* Summer: 20–23.

Christman, J. (1994) *The Myth of Property: Toward an Egalitarian Theory of Ownership*. Oxford University Press, New York.

Community Development Partnership Network (2003) *Re-Shaping the Business of Community Development*. Community Development Partnership Network, Denver.

Dávila, A. (2004) *Barrio Dreams: Puerto Ricans, Latinos and the Neoliberal City*. University of California Press, Berkeley.
Davis, J. (1991) *Contested Ground: Collective Action and the Urban Neighborhood*. Cornell University Press, Ithaca.
DeFilippis, J. (2001) The Myth of Social Capital in Community Development. *Housing Policy Debate* **12** (4): 781–806.
DeFilippis, J. (2002) Equity vs. equity: community control of land and housing in the United States. *Local Economy* 149–153.
DeFilippis, J. (2004) *Unmaking Goliath: Community Control in the Face of Global Capital*. Routledge, New York.
Eichler, M. (1998) Look to the future, learn from the past. *Shelterforce* 101 (September/October).
Epstein, R. (1996) The de-activist. *City Limits* October: 18–23.
Freeman, L. and Braconi, F. (2004) Gentrification and displacement: New York City in the 1990s. *Journal of the American Planning Association* **70** (1): 39–52.
Fyfe, N. (2005) Making space for 'neo-communitarianism'? The third sector, state and civil society in the UK. *Antipode* **37** (4): 536–57.
Gittell, R. and Vidal, A. (1998) *Community Organizing: Building Social Capital as a Development Strategy*. Sage, Thousand Oaks.
Grogan, P. and Proscio, T. (2000) *Comeback Cities*. Westview Press, New York.
Imbroscio, D. (2005) *Shaming the Inside Game: A Critique of the Liberal Expansionist Approach to Addressing Urban Problems*. Unpublished manuscript. Department of Political Science, University of Louisville, Louisville.
Institute for Community Economics (1982) *The Community Land Trust Handbook*. Rodale Press, Emmaus.
Kolodny, R. (1987) The emergence of self-help as a housing strategy for the urban poor. In: Bratt, R., Hartman, C. and Meyerson, A. (eds) *Critical Perspectives on Housing*. Temple University Press, Philadelphia.
Kretzmann, J. and McKnight, J. (1993) *Building Communities from the Inside Out: A Path Toward Finding and Mobilizing a Community's Assets*. ACTA, Chicago.
Krueckeberg, D. (1995) The difficult character of property. To whom do things belong? *Journal of the American Planning Association* Summer: 301–309.
Leavitt, J. and Saegert, S. (1990) *From Abandonment to Hope: Community-Households in Harlem*. Columbia University Press, New York.
Lemann, N. (1994) The myth of community development. *New York Times Magazine*, 9 January: 27–31, 50, 54, 60.
Levitas, R. (2000) Community, utopia and New Labour. *Local Economy* **15** (3): 188–197.
Logan, J. and Molotch, H. (1987) *Urban Fortunes: The Political Economy of Place*. University of California Press, Berkeley.
McKnight, J. (1995) *The Careless Society: Community and its Conterfeits*. Basic Books, New York.
Mollenkopf, J. (1981) Community and accumulation. In: Dear, M. and Scott, A. (eds) *Urbanization and Urban Planning in Capitalist Societies*. Methuen, New York.
Nowak, J. (1998) Expanding the scope of community development. *Shelterforce* 97 (January/February).
Pacenza, M. (2002) East Harlem's bottom line. *Shelterforce* 125 (September/October).
Pierce, N. and Steinbach, C. (1987) *Corrective Capitalism: the Rise of America's Community Development Corporations*. Ford Foundation, New York.

Pitcoff, W. (2003) Has homeownership been oversold. *Shelterforce* 127 (January/February).

PolicyLink (2002) *Equitable Development Toolkit*. PolicyLink, Oakland.

Porter, M. (1997) New Strategies for Inner-City Economic Development. *Economic Development Quarterly* **11** (1): 11–27.

Putnam, R. (1995) Bowling alone: America's declining social capital. *Journal of Democracy* **6**: 65–78.

Putnam, R. (1996) The strange disappearance of civic America. *The American Prospect*, Winter: 34–48.

Putnam, R. (2000) *Bowling Alone: The Collapse and Revival of American Community*. Simon & Schuster, New York.

Rae, R. (1997) *Ownership and Equity: Perceptions of Home Ownership by Low Income Owners of Limited Equity Cooperative Housing*. Ph.D. dissertation Graduate Faculty of Psychology, City University of New York.

Shragge, E. and Fisher, R. (2001) *Community Development Practices: New Forms of Regulation and/or Potential for Social Change*. Paper presented at the annual meeting of the Urban Affairs Association, April, Detroit.

Siegler, R. and Levy, H.J. (1986) Brief history of cooperative housing. *Cooperative Housing Journal* 12–19.

Sites, W. (1998) Communitarian theory and community development in the US. *Community Development Journal* **33** (1): 57–65.

Smith, N. (2002) New globalism, new urbanism: gentrification as global urban strategy. In: Brenner, N. and Theodore, N. (eds) *Spaces of Neoliberalism: Urban Restructuring in North America and Western Europe*. Blackwell Publishing, Oxford.

Stone, M. (1993) *Shelter Poverty*. Temple University Press, Philadelphia.

Weissbourd, R. and Bodini, R. (2005) *Market-Based Community Economic Development*. The Brookings Institution Metropolitan Policy Program, Washington, DC.

Wilson, W.J. (1987) *The Truly Disadvantaged: The Inner City, the Underclass, and Public Policy*. University of Chicago Press, Chicago.

13

Communities, Networks and Power: Is there a Localization of Decision-Making?

Rob Rowlands and Pauline Card

Introduction

Locally based community participation in policy and decision-making processes has become a central tenet of government policy. This has especially been the case since the election of the Labour government at the 1997 general election. The onus on increased participation has been particularly prominent in the areas of housing and regeneration, notably the transfer of council housing stock to not-for-profit bodies and regeneration initiatives such as New Deal for Communities[1], Neighbourhood Renewal Fund[2] and Local Strategic Partnerships[3]. However, the extent and nature of the participation has been criticized. Questions have been raised about the definition of community, the representativeness of those consulted or allowed to participate, and indeed the quality of the participation. The importance of community participation at different scales of decision-making can be seen

[1] New Deal for Communities (NDC) is a key programme in the Government's strategy to tackle multiple deprivation in the most deprived neighbourhoods in England. The aim is to bridge the gap between these neighbourhoods and the rest of England. NDC partnerships are tackling five key themes of: poor job prospects; high levels of crime; educational under-achievement; poor health; and problems with housing and the physical environment.
[2] The Neighbourhood Renewal Fund (NRF) aims to enable England's most deprived local authorities to improve public services, narrowing the gap between deprived areas and the rest of the country.
[3] Local Strategic Partnerships (LSPs) are single non-statutory, multi-agency bodies, which match local authority boundaries, and aim to bring together at a local level the different parts of the public, private, community and voluntary sectors.

as a reflection of the move by central government, at least in rhetorical terms, from a centralist approach to policy and decision making to a 'new localism' that devolves power to regional and local levels and enshrines the participation of local stakeholders and communities in the process (Balls, 2002; Corry and Stoker, 2002; Brown, 2003; Milburn, 2003).

This chapter uses an analytical framework of community, power and social networks to explore the role that communities play in newly localized decision-making processes and analyse the extent to which power and decision making are devolved to lower levels of governance. Essentially our concern is whether the participation of communities in decision-making processes is really aimed at empowering local people and local communities or simply a way of legitimizing policy decisions made more centrally.

We begin by exploring the complex nature of 'community', the definitions applied to the term and the forms that a 'community' can take in practice. This is followed by a brief overview of literature on networks, power and influence to give a theoretical framework for understanding the processes that are occurring at a national and local governance level. The following section sets out the policy context within which the process of stock transfer is being undertaken, firstly at a broad, overarching government policy level and secondly, at the more detailed local authority level, the centrality of community participation will be highlighted. We conclude the chapter by applying the previously identified issues to empirical research on the involvement of black and minority ethnic (BME) tenants and communities in the option appraisal phase of the stock transfer process.

Communities, power and networks

Community

Central to this chapter is the notion of community, a concept that has occupied academics and policy makers for many years and remains contested and imprecisely applied or evoked. In public policy debate in the UK the term community is used in three general senses (Taylor, 2003, p. 34):

- Descriptive: a group or network of people who share something in common or interact with each other.
- Normative: community as a place where solidarity, participation and coherence are found.
- Instrumental: community as an agent acting to maintain or change its circumstances; the location orientation of services and policy interventions.

Communities are often defined or described in terms of common characteristics. These may be personal characteristics, common beliefs, activities and users or providers of services. However, the presence of common characteristics does not necessarily mean that people sharing those characteristics would identify themselves as a community; this requires that they have common interests around which they can mobilize. It is common interests which transform community from a description into an active agent. Common interest may be derived from a number of sources, including cultural heritage, social relationships, economic interests and experience of power or oppression, and a community may have multiple common interests (Taylor, 2003). Individuals may belong to many separate and overlapping communities with different communities and interests taking priority at different times.

The term community has also long been used in a normative sense to comment on the way in which we live our lives and, in particular, contrast the impersonal and contractual nature of social ties to be found in industrial and post-industrial society. Concerns with the 'loss of community' can be traced back to the work of Tonnies (1955) who contrasted the positive attributes of *Gemeinschaft* (village life) with the negative attributes of *Gesellschaft* (urban life). Community in this sense is generally geographically based with common interest arising from social interaction, friendship and kinship ties and common moral values.

Among policy makers there is a tendency to confuse descriptive and normative definitions. Common location or interests are assumed to be accompanied by social and moral cohesion, a sense of security and mutual trust. This assumption is then taken a step further in that it is assumed that norms will be turned into action: that community can be turned into agency. Further confusion can be seen in the use of community to describe the location of service provision, the services provided or the point of contact with the public (Taylor et al., 2000). Such uses can be descriptive but they can also imply beliefs about the way services should be provided or be used symbolically to ascribe values of trust, cohesion and security to policy interventions that may or may not be underpinned or characterized by such values (Taylor, 2003).

In terms of the stock transfer process[4] which is examined in this chapter, the community that is encouraged or expected to participate is the tenant body of the local authority who, it could be argued, have a common interest in their desire for decent homes and a good-quality housing service. However, this would be an over-simplistic definition of the situation. There are many different communities within the tenant body which may have

[4] The transfer of local authority owned and managed housing stock to a not-for-profit housing organization.

different and competing perspectives on the future direction of council housing. The most visible communities are the geographically separate council estates that may or may not be socially cohesive or homogeneous. Within and across the estates there will be other communities that can be identified by their cultural heritage, age, gender, marital status or political ideology, for example, that have very different interests and very different views of the desired outcome for the stock transfer process. Each community may wish their views to be heard but they may not have the influence or power to gain access to the decision-making forums, or the leaders or individuals who are selected to represent particular communities and have access to such forums may not voice their concerns or views.

Definitions of power

If we are to understand the ways in which communities engage in the decision-making process, particularly with regard to who possesses power and influence over whom at the point of decision, we require a way of understanding the nature and application of such power or influence. Early classic studies of organizations identified three dimensions or faces of power. The first dimension, arguably a pluralist view, saw a finite amount of power as being distributed amongst actors, both individuals and organizations, who exercise influence or authority over others (Weber, 1947). Imbalances in power were seen as ensuring that certain interests (generally those of ruling elites) prevail: the capacity to organize or 'mobilize bias' is seen as central to this first dimension (Dahl, 1961). Bachrach and Baratz (1962) identified a second dimension of power which is exerted more surreptitiously than the first dimension by suppressing controversy, restricting opportunities to challenge prevailing norms and institutions. This is important in terms of devolving power down to different levels of governance, because power can be retained at higher levels by controlling the decisions that are to be made and constraining or limiting the options available to the decision-makers, i.e. setting the agenda.

The third dimension of power espoused by Lukes (1974) is seen as protecting the interests of the elite classes through inducing attitudes and practices amongst ordinary people that adhere to accepted norms (Gilchrist, 2004). Once again, this dimension is important in relation to 'community participation' that is invoked through a 'top down' process because of the importance placed on information provision and the ability of the local authority to select which groups or organizations are allowed to participate and those that are excluded.

Using this analytical framework, it can be argued that power and influence can be exercised differentially across, within and between communities, between actors and vertically between different levels of decision making. This differentiation leads to voices which can be heard with different degrees of volume. If power or influence is to be exercised in decision making by the various 'communities', however they are defined, the questions are how and by whom are the views of the community raised. If local authorities, due to time and resource constraints, work with the 'usual' suspects, how are such individuals identified and designated as 'community leaders' by those seeking community views? How do they become part of the 'networks' of participation?

The capacity to participate

Over recent times, in policy arenas, ideas of community have been linked with the presence or absence of social capital. Social capital is seen normatively as the ability or capacity of 'communities' to engage with each other and with groups, organizations, institutions that are involved in a particular policy area. Of particular influence within the British policy arena has been the work of Robert Putnam (1993, 2000), who argues that 'social capital'

> '. . . refers to connections among individuals – social networks and the norms of reciprocity and trustworthiness that arise from them.' (Putnam, 2000, p. 19)

However, in the context of the localization of decision making and community participation and representation, it is more enlightening to consider Bourdieu's notion of 'social capital':

> '. . . the aggregate of the actual or potential resources which are linked to possession of a desirable network of more or less institutionalized relationships of mutual acquaintance or recognition – or in other words, to membership in a group – which provides each of its members with the backing of the collectively-owned capital.' (Bourdieu, 1997, p. 51)

For Bourdieu, 'social capital' is one of four types of capital constituting his concept of *habitus*, which is defined as a way of visualizing the world. Many competing versions of *habitus* may exist at any one time. The other forms of capital are economic (i.e. the financial resources available to an individual), cultural (i.e. the skills, knowledge and credentials that are acquired by upbringing and education that enables accumulation of economic capital and other forms of power) and symbolic capital (i.e. the demonstration of the aesthetic taste of an agent).

Habitus provides a view of society as a set of power relations between groups, enacted through differential utilizations of the forms of capital (Flint and Rowlands, 2003, p. 217). Therefore, *habitus* is useful for understanding the formation of 'communities' of interest, whether these are geographically based or otherwise, that share the same world views. It also describes the relationships between individuals who respond to and alter social structures and divisions within society (Dovey, 2002; Hillier and Rooksby, 2002). It does not, however, confer elite status on particular groups nor does it confer power on any of the groups (communities) formed. *Habitus* does, however, create the conditions for the creation of elite groups and their exercise of power over others (Hall and Rowlands, 2004).

The concept of *habitus* and, in particular, the notion of social networks (social capital), that can provide access to groups that hold power over others or have the power to set the agenda in areas of policy making, is pertinent in the case of community participation and empowerment. Such networks can be inclusive or exclusive and some groups, or individuals within groups (or communities), will have the social capital to gain access while others will not have those links and ties. Some groups will share the world view of those groups who control access to the decision-making elites and will be seen as worthy of admission, while groups that have a differing world view may be excluded. For example, community leaders who are already part of pre-existing networks that have engaged with the local authority previously and are known to be non-obstructive are more likely to be invited to take part in the 'elite' decision-making forums, while those known to be antagonistic or oppositional to the local authority's view are less likely to be given access. The representativeness of those 'favoured' individuals of the communities that they represent may also be questioned.

Having outlined the theoretical concepts that will later be used to explore the issues and problems that arise from the devolution of decision-making processes, the following section sets out the policy context that is to be explored. The concept of 'new localism' underpins a growing commitment to community engagement and one policy area in which this has been implemented has been the transfer of council housing in England to not-for-profit landlords.

Policy context

New localism

Since the 1997 Labour election victory there has been a number of phases to the reform of public services (Corry and Stoker, 2002): 'chaotic centralism', 'steering centralism' and, more recently, 'new localism'. Each provided different policy and regulatory frameworks within which local authorities and other service providers have had to develop locally relevant policies and implementation structures to address centrally identified issues. The initial phase, or 'chaotic centralism', was characterized by a two-pronged approach: the development of a plethora of centrally devised initiatives, strategies, plans and zones, that encouraged 'joined-up' or partnership working alongside a heavy-handed inspection regime. The result was 'initiativitis':

> '. . . the syndrome in which public managers end up swamped by the volume of special projects, discretionary funds and demands to produce plans . . . Too often the politicians' demands for "quick wins" can stifle integration: it takes time to build trust between agencies, identify the correct focus and outcome measures, and develop the right strategy.' (Perri 6 et al., 1999)

This approach produced some success in selected policy areas and the inspection regime saw improvements in service provision. However, it was at great cost in terms of management and financial resources, the stifling of innovation and the creation of a culture averse to risk (Davis et al., 2001). There was little opportunity for the involvement of service providers, service users or local communities in both identifying and resolving problems.

The second phase, 'steering centralism', emerged as the New Labour government recognized that it

> '. . . is simply not possible either to run economic policy or deliver strong public services that meet public expectations using top-down one-size-fits-all solutions of the past.' (Balls, 2002, p. 6)

This phase is characterized by the centre steering and facilitating local service providers through the encouragement of local discretion in service delivery alongside a commitment to national standards. 'Earned autonomy' is a key concept (OPSF, 2002, p. 17): where service providers that are rated highly in inspection are given more freedom and flexibility and a 'lighter touch' inspection regime. Examples of 'earned autonomy' can be seen across a number of policy areas: foundation hospitals in the National Health Service and foundation schools in education. The best performing local authorities are able to add financial borrowing and ALMOs[5] to the options available when trying to achieve stock condition standards. Conversely, the choices available to poor performing local authorities, and therefore their tenants, are much more constrained by the regulatory regime.

While 'steering centralism' can be seen as addressing some of the problems of 'chaotic centralism', the term 'earned autonomy', it is argued, can only

---

[5] Arm's length management organization – a vehicle for distancing the management of local authority housing stock from the authority while retaining ownership.

be understood in a system where the centre 'calls the shots' (Wright, 2002). The limitations of both centralist approaches have been acknowledged by the government (Milburn, 2003) leading, it is suggested, to a move towards a 'new localism' (Balls, 2002; Corry and Stoker, 2002; Brown, 2003).

'New localism', as defined by Corry and Stoker (2002), refers to the decentralization of decision making, power and control from central government. While the devolution of power and decision making has occurred at different scales of government and across many areas of policy and service provision, this chapter concentrates on the role of local authorities in the devolution process, particularly in their role as social housing providers. More significantly however, the notion of 'new localism' embraces community participation and leadership. The policy of 'new localism' encourages local ownership of decision-making processes by stakeholders and local accountability for the decisions taken. Some have argued that a move to 'new localism' can be seen as inequitable and non-redistributive. This is because better resourced, higher performing local authorities benefit above less well resourced, poorly performing authorities. There is also a need for central government to defend the rights of some minority 'communities' to local services, as 'local decision-making can be capricious, short-term and selfish' (Walker, 2002). Furthermore, the centrality of local leadership and accountability to the vision and success of 'new localism' raises concerns about the capacity of local leaders to deliver together with the inadequacy of democratic processes in convincing people to trust local government.

The power that local authorities and local stakeholders have been given is not, however, unfettered; there is a requirement to meet nationally set targets in certain areas of policy, a process identified as 'constrained discretion' (Balls, 2002). An area of policy which exemplifies this process has been in the future investment plans and decisions for English council housing. Here, local decisions and priority setting have been overshadowed by the need to meet the nationally determined Decent Homes Standard[6]. The challenge of meeting this target and finding the large-scale capital investment required has obliged local authorities to consider options for their stock that had previously been dismissed. While the large-scale transfer of council housing has been an option available to local authorities since the late 1980s, and therefore predates the move towards 'new localism', the objectives of transfer have changed over time, as have the inducements to transfer provided by central government (Malpass and Mullins, 2002).

---

[6] The Decent Homes Standard is a benchmark of quality to be attained by all local authority and housing association owned housing by 2010.

Stock transfer

The large-scale transfer of local authority owned housing to not-for-profit landlords, most commonly housing associations, was first introduced in the 1988 Housing Act. Within the system it has been the local authorities themselves (initially the officers and then members) who have decided to propose transfer. Where the option has been financially viable, local authorities have proceeded to a ballot of their tenants. This test of opinion was the opportunity for all tenants to accept or reject the proposal – but on the local authority's terms. Stock transfer was designed as both a means of enabling tenants to leave local authority control and at the same time reduce the size of the local authority housing sector (Hickman, 2004). It has been part of a wider policy of introducing choice and competition to public services and underlying this has been the ideological pursuit to reduce inefficiencies of bureaucratic provision of public services.

The transfer of local authority housing to not-for-profit landlords is now part of the public policy landscape and is a major factor in the changing shape of social housing provision in the UK. Since the election of the Labour government in 1997 there has been an increase in the pace of the transfer programme and the nature of those local authorities taking part. The 2000 Housing Green Paper, *Quality & Choice: A Decent Home for All* (DETR & DSS, 2000), outlined the Decent Homes Standard. Many local authorities have been opposed to housing transfer on the basis that it reduces municipal power and control and also is a step towards the privatization of public services. However, meeting the Decent Homes Standard will be financially challenging and increasing numbers of local authorities are now going down the path of stock transfer.

Increasingly, transfers have been in more deprived urban areas. Murie and Nevin (2001) highlight that in the period up to 1998, only two of the 68 transfers to take place were in disadvantaged urban areas. More importantly this represented only 1.1 percent of the total stock transferred. Since 1998 the picture has changed: the extension of the Estate Renewal Challenge Fund[7] in 1998 has been an aid to negative value stock transfers in England, encouraging 32 transfers in 18 local authorities. The majority of these have been in urban areas with high levels of social and economic deprivation, with 19 of these transfers taking place in inner London boroughs alone. In theory four options may be applied to local authority housing stock: voluntary transfer to a not-for-profit landlord; transfer of management to an arm's length management organization (ALMO); the adoption of a

[7] The Estates Renewal Challenge Fund (ERCF) ran from 1996 to 2000. Under the fund the Office of the Deputy Prime Minister (and its predecessors) provided grants to Registered Social Landlords to compensate them for taking over homes that had a negative value (NAO, 2003).

private finance initiative contract; and retention of ownership and management of the stock by the local authority. In reality, local authorities are constrained in their choice, an illustration of the 'constrained discretion' highlighted above. For those authorities who cannot financially meet the obligations of the Decent Homes Standard if their stock is retained, transfer via one of three options is unavoidable. Furthermore, management performance and the availability of government support restrict the application of ALMOs and PFI to high-performing and relatively well resourced authorities to the extent that the government has effectively eliminated the option of stock retention for all but a handful of local authorities (Weaver, 2004b).

Tenant involvement

The government has stated that tenant involvement is at the centre of stock transfer policy (ODPM, 2004, 2005). At its inception, Conservative transfer policy was predicated on tenants articulating *choice* to leave the local authority sector. Tenant involvement was perceived as a catalyst for change rather than a hurdle to negotiate. Equally, there has been an obligation to hold a ballot where councils wish to pursue a large-scale voluntary transfer and this duty has been the cornerstone of tenant involvement. A constant key element of stock transfer has been the need to secure a 'yes' vote from tenants; this has meant that the engagement of tenants in the process has always been a necessity.

However, as transfer policy has evolved under the Labour government, the requirements for tenant involvement have increased. More recently community participation has been encouraged at earlier stages of the decision-making process. Local authorities are now required to demonstrate that reasonable steps have been taken to involve a large number of tenants during the stock options appraisal and that tenants are empowered to participate in the governance of resulting transfer structures. Table 13.1 illustrates the role of tenant involvement in each of the stock options.

Prior to proceeding with any of the available options, ODPM[8] requires that local authorities demonstrate that they have involved tenants in the appraisal of options for stock. Continued dialogue with tenants is encouraged and this is facilitated through the appointment of an independent tenants' adviser, whose role is to provide advice to tenants about the stock transfer process.

Within the transfer process there is a distinction between communication strategies and empowerment strategies (ODPM, 2003a, b). The latter

---

[8] Office of the Deputy Prime Minister is the new and expanded government department which took on some of the functions of DETR following the 2001 Labour election victory.

Table 13.1 Stock options: main characteristics of management and tenant involvement (adapted from Mullins et al., 2004).

Feature	PFI	ALMO	LSVT
Consultation requirements for approval	Must demonstrate that consultation has taken place and residents' views considered	Must consult, evidence of active support	Formal consultation requirement and evidence that majority of tenants not opposed
Ballot	Not anticipated	LA may hold one, but no obligation	Statutory obligation

is primarily focused on the involvement of tenants in high-level governance of the transfer process and subsequent transfer bodies. It inevitably involves only a small number of individuals and therefore can in practice allow for only limited input from the wider tenant body. On the other hand, communication strategies are focused on providing information to all tenants. Whilst these are aimed at reaching a large number of tenants (ideally all tenants), they provide limited scope for input by tenants to decision-making structures.

Formal tests of opinion, most commonly a ballot of tenants, are generally treated as a cornerstone of tenant consultation and involvement. A ballot is the culmination of a longer process of consultation and appraisal and is characterized by a choice of two scenarios. In all cases this is a choice between the status quo of remaining with the local authority or to transfer. To present tenants with a dichotomous decision limits their choice and leads to a problematic outcome – either vote for a transfer or approve the non-improvement of their homes by the current landlord. Question marks have been raised by a number of groups (Defend Council Housing, 2004; Weaver, 2004a) about the actual level of choice that tenants have even within this two-way scenario. With the UK government committed to meeting the Decent Homes Standard there is limited scope for manoeuvre where tenants choose to vote no. The issue of power, the ability to exercise choice (both for local authorities in the options they can feasibly consider and the tenants in their scope within the ballot) and predetermined decisions loom large within this process. If organizations that hold significant power over communities are seen to ignore the very communities they are seeking to involve and consult, there is a danger that they distance themselves from these communities and create 'hard-to-reach' groups.

'Hard-to-reach' communities

As government policy is steered towards localism and the inclusion of both communities and service users in the development and implementation of policy, so there has been discussion of inclusivity. A discourse has developed in which 'hard-to-reach' groups are highlighted both as a concern because of their non-involvement and as a target for increased attention. Local authorities and housing associations are encouraged to identify these groups and develop new and 'innovative' methods to overcome the barriers to their involvement. There is an assumption that organizations can identify such groups and individuals, but by their very nature, 'hard-to-reach' groups are often hidden within the wider population and their identification is problematic. Often organizations have objectives to provide information to and engage with 'hard-to-reach' groups, but fail to demonstrate that they have a local understanding of who these groups are and of whom they are constituted.

One definition of 'hard-to-reach' groups is that they are 'those with a soft voice' (Oxford City Council, 2004), individuals and communities who are under-represented in consultation and who do not readily engage in participatory mechanisms. Other definitions include:

> '. . . those who do not readily attend, or do not have the opportunity to attend, first-tier consultation mechanisms. Although they may be "picked up" on other types of research, it does not give these groups the opportunity to provide an open forum where they can raise their concerns and become involved in the consultation process.' (Gwent Police Authority, 2004)
>
> '. . . those excluded from day-to-day life' (Medway Council, 2004).

This process assumes that there are specific groups within the community who are 'hard-to-reach'. The question is perhaps better raised in terms of who is hard-to-reach. 'Hard-to-reach' groups are those sections of the community that organizations with a consultative role, such as local authorities and housing associations, have difficulty engaging with. The term itself implies that it is the organization which has difficulty in reaching some of its clients/customers. However, at least one local authority in England[9] has abandoned the term in favour of the phrase 'difficult to hear' groups. Although this may appear to acknowledge the organizations' difficulty in communication it may also be interpreted as putting the responsibility on the consumer to shout more loudly.

There are many different groups who may be considered 'hard-to-reach', yet there are notable groups who are 'hard-to-reach' in numerous situ-

---

[9] Anecdotal evidence from a session with tenant participation managers at an English metropolitan local authority during research for ODPM looking at the involvement of black and minority ethnic tenants in housing investment decisions (see Mullins et al., 2004).

ations. These include, but are not restricted to younger people, black and minority ethnic communities, people with disabilities, the elderly and people with literacy problems. There is no generalized list of types of groups which constitute hard to reach in all situations and the breadth of categories encapsulates the difficulties faced in involving tenants. In society very few people attend meetings to discuss matters that affect their communities. It is difficult to encourage citizens to participate in surveys without incentive. Yet organizations still use these traditional and formalized methods to involve those in their communities who have limited resources or abilities to participate. In this context it is clear to see why those with power, such as local authorities, assume that communities are hard to reach. Therefore, the more important question to pose is why are groups hard to reach?

'Hard-to-reach' or 'difficult-to-hear' groups are created as a result of power inequalities, in particular the ability of those with more power to involve only those groups with whom they wish to engage in dialogue and filter out those they consider oppositional. In terms of community consultation, dialogue is based around different groups' and individuals' interpretation of environments, plans and proposals. Ardener (1993) has highlighted that the creation of 'muted' groups is the product of power relationships:

> 'One social group's construction (or map) of the world, or a sub-structure of it, may dominate the constructions generated by others to the point where they fail to be overtly expressed, or even to be substantially recorded, articulated or encoded.' (Ardener, 1993, p. 20)

This explanation for the existence and creation of 'hard-to-reach' groups highlights the influence of the three faces of power we outlined earlier in the chapter. In particular it emphasizes both the role of agenda setting and, in particular, the creation of norms as a means of excluding certain voices. Therefore it is of paramount importance that we understand why such groups are formed.

The use of the phrase 'so-called' in relation to hard-to-reach groups does imply that there is some dispute about the term. There are critics who suggest that nobody is 'hard-to-reach', rather that the organizations that purport to engage with these groups are doing so in an ineffective way. It also recognizes that steps can be taken to increase reach into these groups and make them more audible. A tenant participation officer in an English local authority encapsulated the reasons for this:

> 'The tenant doesn't have a problem communicating with the housing officer. The housing officer and organisation have a problem communicating with the tenant.' (Tenant Participation Manager, cited in Mullins et al., 2004, p. 74)

This highlights the situation that it is often organizations that have difficulty engaging with their customers and communities, rather than communities having difficulty engaging with their organizations. Ultimately it is important to understand why organizations have difficulty engaging with communities. The resource limitations that public service organizations face make engagement with all tenants problematic. This situation is further exacerbated for local authorities considering and pursuing stock transfer where both financial resources and time scales for involvement are restricted and the engagement of tenants is perhaps of lesser priority than securing investment in the housing stock. Yet the stock transfer process is highly politicized, the pressures to meet centrally set quality standards are considerable and the need to obtain support for initiatives which secure these is an imperative. Therefore there is a strong motive for local authorities to preclude those groups who do not support proposals for transfer.

Issues of participation

Having set out the policy context within which local authorities are currently operating at both a local and national level, it is appropriate at this point to identify the central concerns of the chapter that are explored in greater depth later in this section. First, it has been clearly identified above that, in the context of the stock transfer process, the move towards the localization of decision making as part of the 'new localism' agenda has been more about rhetoric than action. While the final decision of whether to transfer council housing stock to a not-for-profit organization resides with the local authority and their tenants, the choice is highly constrained by centrally set house condition standards and the lack of resources available if the 'wrong' choice is made. This is more a case of 'steering centralism' than 'new localism', raising questions around the issue of power and where power resides at different stages in the process, and at which point do the different stakeholders in the decision-making process have power, if any. When and how do central government, the local authority, the community and even individuals within the community hold power?

This leads us to our second concern, that of community. Central to both 'new localism' and stock transfer is the participation and involvement of the community in the decision-making process. Clearly within the stock transfer process community refers to the tenant body that have a stake in the future of the council stock. However, this is not a homogeneous group but a collection of diverse groups that are differentially connected to the processes of involvement and participation.

Local authorities have to develop both communication and empowerment strategies for the stock option appraisal but questions are raised over who is consulted and particularly over which groups or individuals are potentially empowered and become influential in the decision-making process. This gives rise to our third concern, that of community leadership and representativeness. Highlighted in the policy and good practice literature is the need to access 'hard-to-reach' groups, with emphasis placed on the identification and involvement of BME communities. Some BME groups may contain small numbers of people and, while they may become involved in the consultation process as individuals, it is more difficult for them to get their 'voice' heard in the decision-making forums where issues of representation are problematic. If commitment to understanding and engaging the entire community is lacking and only 'the usual' suspects, i.e. recognized community activists and leaders are involved, questions are raised about the validity of that representation, an issue that goes beyond 'hard-to-reach' groups to cover all claims to representation.

Who holds power?

Within the policy process a number of stakeholders have differential access to power. This power is used to influence agendas as well as implementing decisions. It is essential to examine who holds what power at each of the levels identified above, i.e. central government, local authority and at the community level.

It is clear within urban policy that central government plays the lead role in setting the agenda for other stakeholders to follow. The development of policies and programmes, together with the setting of financial parameters for publicly funded organizations, limits the scope for innovation and variance in the implementation of policy. Stock transfer is a prime example of this and in particular highlights the parameters set by central government in which other networks must operate.

Increasingly, stock transfer is a mechanism which central government is imposing on local authorities and their communities. Although the government has committed itself to localism, in stock transfer the local decision-making process is accompanied with only limited transfer of power to make these decisions and for many of these authorities stock transfer is almost inevitable if the Decent Homes Standard is to be met. Although the 'decision' is seen to be decentralized to the local authority, the control and power with which to make that decision remains predominantly with the centre. Furthermore, as outlined earlier, 'earned autonomy' for local authorities means that the options available to them are linked to performance rather than the suitability of option to their circumstance. Yet within this scenario the missing factor is the tenants and their role in and scope of choice of future for the housing stock. As Murie and Nevin have reflected:

> 'The scale of the proposed transfer programme is planned to accelerate, while plans for arm's length management organisations for council housing represent a major innovation. In other respects the models for the process of transfer and the resulting arrangements reflect a very uncritical view of previous policy in this area.' (Murie and Nevin, 2001, p. 36)

It is unclear what level of 'innovation' can be exercised in meeting the requirements of the Decent Homes Standard when the opportunities remain constrained by central government control of access to alternative stock options.

Although local authorities remain constrained in their choices to a certain degree, in their relationship with their tenants they retain a considerable amount of power. It is the local authority which dictates which of the four options, and in what form they appear, are to be considered under options appraisal, and it is local authorities who dictate the forms of consultation employed to engage tenants and communities. Ultimately it is the local authority which interprets the outcomes of this consultation and, based on consultation and expert knowledge, decides on the next course of action. Within the consultation elements of this process, local authorities fulfil a key role as gatekeeper, controlling who within the community participates in the consultation process. As highlighted above, there is a distinction in the stock transfer process between communication and empowerment of communities. Whilst the local authority will demonstrate wide communication with its tenants, meaningful interaction is often only facilitated through empowerment and only a limited number of tenants can be empowered. Thus the local authority holds some of the reins of power in being able to filter who is empowered within the community and play a significant role in the creation of 'community leaders'. Evidence from regeneration partnerships has identified that in seeking a quick route to consultation the local authority may take a 'godfather' approach. The 'godfather/godmother' is the individual identified 'to represent their (i.e. other user) views in all contexts, assuming a consensus of interests or views' (Williamson, 1993 cited in Mayo, 1997, p. 13). In this situation the local authority can influence the choice of 'representative' to exclude local activists known for 'causing trouble' in favour of the more 'cooperative' (Collins and Lister, 1996). The establishment of community leaders can also be facilitated by the individual themselves through their own skills and acumen, their privileged status within a specific interest group (e.g. faith community) or their prominence amongst those with power, for example local authorities and housing associations. If formalized routes to participation are followed the importance of network membership becomes more significant. Those leaders who are part of pre-existing participation networks, i.e. the 'usual' suspects, are more likely to be consulted or invited to participate.

Ultimately it should be communities themselves who have the power to select the stock option to be taken forward. However, their power is limited. Only one option, large-scale voluntary transfer, affords a compulsory ballot of tenants. Where a ballot is held, tenants hold the ultimate power of veto through a 'no' vote. Because the ballot is a decision between transfer or non-transfer, the choice is not extended to a series of positive options but rather a choice between improvement and the status quo. The collective 'power' of tenants is therefore of limited use if improvement of their homes is the ultimate goal and their collective action prevents this from taking place within the constraints set by central government.

Which community?

As identified above, the concept of community is disputed and challenging. Within the stock transfer process, definition of community is largely geographically based, i.e. areas of council housing. But within that geographical community there are many other communities that are based on shared interest or values. Many such communities have been defined as 'hard-to-reach' for many different reasons, but within policy literature BME communities are most often cited as being 'hard-to-reach'. Yet we should not assume that they are hard to reach in all circumstances. Variations in local contexts make different groups harder to reach than others in different places. Power relationships, representation of groups and the reach of organizations into communities can change at a very local level according to the community's composition. In a recent research report, it was highlighted that in some areas of London the white population makes up the minority ethnic group and feels excluded from decision making (Mullins et al., 2004). However, it is fair to say that in a majority of cases, BME communities are harder to reach than white communities.

The reasons for them being hard-to-reach are the result of various barriers. Mullins et al. (2004) have highlighted the complexity of these barriers for engaging with BME groups. Rather than being uniform, there is a hierarchy of barriers, ranked according to their prominence, visibility and ease to which they can be overcome. Table 13.2 illustrates these barriers.

By illustrating the barriers to involvement, and thus the foundation of why BME communities can be hard to reach, it is possible to outline further the underlying causes of their disengagement. Difficult barriers are the result of structures and dynamics within the community. The identification of fragmentation within BME groups illustrates that there is no single BME community, but rather a series of communities each with different characteristics. These differences include differential access

Table 13.2 Barriers to involvement of BME communities (Mullins et al., 2004, p.19).

Barrier	Key features
Basic	Not enough information about BME tenants, communities and institutions Language and communication
Intermediate	Mistrust of housing organizations Unrepresentative core staff Lack of corporate leadership Inadequate timescale for consultation Unrepresentative existing tenant organizations
Difficult	Fragmented BME communities 'Minorities within minorities'

to power and decision-making forums, a factor in itself which makes some BME communities harder to reach than others. However, even within BME communities there are minorities within minorities, those within communities who have limited voice or who are not spoken for by those with a voice. Clearly there are dynamics at the inter- and intra-community levels which limit some groups' and individuals' ability to be heard.

However, the basic and intermediate barriers identified above are often the result of practices and procedures of the organizations involved in instigating the consultation. Issues such as unrepresentative staff and mistrust of the organizations by their communities are issues which organizations can begin to overcome: for example, increasing the number of staff who are reflective of the communities the organization serves can foster greater trust. So too can the delivery of the service for which the organization is responsible. But unrepresentative existing tenant structures and time constraints can be used by organizations as an excuse for ineffectual engagement with more difficult-to-reach groups and individuals. Therefore it is difficult to justify that it is the fault of the community alone.

Most importantly, many organizations do not understand who constitutes their community. In the case of housing organizations, supermarket chains that operate loyalty card schemes have greater information about their customers than housing organizations have about their tenants. In a survey of authorities who were proposing to undertake stock transfer (Niner and Rowlands, 2003), most organizations made an assumption about their BME communities based on 1991 census data. Few organizations were able to identify with any certainty the ethnicity of their tenants or where tenants from different ethnic groups lived. If organizations cannot identify who their communities are in their entirety, how can they demonstrate and then seek to enhance engagement with those people who are hard to reach?

Even where there is engagement with BME communities there are question marks as to the representativeness of those who are consulted. A recent survey for the ODPM found that 44 percent of stock transfer authorities used BME community organizations as an access point for consultation (Niner and Rowlands, 2003, p. 23). Whilst such organizations can be a useful access point into communities, they do not necessarily offer unlimited contact with all community members. Firstly, the purpose of some organizations may not be suited for engaging with communities about issues such as stock transfer. Secondly, there is an issue about the reach of these organizations. In one respect, BME community organizations serve a wide membership of these communities and not just those who are council tenants. But equally, and of particular concern to some, is the representativeness of those who purport to speak for the community. In some cases, so-called community leaders may claim to represent the views of all members of the community but in reality have an obscured understanding of opinions and aspirations. While they may be embedded in elite participate networks that are outward looking, they may be less well embedded in community networks that they are seen as representing. For example, an elderly, male spokesperson who has gained community leadership status through his position within a faith hierarchy may be unable to represent the views of young people and women within that community. Issues of power within the community are clearly at play in this respect and highlight how housing organizations and internal community politics can inadvertently collude to create hard-to-reach groups.

Which representative?

It is critical to identify who takes part in the 'leadership' roles afforded in community participation. Whilst the spirit of the policy is to encourage the empowerment of a wide range of members of the community, often the outcome is that the 'usual' suspects are involved. In part this is the product of the process by which the local authority who, constrained by limited time and financial resources, seek the most efficient method of consultation. But there is scope for those who already hold preferential access to power to ensure that they protect their position.

Within the stock transfer process there are opportunities for tenant empowerment. As described earlier, there are different roles for communication and empowerment strategies. Whilst empowerment strategies are aimed at reaching a large number of tenants (ideally all tenants), they provide limited scope for input by tenants to decision-making structures. In essence, the two represent two extremes of a pyramid, as illustrated in Figure 13.1.

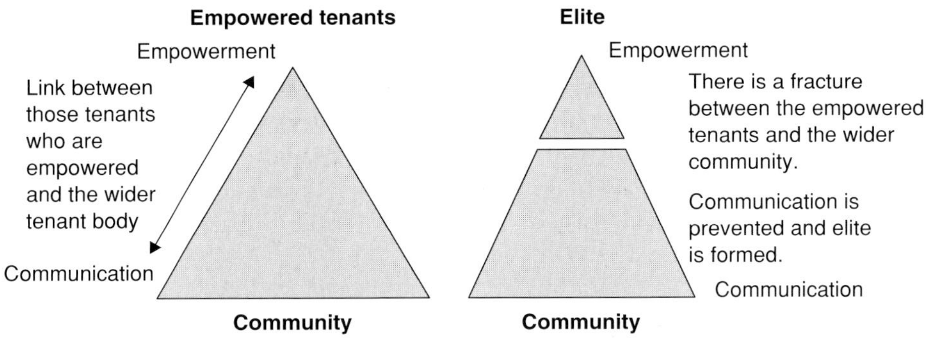

(a) The operation of empowerment and communication strategies should be linked.

(b) In practice, empowerment strategies can assist in developing an 'elite' who are disconnected from the wider community.

Figure 13.1 Pyramid of power in community consultation.

As we have highlighted above, the level of input which tenants have in the stock transfer process is limited. Empowerment strategies are aimed at promoting high-level involvement in the governance of steering groups, shadow boards and boards of management. However, they provide opportunity for only a small number of tenants to become involved and to 'represent' the views of the wider tenant body. At the opposite end of the scale, communication strategies are designed to facilitate the provision of information to all tenants but provide no, or limited, opportunity for wider engagement. Figure 13.1a illustrates the pyramid which is created within society by this. However, in practice two issues arise from the implementation of these strands of policy. First, there is a clear separation between the wider community and those who are supposed to 'represent their views'. Arnstein's ladder of participation (1969) assumes that there is a progression between information provision at the lower end and citizen control at the upper end. However, in the stock transfer process there are notable rungs missing from this ladder which deprive individuals at the lower end of an opportunity from becoming empowered (Rowlands et al., 2003). Secondly, the disconnection created by the lack of opportunities to 'move up the ladder' may help contribute to the establishment of an 'elite' group of tenant activists who are disengaged from the wider tenant body. The outcome of this can be that the apex of the pyramid becomes divorced from the much larger base (Figure 13.1b). Empowerment is not connected to communication and communication becomes a one-way process. Overall this creates imperfect involvement of communities. It may meet the criteria required by government but it does not enable the voices of all the community to be heard. What does emerge is the presence of 'community leaders' who 'represent' the views of other tenants and residents. Often there is no check made to ensure that these individuals are representative and furthermore there is not necessarily any commitment to ensuring that they reflect the range of views which may be expressed at grassroots. There is a dysfunction between communication and empowerment which requires commitment and innovation on the part of the local authority to overcome.

Conclusion

Central to the rhetoric of 'new localism' is the devolution of power and decision making to the local level, tempered with a requirement to meet nationally set standards in certain policy areas. We have demonstrated in this chapter that the decision-making process has been devolved to the local level but the power to make unconstrained decisions remains firmly vested in central government. Central government's control of the decision-making agenda and the constraining of the options available to local authorities, particularly in their attempt to meet the nationally set Decent Homes Standards, illustrate this.

Issues of power hinder further devolution of power to the community level. The process at the local level remains 'top-down' from the local authority. This has a number of implications: once again the options made available to the local communities are limited by the discretion of the local authority and the local authority identifies and chooses who to engage within the process. Part of the latter issue is connected with time and resource constraints whereby organizations engage with easily identified contact points. Furthermore, the chosen representatives may not reach into the communities they are meant to represent nor may they have an understanding of the housing aspirations and choices of their constituents. Choice of representative groups or organizations may also be governed by the local authority's perceptions of the groups' willingness to positively support the local authority's proposals: that is, adhere to the commonly accepted norms and values.

At the local level it is a prerequisite that communities will participate in the decision-making process. However, local authorities often fail to understand the complex nature and composition of 'communities' within their areas. In particular, few local authorities are able to demonstrate a detailed knowledge of the demographics and socio-economic characteristics of their tenants. This makes the identification of 'hard-to-reach' groups problematic, especially if such groups are small, diverse and isolated. Lack of information can lead to assumptions being made about the 'communities' that exist which in turn may lead to representative structures that reflect the local authority's construction of their target audience rather than the reality of the communities. Also, in involving more established, larger networks or groups in the participation process, small groups either have to join larger networks and risk losing their voice, or 'punch above their weight' in getting their voice heard.

Communities hold a wide variety of views on their housing choices and aspirations. In theory the basis for involving communities in these decision-making processes is to reach a consensus on the direction of policy and the outcome for all stakeholders. However, in reality the involvement of communities can be seen as a legitimizing process for decisions predetermined at higher levels. Our contention is that community participation is being used to rubber stamp policy decisions rather than as a tool to empower the communities involved.

References

Ardener, S. (1993) Ground rules and social maps for women: an introduction. In: Ardener, S. (ed.) *Ground Rules & Social Maps for Women*, pp. 1–30. Berg, Oxford.

Arnstein, S. (1969) A ladder of citizen participation. *Journal of American Institute of Planners* **35** (4): 214–24.

Bachrach, P. and Baratz, M.S. (1962) Two faces of power. *American Political Science Review* **56**: 947–52.

Balls, E. (2002) Foreword. In: Corry, D. and Stoker, G. (eds) *New Localism*. New Local Government Network, London.

Bourdieu, P. (1997) The forms of capital. In: Halsey, A.H., Lauder, H., Brown, P. and Stuart Wells, A. (eds) *Education: Culture, Economy, Society*, pp. 46–58. Oxford University Press, Oxford.

Brown, G. (2003) *New Localism – Empowering Local Communities*. Speech made at the University of Wolverhampton, 7 February.

Collins, C. and Lister, J. (1996) Hands up or heads up? Community work, democracy and the language of 'partnership'. In: Cooke, I. and Shaw, M. (eds) *Radical Community Work*. Moray House Publications, Edinburgh.

Corry, D. and Stoker, G. (2002) *New Localism*. New Local Government Network, London.

Dahl, R.A. (1961) *Who Governs? Democracy and Power in an American City*. Yale University Press, New Haven.

Davis, H., Downe, J. and Martin, S. (2001) *External Inspection of Local Government*. Joseph Rowntree Foundation, York.

Defend Council Housing (2004) *A Manifesto for Council Housing*, http://www.defendcouncilhousing.org.uk/dch/resources/manifesto.pdf (accessed 24 June 2004).

DETR & DSS (2000) *Quality & Choice: A Decent Home for All*. DETR, London.

Dovey, K. (2002) The silent complicity of architecture. In: Hillier, J. and Rooksby, E. (eds) *Habitus: a Sense of Place*, pp. 267–80. Ashgate, Aldershot.

Flint, J. and Rowlands, R. (2003) Commodification, normalisation and intervention: cultural, social and symbolic capital in housing consumption and governance. *Journal of Housing and the Built Environment* **18** (3): 213–32.

Gilchrist, A. (2004) *The Well-Connected Community: A Networking Approach to Community Development*. Policy Press, Bristol.

Gwent Police Authority (2004) http://www.gwent.police.uk/pa/consultation.htm (accessed 22 June 2004).

Hall, S. and Rowlands, R. (2004) Place making in theory and practice: collaborative planning, power, governance and large housing estates. In: Dekker, K., Hall, S., Tosics, I. and van Kempen, R. (eds) *Restructuring Large Housing Estates in European Cities*. University of Utrecht, Utrecht.

Hickman, P. (2004) *The Changing Nature of Tenant Participation in the English Local Authority Housing Sector*. An unpublished paper for CRESR, Sheffield Hallam University.

Hillier, J. and Rooksby, E. (eds) (2002) *Habitus: a Sense of Place*. Ashgate, Aldershot.

Hovik, S. (2003) *Coastal Zone Planning as Multilevel Governance?* Paper to AESOP-ACSP Congress, Leuven, 8–12 July.

Lukes, S. (1974) *Power: a Radical View*. Macmillan, Basingstoke.

Malpass, P. and Mullins, D. (2002) Local authority housing stock transfer in the UK: from local initiative to national policy. *Housing Studies* **17** (4): 673–86.

Mayo, M. (1997) Partnerships for regeneration and community development. *Critical Social Policy* **52** (17): 3–26.

Medway Council (2004) *Medway Council Community Plan* http://www.medway.gov.uk/index/council/communityplan/24031/24031-2/24031-5.htm) (accessed 22 June 2004).

Milburn, A. (2003) *Localism: from Rhetoric to Reality*. Speech made to the New Health Network and the New Local Government Network, 5 February.

Mullins, D., Beider, H. and Rowlands, R. (2004) *Empowering Communities, Improving Housing: Involving Black & Minority Ethnic Tenants and Communities*. Office of the Deputy Prime Minister, London.

Murie, A. and Nevin, B. (2001) New labour transfers. In: Cowan, D. and Marsh, A. (eds) *Two Steps Forward: Housing Policy into the New Millennium*. The Policy Press, Bristol.

National Audit Office (NAO) (2003) *Improving Social Housing Through Transfer*. SO Report to the Comptroller and Auditor General HC 496. The Stationery Office, London.

Niner, P. and Rowlands, R. (2003) *Involving Black and Minority Ethnic Tenants in Decisions on Housing Investment Options: Report on Postal Surveys*. CURS, Birmingham.

Office for Public Services Reform (OPSR) (2002) *Reforming our Public Services: Principles into Practice*. Office for Public Services Reform, London.

Office of the Deputy Prime Minister (ODPM) (2003a) *Communication and Consultation Strategies in Option Appraisal*. ODPM, London.

Office of the Deputy Prime Minister (ODPM) (2003b) *Tenant Empowerment Strategies in Option Appraisal*. ODPM, London.

Office of the Deputy Prime Minister (ODPM) (2004) *Housing Transfer Manual* 2005. ODPM, London.

Office of the Deputy Prime Minister (ODPM) (2005) *National Framework for Tenant Participation Compacts*. ODPM, London.

Oxford City Council (2004) *Regeneration in Oxford*. Oxford City Council, Oxford. http://www.oxford.gov.uk/oxford/yourcity.nsf/pages/Howitwas161942.html (accessed 22 June 2004).

Perri 6, Leat, D., Seltzer, K. and Stoker, G. (1999) *Governing in the Round.* DEMOS, London.

Putnam, R. (1993) *Making Democracy Work.* Princeton University Press, Princeton.

Putnam, R. (2000) *Bowling Alone: the Collapse and Revival of American Community.* Simon and Schuster, New York.

Rowlands, R., Mullins, D. and Beider, H. (2003) *'You're listening to the streets?' The Role of Informal Networks in Community Consultation.* Paper presented to Housing Studies Association Conference, Autumn 2003, University of Bristol.

Taylor, M. (2003) *Public Policy in the Community.* Palgrave Macmillan, Basingstoke.

Taylor, M., Barr, A. and West, A. (2000) *Signposts to Community Development.* Community Development Foundation, London.

Tonnies, F. (1955) *Community and Association.* Routledge and Kegan Paul, London.

Walker, D. (2002) *In Praise of Centralism – A Critique of the New Localism.* The Catalyst Forum, London.

Weaver, M. (2004a) Coalition to press ministers to listen to tenants. *Guardian Society* 2 June.

Weaver, M. (2004b) Minister refuses to back down over unpopular transfer policy. *Guardian Society* 28 January.

Weber, M. (1947) *The Theory of Social and Economic Organisation.* Free Press, New York.

Wright, A. (2002) The curse of centralism. *Public Finance* 28 June–4 July.

Conclusion

14

Towards a New Policy Agenda?

Harris Beider

Introduction

The previous chapters demonstrate that US and UK approaches to housing, neighbourhood renewal and community engagement are beyond debates about convergence or divergence of systems. The contributors comment on and analyse city and neighbourhoods, the role and capacity of housing-based organizations to generate renewal and meet needs, and reflect upon wider policy concerns on race, renewal and revival of cities. Different places in the US and the UK are facing similar problems and sometimes coming up with similar solutions to these problems. Moreover the processes and politics applied to issues such as urban decline, community engagement and managing complex and diverse neighbourhoods appear to show areas of overlap. This does not mean that the US and UK housing systems are the same.

We explained at the start of the book that housing policy had very different origins after World War II. The scale, tradition and importance of social housing are much greater in the UK than in the US where home ownership has long held a supreme position. Though process and problems are similar, the housing systems, cities and neighbourhoods reflect the legacy of national policy practised by governments. Differences persist in the role and scale of housing and community organizations and strategic priorities of national and local governments.

In this final chapter we will attempt to draw together the disparate strands from previous discussion to construct a new policy agenda. This will seek to challenge academics and policy makers to develop new ways to address common problems that are to be found in cities on both sides of the Atlantic. In doing so we make no apologies for the blunt language used. The fact that people who live in cities and neighbourhoods have had to endure persistent levels of housing disadvantage, crime and racism is cause enough to motivate ourselves to do much more in the future.

The story so far: key themes for cities, neighbourhoods and communities

The debate in previous chapters seems to indicate patterns of convergence on policy and practice for housing and neighbourhoods in the UK and US but with caveats. The framework for the discussion of urban problems is shaped by the origins, and continuing importance, of the welfare state in the UK compared to the US. In the former, social housing (that is housing built and managed by local authorities and housing associations) accounts for more than 18 percent of all housing stock. Social housing still has the capacity to stir political campaigns at local and national levels against perceived attempts to sell it off to private interests. This could not be said to reflect the situation in the US. Here social housing makes up just 3 percent of all housing stock and has never been higher then 5 percent of the total. Private interests dominate even in cities such as New York that have led the country in developing innovate, subsidized housing projects. The role and important legacy of the welfare state in the UK makes total convergence with the US unlikely despite similarities in social policies and problems associated with housing and communities. At best, housing and social policies in the two countries have aligned but are running parallel rather than crossing over.

Several contributors have commented on problems of racial polarization in cities and neighbourhoods in the US and UK. Clearly the nexus of racism and poverty help to shape residential patterns across many of the cities studied in this volume. Active promotion of home ownership, racism within public housing markets, the importance of maintaining links to places of employment as well as family and social networks distorted housing, race and location. Policy and practice have largely failed to resolve the issue of segregation in both countries. Whilst it can be claimed that racial segregation is much more entrenched in US cities than is the case in the UK, there is a danger that such a sweeping statement misses important sub-messages. In fact, in both countries, minorities are showing a tendency to move out from traditional areas of settlement to live in towns and cities that have had little experience of immigration. This will challenge local authorities and housing organizations to come up with new ideas and solutions to meet the housing and neighbourhood aspirations of these secondary pioneer communities. This issue and associated problems are explored later in the chapter. Indexes of segregation show that, contrary to government pronouncements, the UK is actually becoming more diverse and thus less segregated. The problems of racial polarization appear to be much greater in the US than the UK (according to the index of segregation), reflecting decades of housing and public policy that restricted the choices open to minorities, especially African-Americans.

Institutions and change agents operating in neighbourhoods demonstrate levels of entrepeneurism to meet housing and related needs of communities. However they face a number of challenges to their role. Increasingly in both countries government emphasizes the importance of regional, national and, indeed, global solutions to the very local problems of neighbourhood disadvantage, housing choice, segregation. In the UK, housing associations are forming larger organizations to reap the benefits of economies of scale. Regional and sometimes national housing associations are becoming the norm to access the most efficient levels of operation. To what extent does this squeeze out housing organizations that advocate on behalf of and serve poor and minority communities? Clearly these bodies cannot compete in terms of economic efficiency and the regeneration and renewal of cities and neighbourhoods. The emphasis on market solutions to social problems simply restates the problem.

To make matters worse, increased diversity within neighbourhoods and cities makes the future appear bleak for those housing and community organizations that have based their raison d'etre on representation and advocacy of specific groups. Despite all this, community organizations can still play an important role in advocacy and building social capital in neighbourhoods that have been in economic decline. Neighbourhoods and community organizations are not dead.

Increasingly complex and diverse communities in the US and UK mean that searching for common American or British identities and building community cohesion are problematic. American identity has been complicated by mass Latino migration and the increasing use of Spanish as the language of common currency. The UK has based common identity on building shared values through community cohesion. This is a false premise. Increasing fragmentation pulls against an easily recognizable identity. Moreover, minority communities may not want to trade in powerful and clear ideas on ethnic identity for unclear and weak notions of British identity in an increasingly diverse society.

It could be argued that race relations legislation has encouraged the preservation of different groups in the UK. Now it appears that government is using equal opportunities legislation to help managed societal change. This could be viewed as mediating a position between the cultural melting pot and the homogenous society. Rapid societal change and growing prosperity in the UK could mean that neither is a realistic policy objective. A new approach could be considered: this should be predicated on promoting the economic, social and cultural benefits of immigration, recognising and addressing the damaging impact of racism and disadvantage and encouraging Britons to support a new and modern multicultural society based on civic rights, shifting identities and responsibilities.

In this concluding chapter we will resist the temptation to review previous material. The emphasis will be on looking ahead. Recommendations will be put forward to address concerns of policy makers and suggest productive areas for future research. This is then not the end of the process but merely the start.

Comparative policy responses

The evidence presented here suggests that cities in the US and UK are confronting similar types of problems but are still characterized by different responses, reflecting differences in history and responses to immigration, the role of the public and private sectors, and in the assumptions that underlie policy interventions. It is clear that serious debates at both policy and academic levels are taking place on matters that include increasing housing supply, quality and choice, de-concentrating poverty and racial polarization, managing changing markets and working with community groups as well as private developers. This cocktail of issues is not so dissimilar to debates in other cities across Europe. In this section we review and assess approaches to some of these issues in the US and UK.

Reactions to new migration

The context for differing responses in the US and UK is rooted in historical reactions to immigration. Generally the US has viewed immigration as a basis to build a nation and an important component in wealth creation and widening opportunities. Despite an increasingly conservative economic and social agenda it could be argued that the US government still has a relatively constructive and welcoming approach to migration compared to the UK.

Latino communities form a substantial electoral block that both Republicans and Democrats court at local and national elections. Census figures show that there are now over 35 million Hispanics/Latinos in the US, and the population is growing especially fast in the important electoral states of the South and West. Recent announcements by President Bush on immigration reforms have been seen as a palliative to minority groups. The key feature of this has been an offer of legal status for new migrants or those who entered the US illegally. This has been viewed as a step against the exploitation of migrants working in the unregulated economy and in low-skilled, high-risk jobs. In addition, cities such as Baltimore are marketing themselves as 'immigration gateways' in the hope that new migration can boost flagging economic growth and start the process of regeneration.

In contrast, race and immigration have been uneasy bedfellows in UK social policy. The two waves of immigration in the post-1945 period (broadly speaking encompassing migrants from the Caribbean, India and Pakistan during the 1950s and 1960s and from Africa, Eastern Europe and the Middle East since the 1990s) have culminated in restrictive immigration controls, despite pressing economic needs, as well as a frantic debate on the politics of race and community cohesion. Compared to the general receptivity in the US towards migrants, UK policy across the political spectrum has construed immigration as a problem rather than a solution. Its impact is often linked to societal breakdown, and a threat to what is portrayed as a relatively homogeneous society. It could be argued that there has been a conspicuous failure to resolve, or at best mediate, between cohesion, stability and politics. Often, short-term political interests have triumphed over long-term economic goals.

Political responses towards minorities have shaped policy interventions in housing and renewal. In the UK the result of exclusionary policies that prevented access to public sector housing led to the concentration of black and minority ethnic communities in poor private sector housing located in the inner urban areas of major cities and towns. Here too are substandard schools, health services and local environment. Whilst it could be argued that new migrants in the US have been viewed as a key component of economic growth, the legacy of slavery and racism has left entrenched patterns of segregation in the US. In both countries, poor people living in substandard housing in areas of social disadvantage tend to be drawn from black and minority ethnic communities.

Directions in US and UK housing policy

American housing policy immediately after 1945 was driven by supply-side inducements to local government and private developers to build high-quality public sector housing for middle-income groups. More recently, public sector housing developments have become associated with entrenched poverty, racial segregation, high crime rates and the selling of drugs. Economically mobile tenants have moved out to the suburbs and the private sector, leaving a residualized tenant population cut off from decent jobs, schools and transport.

The response of the US Government was to develop the HOPE VI Program (Housing Opportunities for People Everywhere). This is a \$6 billion national programme based on the demolition of substandard public sector housing and its replacement with mixed-income developments.

Though Congress has provided funding since 1992, authorization for the demolition and rebuild was only granted when the Quality Housing and Work Responsibility Act was passed in 1998. Since 1992, more than 166 housing authorities have received grants through this programme, leading to 60,000 public housing units being demolished (see Chapter 6).

HOPE VI now includes demand-side inducements in the form of housing vouchers to assist people on low incomes to improve their housing choice. The voucher scheme, known as Section 8 after the passage in the Act, was first authorized by Congress in 1974 and was developed by the US Department of Housing and Urban Development (HUD) to provide rental subsidies for eligible tenant families residing in newly constructed, rehabilitated and existing rental and cooperative apartment projects. The scheme now provides low-income families with an opportunity to access decent quality private sector housing through vouchers to supplement what they would otherwise be able to afford. Public housing authorities apply to HUD for Section 8 funds, which are then given to eligible families. Each housing authority will set different limits on income and eligibility for applicants according to local need. Housing vouchers are cheaper to administer than addressing housing supply issues and help to maximize choice, and a pathway to home ownership, for even the lowest-income groups.

The results of this policy are only just starting to emerge. The suggestion is that although the federal government was keen to encourage public sector voucher recipients to move to low-poverty neighbourhoods through programmes such as Movement to Opportunity (MTO), the outcomes did not lead to de-segregation. The problem was, and continues to be, that the voucher programme is based on reducing economic poverty and not racial segregation. Either way, dispersal of families using vouchers has led to opposition from neighbourhood-based politicians and resident associations who argue that the programme leads to unstable communities.

More generally, there are mixed outcomes on the HOPE VI Program. There is clear evidence that demolition of the worst public sector housing in the US has led to reductions in poverty and crime in these areas. However, relatively few tenants return to mixed-income and mixed-tenure neighbourhoods after redevelopment (less than 11 percent across the US) and it is often higher-income groups who tend to benefit from the better housing that has been developed (see Chapter 6).

In the introduction we stated that social housing in the US is perceived as an entitlement and not a right. Both Republicans and New Democrats stress the importance of linking social housing with welfare reforms. Tenants have responsibilities. These are seen in rental contributions required from low-income households, the importance of participating in local community and voluntary sectors (almost a form of compulsory tenant participation), criminal records checks and tough rules on anti-social behaviour. The ultimate aim of housing policies is to reframe social housing as the tenure of aspiration rather than stigmatization, to create routes to home ownership and to build sustainable, mixed and thriving neighbourhoods.

In the UK, policies to break up concentrations of poverty and segregation are less well developed. Successive regeneration activity based at neighbourhood level has tried to revive communities. However, it is only recently, following concerns over disturbances and minority disenfranchisement in some parts of the UK, that there has been interest in a more concerted attempt to de-concentrate poverty, address ethnic segregation and build sustainable communities. Race, housing and community have become intertwined and gained a high political profile following recent events, including the publication of the Macpherson Report into the murder of Stephen Lawrence in 1999 and the Cantle Report that investigated the disturbances in Burnley, Bradford and Oldham in 2001.

Following the riots in these northern cities, there was wide-ranging debate on the role of housing improvement in relation to ethnically segregated neighbourhoods in many urban areas. The Cantle Report set out the depths of polarization in these areas in terms of education, housing, social and cultural networks. The report painted a bleak picture of divided communities, which had become ethnically segregated in different areas and in different forms of housing tenure. In effect, the separation of communities in the same towns and cities was akin to people living 'a series of parallel lives' with real differences in education, employment and housing. Cantle made links between place, identity and community cohesion. One of the major recommendations of the report was to urge housing agencies to review their policies and strategies to provide more mixed housing areas, together with supportive mechanisms for residents who faced intimidation and harassment. The Cantle Report also noted that funding for housing improvement could distort regeneration programmes as it is capital intensive, and suggested that some change of emphasis may be needed to develop a *people-focused*, rather than what they termed a *property-needs*, approach to areas.

In considering the overall policy directions in the US and the UK, it is evident that there are a number of problems with breaking up poverty and ethnic segregation. One of the most significant barriers is racism within some communities which may resist measures that encourage mixed-tenure, mixed-income and diverse racial neighbourhoods. The negative reaction to the dispersal of asylum seekers in the UK gives an insight into potential problems. Taken together with the waves of 'white flight' from cities, this is a significant barrier to changing neighbourhoods.

Minority communities may also lose out with housing mobility programmes because they reduce their bonding capital and erode local political power. Moreover, some minorities may believe that the benefits of living in mixed neighbourhoods have been oversold. Recent research shows that housing mobility schemes in the US, such as housing vouchers, do not in themselves break up poverty and racial segregation. Many claimants move to contiguous areas that suffer from poor socio-economic indicators and remain racially polarized. This is because of the need to remain connected to social networks and also exclusionary practices by housing organizations and opposition from resident groups.

Gentrification, communities and race

There has been much debate on both sides of the Atlantic over housing policies encouraging middle- and high-income communities back to inner-city environments as part of a package of neighbourhood renewal measures. The process of gentrification has been described as taking 'an interest in working class and cheaper neighbourhoods by middle class and professional households' (Atkinson, 2002). Our concern here is to focus on the impact of this process on minority communities and changing neighbourhoods.

There is sharp division on the impact of gentrification. Many commentators remark on the adverse effects of rising house prices in disadvantaged communities. These concerns focus on the displacement of poor – and often minority – communities as a result of increased property and rental values. In this way gentrification can have a racial dimension. Some comment that displacement of inner-city communities is a local authority driven policy marked by encouraging private developers to buy discounted land and property for resale to middle class and white communities. Gentrification destroys social relations and community capital.

There is an alternative view to the conflict model. Gentrification can bring increased tax revenues (based on physical re-development and the break-up of concentrations of poverty and racial segregation) and it can help to create diverse, mixed-tenure neighbourhoods. In New York City, high property values have forced middle-income professionals to consider neighbourhoods which were previously seen as 'no go areas'. The US Housing Department's Ten Year Housing Plan has also supported housing market change (see Chapter 4). For example, between 1971 and 1981, New York City lost more than 320,000 housing units through arson and abandonment, equivalent to the housing stock of a city the size of Baltimore and three times the size of the recent Birmingham stock transfer proposal. Most of this was concentrated in the poor African-American communities of South Bronx, Brooklyn and Harlem.

The policy community was concerned about 'planned shrinkage', convinced that New York would never recover from the loss of people, business and tax revenues. The Housing Department responded by raising revenue from selling municipal bonds, buying abandoned properties and offering them to property developers in the not-for-profit and private sectors. The objective of the process was to stabilize communities and to change the housing trajectories of poor neighbourhoods, by limiting homeless household tenancies to less than 15 percent of new housing developments and actively encouraging middle-income earners to buy and rent property. The determined lobbying of big business was also designed to help rebuild economies. For example, private sector companies such as Starbucks, Old Navy and Gap were offered inducements to locate flagship stores in Harlem.

The impact of this intervention, which can be seen as a planned gentrification in a city with a tight housing market, has been profound in some communities. Harlem has witnessed dramatic change. Gentrification and housing transformation has swept upwards from 79th Street to 125th Street, often cited as the frontline of Harlem. It is noticeable that people going to and from work in this neighbourhood are young, professional and mainly white. Moreover, housing market change has been supported by the Housing Department as 'fast tracking' regeneration. As one senior official stated, 'gentrification is good and we want [the ambience] of 79th Street to be replicated elsewhere'.

Minority communities should not be viewed as passive actors in the process of gentrification. Housing markets that were deemed to be low price and low quality, and the only areas that migrants could access over 30 years ago, have in some cases seen rapid price hikes as a result of demand exceeding supply. In the UK, Brixton, an area long associated with Caribbean settlement, has boomed in the last two decades. Houses bought for less than £5000 in 1964 were selling for £500,000 in 2004. The question is whether minority communities have gained from this. Rising property prices need not spell the end of inner-city neighbourhoods but may simply provide opportunities for social mobility for those in a position to benefit.

The process of housing market change and displacement may not be as far advanced in the UK but there are nevertheless examples of planned change. In Birmingham, new housing markets in the city centre have been created and marketed to middle- and high-income professionals. This was part of the policy by the local authority to revive run-down neighbourhoods and bring renewed life and vigour to the city centre to go alongside flagship schemes such as the International Convention Centre, Brindley Place and the new Bull Ring (see Chapter 5). Early indications are that city centre living has worked in Birmingham. Property prices and demand are high. During recent years there has been increased consumption of goods and services in the city as a result of people being drawn to jobs and access to amenities. Interestingly, this type of housing market renewal has been attractive to black and minority ethnic communities as well. More than

15 percent of these new housing markets are drawn from minority communities. Indeed, it has been argued that multiculturism has been used as a marketing ploy and become an integral part of selling 'new urban' living.

Fresh from the success of this initiative, the local authority is addressing the problem of de-concentrating poverty by demolishing public sector housing in contiguous areas. Private developers are re-building communities and replacing public sector housing with mixed developments, with an emphasis on marketing to middle-income owner-occupiers and key public sector workers. Some previous occupants think that neighbourhoods have changed substantially and decided not to return to the new housing. The reasons vary but link to the transformation of the area from public to mixed-tenure housing, perceived problems of affordability, the loss of established residents and community networks and the creation of new amenities and shopping facilities that previous occupants do not want and cannot afford.

In both the US and UK, policies to change housing markets and attract investment have led to increased property prices, tax revenues and a revived culture of urban living. The question is whether the New York and Birmingham examples of planned market change will lead to new patterns of segregation by income and ethnicity or whether they are simply providing the tenure and neighbourhoods of choice demanded by upwardly mobile young communities – both white and from minority groups.

Policy implications: a UK perspective

Introduction

In the UK, recent events and policy developments have served to highlight the importance of race, communities and housing issues at a national level. These include the concerns about race and segregation identified in the Home Office's Community Cohesion Report, the evolving agenda to promote sustainable communities set out in the ODPM's Communities Plan (£22 billion initiative to address the problems of housing supply in the south-east and low demand in the north by delivering sustainable communities) (OPDM, 2003) and the renewed interest in investment in housing and changes to the planning system. In America the new approaches being adapted to housing, with HOPE VI, the Voucher Program and the further move away from public housing, also show the importance of these issues nationally. Changing patterns of race and residence, and new approaches to community involvement and empowerment are evident in both countries. Given this context, and the previous discussion, we will now focus on the implications for policy makers looking at the role of housing organizations working within neighbourhoods, the relative importance of connecting communities to regional and national policy currents and critically appraising social capital as applied to race.

Housing products and neighbourhood choice

Minority communities in the UK are changing, spatially, educationally and also regarding preferences for neighbourhood choice and tenure (see Chapter 2). Much more work needs to be undertaken to understand the contours of change. As noted earlier, the speed of change is faster for some groups than others and the first step for public agencies is to map out existing communities, their preferences and aspirations before designing and building new housing and planning future services. These changes go beyond a binary debate between white and black communities and even new groupings such as white, Caribbean and Asian. Discussion on race must be placed in a context of valuing differences between 'new' and 'old' migrants, class, income and age which has flowed from demographic, economic, social and spatial change.

The task for housing organizations in the US and UK is to proactively respond to change. At a micro level, housing organizations need to develop more diverse policies to meet the needs of increasingly heterogeneous communities. Importantly, housing organizations also need to extend the range of housing products and choice of neighbourhoods available to communities. Given demographic changes in minority communities, housing organizations need to consider attracting communities to new housing developments contiguous to previous areas of settlement. Moreover, housing preference for minority communities should be linked to a greater understanding of wider 'lifestyle' choices concerning schools, amenities and access to retail outlets, as well as proximity to social networks and places of worship.

There is a consensus within the academic and policy communities about the urgent need to break up concentrations of poverty and, to a lesser extent, racially segregated communities. A route out of high-density, poor-quality housing in the inner cities to better-quality and low-density housing in the suburbs is a policy that has been tested in the US. The results are mixed. However, a pathway out of poverty through increased housing choice could be reconsidered. Schemes such as vouchers, or subsidies, to access decent housing in neighbourhoods of choice may lead to housing change but also opposition by groups on the grounds of unfair allocation of housing goods. Enforcement of any scheme and the active promotion of diversity are required to build sustainable and diverse neighbourhoods.

Just as interesting are the attempts to revive housing markets located within inner-city neighbourhoods in both countries. We have already shown how neighbourhood trajectories have been changed through intervention by local authorities and private developers. Making the most of historic districts, waterfronts and easy access to the main employment core may attract middle- and high-income earners to these markets. In addition, housing mobility schemes operating in the public but especially the private sector could also lure key workers and young families. Local authorities in the UK could replicate (on a much smaller scale) the New York approach to buying derelict property and marketing this to communities whose incomes are likely to rise. This may also contribute to breaking up concentrations of poverty and racial segregation, increasing investment and encouraging diverse communities and improved neighbourhood trajectories.

Housing organizations need to develop projects and initiatives that are based on meeting the concerns of tenants in terms of routes into home ownership, increased quality of social housing and providing access to neighbourhoods of choice. Housing mobility schemes together with proactive intervention in declining housing markets may contribute to creating mixed communities in both the inner city and outer suburbs. Vested interests in both types of areas need to be addressed to ensure that policies can work in practice. In this way, access to new housing markets can be made much easier than is currently the case.

Connection to 'geographies of opportunity'

Communities in themselves cannot transform neighbourhoods. Connections need to be made with wider economic and public policy at regional and sub-regional levels. The task is to link disadvantaged minority communities to the wider 'geography of opportunity'. Sometimes there is a housing/jobs mismatch with an oversupply of poor, unemployed and under-skilled people living in inner-city housing markets and a surplus of middle- to high-skilled jobs in the wider region. Matching housing and jobs cannot be achieved at a neighbourhood level. The challenge has to be taken up at regional and national levels. The regional task is to make both inner-city and suburban housing markets work.

Conceiving race as part of a wider geography of opportunity in the UK could be problematic. The fact remains that minority communities (despite new population trajectories) reside in concentrated neighbourhoods in the conurbations of London, the Midlands, north-west and west Yorkshire. There may be a reluctance to relocate for new jobs and housing. However, people first of all need be made aware that there is a *choice* and then be given the *tools* to access opportunities. For example, large regeneration projects in Stratford and Docklands provide job opportunities for minority communities in Newham, Tower Hamlets and Hackney. Government has a key role to connect people on low incomes to jobs at a regional and local level.

Does this mean we can now discard the concept of neighbourhood? Community-based organizations operating at local levels do have an important role in campaigning for new investment and better amenities. Indeed, Birmingham has seen two celebrated examples of community activism; residents in Balsall Heath (protesting against prostitution and crime) and Lea Bank (fighting declining housing and rising crime) activated change at both local and national level. These are exceptions rather than the rule. It should be noted that Balsall Heath could be viewed as conflict over space between two marginal groups – low-income Pakistanis and prostitutes and their pimps. The conflict was not characterized by discussion and debate. It was akin on some occasions to 'hand-to-hand' combat as one group tried to reclaim a community. The results have been impressive in the form of neighbourhood development, housing popularity and reduced crime, but the prostitutes have simply decamped to the neighbouring area. The expectation is that neighbourhoods, politics and community development will be shaped by factors and decisions taken at regional, national and local level. The task for activists is to link into this new geography and economy of opportunity and ensure integration between neighbourhood and housing market renewal agendas works for the benefit of local people.

Social capital and race

The debates on social capital have influenced much public policy making in the US and UK. Social capital is associated with Putnam who defined it as 'connections among individuals – social networks and the norms of reciprocity and trustworthiness that arise from them' (Putnam, 2000). There are two further dimensions of the debate: *bonding* and *bridging* capital. Bonding gives people a sense of identity and purpose (such as ethnicity) and holds people together in groups. Bridging provides connections between people who may live in the same neighbourhood, city or country. Bridging capital can be conceptualized as the key towards building community cohesion and greater understanding between communities.

There are some problems with the social capital debate when linked to the issues of race. For example minority communities may be less likely to join organizations, such as tenant associations, where they are not perceived to be relevant to their lives and aspirations (Mullins et al., 2004). Formal and informal mechanisms such as black-led churches, mosques and carnivals provide a basis for building social capital. Not all forms of social capital have positive consequences – for instance inner-city gangs may have a corrosive impact on the wider community. As we have mentioned elsewhere, minorities may not wish to belong to an integrated neighbourhood that dilutes capital.

There is a view that new policy drivers, such as neighbourhood renewal and establishing consultative mechanisms, neutralize debate and allow planned and market-led displacement to take place (see Chapter 12). Justification is provided by social capital theorists who emphasize the lack of 'connections among individuals' (Putnam, 2000). The assumption is that poor and minority communities have a social capital deficit, which needs to be accumulated through the creation of new businesses and mixed neighbourhoods. Apart from the inherent dangers of developing pathologies on poor and minority communities, it neglects the importance of social networks, the role of grassroots organizations and the important collectivist ideology within poor communities and families. Often this type of social capital holds communities together when local authority and other stakeholders have long exited from a neighbourhood (Venkatesh, 2000).

Policy based on housing market change either through demolition and the creation of mixed-tenure communities, or the work of the market through gentrification, should consider the importance of building on the social capital, networks and culture existing in poor, minority neighbourhoods. The process of 'washing out' or emptying communities should be resisted. It is argued that 'gentrification has left residents of low-income neighbourhoods in a situation where, since they exert little control over either investment capital or their homes, they are facing the "choice" of either continued disinvestments and the decline of the quality of their neighbourhood, or reinvestment that results in their displacement' (see Chapter 12).

There is a danger that social capital may be associated with groups, households and spatial areas that are white, owner-occupied and suburban communities. Reducing diversification in terms of ethnicity and tenure may ultimately lead to neighbourhoods and communities that are devoid of ideas and tools required to meet the challenges of an increasingly complex world. The premise is that new communities can be built on existing social capital, organizations need to meet their obligations on representation of issues and there has to be a menu of housing choice and ownership for increasingly diverse communities. The implications for policy makers are that poor minority communities and inner-city areas need investment and support to rebuild neighbourhoods. Demolition, displacement and replacement may erode social capital rather than help to create it.

Social capital is a useful tool to discuss the renewal of civic engagement. On its own it does not explain the problems of cities, segregation and sustainability. Complex relationships, ties and bonds within minority communities are sometimes simply not recognized and valued by policy makers and researchers. The task is to develop a much more inclusive social capital model.

Increased diversity, community cohesion and shifting identities

We have already noted that the UK and US are becoming increasingly diverse, fragmented and polarized (see Chapter 2). This is even more the case for communities residing in cities. Seismic social movements, such as immigration and aging, combined with global economic change and technological advances are re-shaping countries, cities and neighbourhoods (Katz, 2004).

The discussion of race relations cannot be adequately referenced through binary analysis of 'black' and 'white'. Diverse societies need diverse solutions, which sometimes rest uneasily with equal opportunities legislation and fair housing policies. Census material in both countries shows that some minority groups have become less polarized and moved to outer suburbs from inner-city locations. This may be as much to do with educational and income prosperity as opposed to any direct policy interventions. We must recognize the mosaic of different communities that make up many urban neighbourhoods. A variety of responses is required to ensure that all communities have access to and choice in education, housing and welfare. This may have a resource implication. For example, it could be argued that much more should be done to support new migrants than established minority communities. Of course, the response from some minority organizations could be that new and old migrants still need help and support. However, a finite budget and infinite choices mean hard decisions have to be taken. Positive action becomes more complicated when the decision to appoint is not based on a white or black applicant but on a Pakistani or Somali person. For a society that has not resolved the binary problems on race equality, dealing with a much more complicated picture becomes highly problematic. Society as a whole has to deal with not only primary racism but also secondary racism as well.

Mediation on race and creation of stable communities has been a concern on both sides of the Atlantic. In the US, directive policies restricting low-income housing, prescribed rights and responsibilities and housing vouchers are tools at the disposal of government. American identity is promoted and citizenship is enshrined in the Bill of Rights. As we have discussed above, the UK is struggling to come to terms with transforming itself to accommodate increasing ethnic diversity. Responses to the disturbances in Oldham, Bradford and Burnley in 2001, together with the negative debates on asylum and immigration, could be seen as counter-productive to the goals of shared identity and citizenship. New and old migrants are less likely to feel any obligation to contribute to community cohesion when they are being identified as part of the problem.

The fact remains that Britain has seen successive waves of migration into towns and cities. The difference in post-1945 migration is that newcomers are more visible and different. In reality, community cohesion based on a mix of communities, tenure and income is the exception rather than the rule. Social intervention is unlikely to speed up the process. Would it not be better to accept that the path to community cohesion be based on choice, celebrating differences and recognizing that migrant communities, new and old, have stabilized neighbourhoods rendered weak by 'white flight'? Connecting to neighbourhoods of choice, providing options for key public services, such as education and health and re-modelling transport, show another way to cohesion. Moreover, some neighbourhoods have benefited from the impact of new migration. For example, the Sparkbrook area of Birmingham had seen an outflow of whites, Irish and Pakistani communities and been associated with decline and abandonment. In recent years, Somali communities have concentrated in the area and this has led to the flourishing of shops and cultural amenities and greater cohesion. In short, the debate on race and immigration must shift to the value and benefits migrants bring rather than the problems they are believed to create.

The search for common identity in the UK can also be viewed as problematic. Increasing fragmentation pulls against an easily identifiable identity. In truth, established and new migrant communities have composite identities that provide a framework and reference point in day-to-day life. This will vary in different contexts and locations – at work, religious festivals and cultural events. Trading ethnic identity for unclear notions of common identity leads to a zero sum scenario. Britons are defined by a number of different identities including ethnicity, class, region, city and neighbourhood. Race relations legislation has encouraged the preservation of different groups and communities. Now government has to use equal opportunities legislation to manage societal change. It could be viewed as mediating a position between the cultural melting pot and the homogeneous society. Growing prosperity and change in Britain mean that neither will be a realistic policy objective that is worth pursuing. A third way should be predicated on promoting the economic and social benefit of immigration, addressing the damaging impact of racism and encouraging all Britons to buy into a new and modern multicultural society based on civic rights and responsibilities.

Key issues for research

Are we heading towards an urban 'melting pot' or an urban 'meltdown'?

Immigration and asylum are viewed as two of the most important issues facing the UK. There have been outbreaks of 'moral panic' on immigration and asylum during recent years. This, together with the urban disturbances in Bradford, Burnley and Oldham in 2001, may suggest that we are heading towards an urban meltdown, where the factors which hold communities together are tested and broken. The events following the London bombings on 7 July 2005 and the realization that the bombers were 'home-made terrorists' who were either born or grew up in the UK, further escalated collective soul searching on whether the country, cities and neighbourhoods can rally around a set of societal norms. This includes shared identity, an acceptance of tolerance and trust in legal and political institutions to resolve problems. The vision of atomized and conflicted communities may be apocalyptic but this was partly the conclusion of the Community Cohesion Report. In contrast, some parts of the UK demonstrate a 'melting pot' of cultures and communities that live and co-exist alongside each other. Moreover, it should be noted that the country has the highest proportion of dual heritage couples in Europe.

In the US these types of debates are different and have been going on for much longer. They are different because citizenship is enshrined in the Bill of Rights and to an extent is much more clear-cut than in the UK. Unlike the UK, the US has never had to manage the implications of the end of Empire and its position in the world. It could be argued that this goes to the heart of the problem that the UK has on the issue of race, neighbourhoods and cities. Black and minority ethnic communities have been portrayed as a problem. This has structured responses by local authorities to housing and community interventions. These communities were shut out from public sector housing and located in the poorest neighbourhoods. The UK still has difficulty in conceiving diverse neighbourhoods and people as an asset rather than a liability, a strength as opposed to a weakness. The poorest housing in cities in the US has entrenched levels of segregation. African-Americans also continue to be on the receiving end of dilapidated housing and poor public services, live in areas of high crime and experience educational underachievement.

In parts of the US and UK, communities are in meltdown. The policy vision is to transform these areas from meltdown to melting pot similar to the most vigorous and interesting parts of London and New York where a mix of people, income and communities co-exist side by side. How can this be achieved? What is the respective role of institutions and citizens? We do not seek to prescribe policy solutions to deep-seated problems here. There has been enough of the 'quick fix' mentality for policy makers in the US and UK. Suffice to say that sometimes the ideal of melting pot and diversity have been oversold as concepts to communities who distrust institutions and politicians in equal measure. Perhaps the best way forward is a combination of understanding the root causes of city and neighbourhood decline (connected to regional, national and international policies) and providing people a real choice in terms of both housing and neighbourhoods of aspiration.

Can private sector rather than government sponsored renewal lead to better economic and social outcomes for communities?

There could be a review of the relative impacts of government sponsored regeneration and market driven change on minorities, especially in the housing sector. Do social regeneration programmes simply pathologize areas and residents whilst failing to link into their housing aspirations and trajectories? Gentrification is one aspect of change but does there need to be more consideration of choice-based lettings, shared ownership and low-cost loans to different communities? The notion of the private sector delivering to poor communities also needs to be scrutinized. Research could also see how different agendas for regeneration, such as housing market renewal and neighbourhood renewal, come together to benefit local communities.

Applying market principles to community renewal and engagement seems at one level to be inviting problems. After all were not some of these dedicated followers of the market calling for 'planned shrinkage' of US cities such as New York? There are some very real dangers of applying profit maximization to services that are needed in disadvantaged communities, such as affordable childcare, decent housing and cheap transport. As Defilippis stated in Chapter 12, the market riptide could wash out communities, support and structures and deposit them in places and circumstances from which they may not recover. Balanced against this is the importance of securing private investment and project management of interventions that help to regenerate neighbourhoods.

A critical problem with renewal and regeneration is that much of it is centrally imposed. There is little consideration of the nuances that hold communities together despite many seeing an outflow of capital, people and services over many years. Recognizing the capital that already exists in communities is important to their regeneration and renewal. At various points in this book communities have played a critical role in finding their own solutions to housing and social crisis that afflict neighbourhoods. Renewal does not work when it is simply concerned with erecting buildings. In this context, a people- rather than a property-based focus is needed.

Can housing and community institutions play a role in the renewal of neighbourhoods?

Some of the contributors have concerned themselves in the role of institutions to meet neighbourhood needs. We have already recounted that following World War II the government decided that local authorities rather than private developers had the skills, capacity and means to volume build the housing that was required in major cities after bomb damage. Despite the growth of private sector housing to become the major form of tenure in both the US and UK, community organizations continue to play an important role in engaging with local people. Housing associations in the UK have become leading suppliers of social housing during the last 20 years and many have played a key role in renewal and regeneration. In some respects, housing associations are very well equipped because they work with low-income communities, plan for the long term, attract private sector investment and have a range of project management and financial skills that are required to make successful interventions at the local level. In addition, it could be argued that because housing associations are regulated by the Housing Corporation, tenants have more security than people in the private sector.

Housing organizations and community-based organizations in the US and UK have demonstrated that they can be agents of change in neighbourhoods. In New York a faith-based community organization has played an important role in a number of different policy areas including housing, retail, childcare and increasing the skills base of local people through employment-related training. Likewise specialist housing associations in the UK have undertaken a key role in advocating and providing leadership and ultimately increasing investment for poor communities.

Yet the role of community organizations has been questioned by government and key stakeholders on both sides of the Atlantic. First, the complaint is that community organizations do not always have the skills and capacity to deliver on problems that have become entrenched over many years. Far better, this argument goes, to contract with a larger organization, which has more capacity, skills and knowledge to complete the task on time. Second, conversely, is the view that these housing and community organizations have become too professionalized and remote from the communities that they seek to serve. They are merely contractors to deliver centrally generated policies at a local level. Black and minority ethnic housing associations and CDCs in particular are left in a very difficult position of being advocates and service providers that may sometimes be in direct contradiction to each other. Finally, organizations that have historically been based on representing the needs of sectional community interests may have problems advocating on behalf of more diverse groups. Competition for scarce resources within neighbourhoods leads to sharp community politics. It is sometimes very difficult for housing and community organizations to reconcile increasingly disparate groups.

On a more positive note some of the best examples of community action are located in the work of neighbourhood-based organizations. These are people and organizations that have developed new responses to old problems, showing government how to engage and involve excluded communities and working alongside the private and public sectors for the common good.

To what extent is the concept of neighbourhood dead?

Recent policy in the UK and the US has emphasized neighbourhood-based intervention to address economic and social problems. Neighbourhood management, economic projects and community participation will lead to flourishing communities, cities and regions. This now has to be questioned. Neighbourhood problems are not born in a local area but have regional, national and international dimensions. Necessarily this means that a policy focus at these levels is required to resolve problems. The Communities Plan and Housing Market Renewal Areas in the UK perhaps signify a shift in thinking but many still regard their neighbourhood as the main point of reference. Do we consider that the neighbourhood is dead and give it a decent burial? Or should we review how the neighbourhood can play a role within a wider frame of reference?

The problems of cities and neighbourhoods are located in national and international perspectives. However, rather than burying neighbourhoods, it is important to connect people to new opportunities that exist in contiguous areas. Housing has an important role to play. Recent research in Birmingham has shown that the range of housing products and neighbourhood is restricted (Beider and Goodson, 2006). People want housing organizations to meet their own aspirations in terms of dwelling and neighbourhood rather than coming to default conclusions about what is required. Neighbourhood management will contribute to resolving local issues but will not in itself solve the problems facing cities. This needs a regional, national and international focus.

To what extent does migration help to build community cohesion and social capital in neighbourhoods that had previously been in economic decline?

Research should consider the alternative view of social capital that positions migrant communities as asset builders and a force to stabilize housing markets. This could then be a basis for constructing a new social cohesion model and also to question the wisdom of striving towards a single shared identity.

Does the idea of citizenship need to be reviewed in a diverse and fragmented society?

Are we moving to a post equal opportunities era? It could be argued that equal opportunities legislation cannot function in an increasingly diverse society. Representation of new migrants may conflict with the needs and interests of established white and minority communities. Managing diversity is now much more about the pragmatic concerns of diverse neighbourhoods and people rather than advocating for specific minority groups. The challenge for organizations is to mediate between new tensions within diverse minority ethnic communities as well as service the needs of well established residents. This could lead to a new framework for discussing race relations, choice and responsibilities. Research into how service providers can respond to diversity may be needed to push this agenda forward.

Summary

This book has been about comparing housing markets, neighbourhood renewal and community engagement in the US and UK. The chapters have covered the key areas of mapping and spatial change, housing interventions in changing cities and neighbourhoods, the role of organisational change agents working with diverse communities and the contested issues of power and conflict. So what can be said about this heady mix of ideas, issues and analysis? Change and dynamism is the theme that permeates across issues discussed in the chapters: population driven by global immigration, economic and social change leading to attempts to revive housing markets and neighbourhoods, organisations needing to transform themselves because of consumer preferences, and the jostling for power and access between shifting elites and communities within cities.

Managing change at neighbourhood, city and regional level requires flexible and innovative responses. This book has demonstrated policy interventions by housing agencies and organisations responding to changing circumstances. Discussion has also focused on new models of intervention and community engagement with diverse communities. Despite these interesting examples there is still a sense that academics and practitioners need to embrace new ideas on housing change, market renewal and especially managing increasingly diverse communities. We still have a tendency to view cities, neighbourhoods and communities from a narrow rather than wide focus. By inviting contributions from the US and UK the book has shown that both sides of the Atlantic face similar problems and issues but policy and interventions are shaped by differences in legal frameworks, politics and organisational culture.

The book should be seen as part of the process to widen a discussion and debate that is too easily closed down. The hope is that it will spark new ideas and thinking which lead to public policy change for those who need most support in our cities and neighbourhoods. To my mind this is what research should be about.

References

Atkinson, R. (2002) Does gentrification help or harm urban neighbourhoods? An assessment of the evidence-base in the context of the new urban agenda. ESRC Centre for Neighbourhood Research, Glasgow/Bristol.

Beider, H. and Goodson, L. (2006) *BME Communities in the Eastern Corridor: Aspirations, Neighbourhood 'Choice' and Tenure.* Birmingham City Council, Birmingham.

Katz, B. (2004) *Neighbourhoods of Choice and Connection: the Evolution of American Neighbourhood Policy and What it Means for the United Kingdom.* Joseph Rowntree Foundation, York.

Mullins, D., Beider, H. and Rowlands, R. (2004) *Empowering Communities, Improving Housing: Involving Black and Minority Ethnic Tenants and Communities.* Office of the Deputy Prime Minister, London.

Office of the Deputy Prime Minister (2003) *Sustainable Communities: Building for the Future.* Office of the Deputy Prime Minister, London.

Putnam, R. (2000) *Bowling Alone. The Collapse and Revival of American Community.* Simon & Schuster, New York.

Venkatesh, S. (2000) *American Project: The Rise and Fall of a Modern Ghetto.* Harvard University Press, Cambridge.

Glossary

BME Housing Associations Black and minority ethnic housing associations are defined as those that have 80 percent or more of their board or management committee drawn from black and minority ethnic communities. This definition is currently under review. Black and minority ethnic housing associations were established from the 1970s onwards to meet the specific housing needs of black and minority ethnic communities. The Housing Corporation sponsored the development of BME housing associations in various policies commencing in 1986 with a view to meeting housing needs but also developing black and minority leadership and capacity in the housing sector. However, recent policy interventions and focus on financial viability has reduced the number of BME housing associations from a peak of over 50 during the 1990s and the future is uncertain.

CDCs Community Development Corporations are small, not-for-profit organizations that generally operate in distressed neighbourhoods in the US and provide housing, social and economic services to low-income communities. Most CDCs are based in the east and west coasts and annually manage less then 30 homes. Estimates place the number of CDCs operating in the US at between 2000 and 2500.

CIH The Chartered Institute of Housing is the professional body for people who work in the housing sector in the UK and was founded in 1938 as the Institute of Housing but changed its name in 1994 after receiving a Royal Charter. More than 19,000 people are members of the CIH. Each year the CIH organises its Annual Conference and Exhibition, which is the largest housing event in Europe.

CRE The Commission for Racial Equality is a national organization that works in 'both the public and private sectors to encourage fair treatment and to promote equal opportunities for everyone, regardless of their race, colour, nationality, or national or ethnic origin'. It has statutory powers under the 1976 Race Relations Act which include helping people who have suffered discrimination, harassment and abuse; conducting formal investigations into organizations who may be operating in a discriminatory way and oblige them to change; taking action against organizations or adverts that are discriminatory; and helping individuals with judicial review action against the decisions by public funded organizations such as local authorities. In 2007 a new Commission for Equalities and Human Rights will be established to provide a single point of access for equalities matters.

For the time being, the CRE has decided that it will not be part of this new body, believing that, amongst other issues, race equality will be diluted.

Council housing Council housing was built and managed by local authorities in the UK during the twentieth century to meet the needs of local people. After World War II there was a major boom in council housing given that one out of every three houses was destroyed. In addition, slum clearance in cities was a key policy priority. Many developments during this period were characterized by generous standards of space and layout together with gardens and inside toilets. Council housing from the 1960s became increasingly characterized by high-rise tower blocks that were deemed to provide better housing density. The rise of home ownership, and especially the introduction of Right to Buy in the 1980 Housing Act (see below) by the Conservative government, together with design faults and reduced investment led to sharp reductions in popularity and tenure. Council housing became associated with poor quality standards inhabited by poor households. The housing stock transfer programme provides an opportunity for councils to attract inward investment into council housing through establishing different types of management arrangement. Both Labour and Conservative governments have encouraged this.

Decent Homes Standard This is commitment by government that all public sector (including housing associations') dwellings to be of decent standard by 2010. This means that the onus is on local authorities and housing associations to undertake stock condition surveys, demolish housing that is low demand or unfit and establish how repairs could be funded.

Fannie May Federal Government established Fannie May in 1938 to help low- and middle-income Americans access loans to buy their homes. Fannie May does not lend directly to home buyers but instead works with lenders to make sure that they have a supply of mortgage funds available to increase the goal of home ownership. In 1968 Fannie May became a private company operating with private capital and on a self-sustaining basis. Though it operates under Congressional Charter, Fannie May does not receive any funding or indeed backing from government.

Freddie Mac Established by federal government in 1970 and operating under Congressional Charter, Freddie Mac operates in the same way as Fannie May. It helps to provide a stable supply of mortgages that lenders offer to aspiring homeowners and renters. In effect, Freddie Mac is the supplier of capital to lenders by operating in the secondary mortgage market.

HMRP There are nine Housing Market Renewal Pathfinders that were established by the government in 2003 to address the problems of low housing demand and abandonment in the Midlands and northern England. More than £1.2 billion will be invested in these sub-regional housing markets until 2008. This will help to deal with housing market problems as well as help with attempts at neighbourhood renewal and regeneration. Each HMRP is expected to work with public and private sector housing partners to develop a strategic approach and also draw down additional investment.

HOPE VI HOPE VI stands for 'Housing Opportunities for People Everywhere'. Initially labelled the Urban Revitalization Demonstration Program, the objectives of HOPE VI were to comprehensively address the physical and social problems of very distressed public housing communities. The programme was first funded in 1993 but did not receive Congressional approval until the passage of the Quality Housing and Work Responsibility Act of 1998.

Housing associations Housing associations (sometimes referred to as registered social landlords) are the main providers of new and affordable homes to rent and buy in England. They operate on a not-for-profit basis and any surplus generated is invested into housing management and maintenance and helping to finance new housing developments. There are over 1500 housing associations in England managing over two million dwellings and housing over four million people. Some housing associations were first established over 100 years ago but many were founded in the 1960s and 1970s. The Housing Corporation is the regulatory agency that supports, provides investment and measures performance of housing associations.

Housing stock transfer Housing stock transfer is when a local council transfers some or part of its stock to a housing association. This enables local authorities to access investment to repair housing stock and reach Decent Homes Standard. Councils that need additional investment to improve stock can choose from arm's length management organizations (ALMO), private finance initiative (PFI) and housing stock transfer. Before the process begins, the council and key stakeholders have to undertake an option appraisal and then eventually take the preferred route for a tenants' vote. Provided that a majority of tenants vote in favour, the transfer can then take place.

HUD The US Department of Housing and Urban Development was established in 1965. It seeks to create decent and affordable homes for

Americans by addressing housing needs, investing in neighbourhoods and enforcing fair housing laws. The agency regulates public housing authorities.

Low-income housing tax credits This is arguably the most successful national production programme for low-income housing. It works by providing a tax liability and incentives for large private sector developers and indeed individuals to support housing programmes designed for low-income communities. This initiative was created by the 1986 Tax Reform Act. More than 1 million housing units have been built and rehabilitated. Public housing authorities have used this tax break to support and fund the supply of affordable housing outside the scope of HUD.

Macpherson Report The Macpherson Report is sometimes known as the Lawrence Report and was published in 1999. Chaired by Judge William Macpherson it was the culmination of an inquiry into the unlawful and racist killing of Stephen Lawrence in south London. The report was important because it highlighted the institutional racism in the failure of the Metropolitan Police successfully to complete their investigation, leading to the non-conviction of the alleged perpetrators. The Macpherson Report was hugely influential in establishing a framework for race equality in the UK.

MTO Movement to Opportunity is a HUD demonstration programme that helps very low-income families move to low-poverty neighbourhoods by using vouchers and counselling. The MTO objective is to develop effective mobility strategies for low-income tenants across metropolitan areas.

NASS The National Asylum Support Service is the government agency that has responsibility for people who are seeking asylum in the UK. Limited support is provided with accommodation and other areas of support. Importantly, once a decision has been made on status, people who have been seeking asylum have only 28 days to leave NASS-supported accommodation, which may increase the risk of homelessness.

New urbanism This is a movement that promotes sustainable and compact, multi-layered cities that mix living, work and leisure with well connected and designed urban spaces. It is sometimes seen as the blueprint for the regeneration and renewal of cities in both the US and UK.

NHF The National Housing Federation is the trade body for 1400 independent and not-for-profit housing associations in England. It promotes the needs of members to government and other key agencies, influences the public policy agenda on housing and supports its members through briefings, conferences and workshops.

ODPM The Office of the Deputy Prime Minister is the government department responsible, amongst other areas, for creating sustainable communities. High-level objectives include increasing affordability of housing, helping local authorities to deliver high-quality public services and cohesive communities. The ODPM has responsibility for managing a number of national programmes that help organizations meet these high-level goals.

Public housing authorities Public housing authorities are independent organizations established by local authorities to develop, own and manage low-rent housing for disadvantaged communities. The public housing programme has grown to more than 1.3 million houses, providing housing to 3 million people. Public housing authorities are regulated by HUD.

Right to Buy The 1980 Housing Act first introduced the Right to Buy Scheme. It enables local authority tenants with at least 2 years' tenancy to buy their homes at below market rates. More than 1.6 million tenants have exercised their Right to Buy since 1980. For some the policy has helped to create mixed-income and mixed-tenure neighbourhoods but others suggest that it has also increased polarization and led to further erosion of council housing stock.

Sustainable Communities Plan Launched in 2003 the Communities Plan is the major government policy intervention to address the problems of low demand in the Midlands and northern England, increasing the housing supply and affordability in the south and quality of public services across the country. This will be delivered within a framework to protect the countryside and backed up with considerable resources.

The Housing Corporation The Housing Corporation is the government agency that funds new affordable housing and regulates housing associations across England. This is achieved by investment through the approved development programme to housing associations, being the statutory regulator for housing associations and helping to influence housing policy nationally, regionally and locally.

Urban renaissance Framed by the government White Paper in 2000, urban renaissance is the process of improving cities and towns and making them places that people want to live, work and play.

Voucher Program A Federal supported program that helps two million families receive a rent subsidy to offset the costs of renting in the private sector. The voucher program ensures that people pay not more than 30 percent of their income on rent and 75 percent of those eligible must qualify as 'extremely poor'.

Index